Luca Bertolini, Bernhard
Elsener, Pietro Pedeferri,
Elena Redaelli, and
Rob Polder

**Corrosion of Steel
in Concrete**

Related Titles

Sharma, S. K. (ed.)

Green Corrosion Chemistry and Engineering

Opportunities and Challenges With a Foreword by Nabuk Okon Eddy

2012
ISBN: 978-3-527-32930-4

Kreysa, G., Schütze, M. (eds.)

Corrosion Handbook – Corrosive Agents and Their Interaction with Materials

13 Volume Set

2009
ISBN: 978-3-527-31217-7

Schütze, M., Wieser, D., Bender, R. (eds.)

Corrosion Resistance of Aluminium and Aluminium Alloys

2010
ISBN: 978-3-527-33001-0

Galambos, T. V., Surovek, A. E.

Structural Stability of Steel

Concepts and Applications for Structural Engineers

2009
ISBN: 978-0-470-03778-2

Bernold, L. E., AbouRizk, S. M.

Managing Performance in Construction

2010
ISBN: 978-0-470-17164-6

Heimann, R. B.

Plasma Spray Coating

Principles and Applications

2008
ISBN: 978-3-527-32050-9

Krzyzanowski, M., Beynon, J. H., Farrugia, D. C. J.

Oxide Scale Behavior in High Temperature Metal Processing

2010
ISBN: 978-3-527-32518-4

Geschwindner, L. F.

Unified Design of Steel Structures

2008
ISBN: 978-0-471-47558-3

Revie, R. W.

Corrosion and Corrosion Control, Fourth Edition

2010
ISBN: 978-0-470-65366-1

Roberge, P. R., Revie, R. W.

Corrosion Inspection and Monitoring

2007
ISBN: 978-0-471-74248-7

Luca Bertolini, Bernhard Elsener, Pietro Pedeferri,
Elena Redaelli, and Rob Polder

Corrosion of Steel in Concrete

Prevention, Diagnosis, Repair

Second, completely revised and enlarged edition

WILEY-VCH Verlag GmbH & Co. KGaA

The Authors

Prof. Luca Bertolini
Politecnico di Milano
Department of Chemistry,
Materials, and Chemical
Engineering "G. Natta"
Piazza Leonardo da Vinci 32
20133 Milano
Italy

Prof. Bernhard Elsener
ETH Zürich
Institute for Building Materials
ETH Hönggerberg
8093 Zürich
Switzerland

Dr. Elena Redaelli
Politecnico di Milano
Department of Chemistry,
Materials, and Chemical
Engineering "G. Natta"
Piazza Leonardo da Vinci 32
20133 Milano
Italy

Prof. Rob Polder
TNO Technical Sciences/
Built Environment
P.O. Box 49
2600 AA Delft
The Netherlands

Delft University of Technology
P.O. Box 5048
2600 CA Delft
The Netherlands

Cover:
The picture used on the cover is a painting on a titanium surface, created by Prof. Pietro Pedeferri. We thank his heirs for the kind permission to use this work of art.

■ All books published by **Wiley-VCH** are carefully produced. Nevertheless, authors, editors, and publisher do not warrant the information contained in these books, including this book, to be free of errors. Readers are advised to keep in mind that statements, data, illustrations, procedural details or other items may inadvertently be inaccurate.

Library of Congress Card No.: applied for

British Library Cataloguing-in-Publication Data
A catalogue record for this book is available from the British Library.

Bibliographic information published by the Deutsche Nationalbibliothek
The Deutsche Nationalbibliothek lists this publication in the Deutsche Nationalbibliografie; detailed bibliographic data are available on the Internet at <http://dnb.d-nb.de>.

© 2013 Wiley-VCH Verlag GmbH & Co. KGaA, Boschstr. 12, 69469 Weinheim, Germany

All rights reserved (including those of translation into other languages). No part of this book may be reproduced in any form – by photoprinting, microfilm, or any other means – nor transmitted or translated into a machine language without written permission from the publishers. Registered names, trademarks, etc. used in this book, even when not specifically marked as such, are not to be considered unprotected by law.

Composition   Toppan Best-set Premedia Ltd., Hong Kong
Printing and Binding   Markono Print Media Pte Ltd, Singapore
Cover Design   Bluesea Design, Simone Benjamin, McLeese Lake, Canada

Print ISBN: 978-3-527-33146-8
ePDF ISBN: 978-3-527-65172-6
ePub ISBN: 978-3-527-65171-9
mobi ISBN: 978-3-527-65170-2
oBook ISBN: 978-3-527-65169-6

Printed in Singapore
Printed on acid-free paper

Contents

Preface to the Second Edition XV
Preface to the First Edition XVII

1   Cements and Cement Paste 1
1.1   Portland Cement and Hydration Reactions   1
1.2   Porosity and Transport Processes   3
1.2.1 Water/Cement Ratio and Curing   4
1.2.2 Porosity, Permeability and Percolation   7
1.3   Blended Cements   8
1.3.1 Pozzolanic Materials   9
       Natural Pozzolana   9
       Fly Ash   10
       Silica Fume   10
1.3.2 Ground Granulated Blast Furnace Slag   10
1.3.3 Ground Limestone   11
1.3.4 Other Additions   11
1.3.5 Properties of Blended Cements   11
1.4   Common Cements   13
1.5   Other Types of Cement   15
       High Alumina Cement (HAC)   18
       Calcium Sulfoaluminate Cements (CSA)   19
       References   19

2   Transport Processes in Concrete 21
2.1   Composition of Pore Solution and Water Content   22
2.1.1 Composition of Pore Solution   22
2.1.2 Water in Concrete   23
       Capillary Water   23
       Adsorbed Water   26
       Interlayer Water   26
       Chemically Combined Water   26
2.1.3 Water Content and Transport Processes   26

2.2	Diffusion 27
2.2.1	Stationary Diffusion 28
2.2.2	Nonstationary Diffusion 29
2.2.3	Diffusion and Binding 30
2.3	Capillary Suction 32
2.4	Permeation 33
2.4.1	Water Permeability Coefficient 34
2.4.2	Gas Permeability Coefficient 35
2.5	Migration 35
2.5.1	Ion Transport in Solution 35
2.5.2	Ion Transport in Concrete 36
2.5.3	Resistivity of Concrete 37
	Temperature Dependence 38
	Concrete Resistivity and Corrosion Rate 39
	Measuring Concrete Resistivity 39
2.6	Mechanisms and Significant Parameters 40
	Correlations 40
	Presence of More Than One Transport Mechanism 43
	References 45

3 Degradation of Concrete 49

3.1	Freeze–Thaw Attack 50
3.1.1	Mechanism 51
3.1.2	Factors Influencing Frost Resistance 52
3.1.3	Air-Entrained Concrete 53
3.2	Attack by Acids and Pure Water 54
3.2.1	Acid Attack 54
3.2.2	Biogenic Sulfuric Acid Attack 56
3.2.3	Attack by Pure Water 58
3.2.4	Ammonium Attack 58
3.3	Sulfate Attack 59
3.3.1	External Sulfate Attack 59
	Protection 60
3.3.2	Internal Sulfate Attack 60
	Prevention 61
3.4	Alkali Silica Reaction 61
3.4.1	Alkali Content in Cement and Pore Solution 62
3.4.2	Alkali Silica Reaction (*ASR*) 63
	Presence and Quantity of Reactive Aggregate 64
	Alkali Content in the Pore Liquid of Concrete 64
	Type and Quantity of Cement 64
	Environment 65
	Prevention 65
3.5	Attack by Seawater 66
	References 67

4	**General Aspects**　*71*	
4.1	Initiation and Propagation of Corrosion　*71*	
4.1.1	Initiation Phase　*71*	
4.1.2	Propagation Phase　*73*	
4.2	Corrosion Rate　*73*	
4.3	Consequences　*74*	
4.4	Behavior of Other Metals　*75*	
	References　*77*	
5	**Carbonation-Induced Corrosion**　*79*	
5.1	Carbonation of Concrete　*79*	
5.1.1	Penetration of Carbonation　*80*	
5.1.2	Factors That Influence the Carbonation Rate　*81*	
	Humidity　*81*	
	CO_2 Concentration　*83*	
	Temperature　*84*	
	Concrete Composition　*84*	
5.2	Initiation Time　*85*	
5.2.1	Parabolic Formula　*86*	
5.2.2	Other Formulas　*86*	
5.3	Corrosion Rate　*87*	
5.3.1	Carbonated Concrete without Chlorides　*87*	
5.3.2	Carbonated and Chloride-Contaminated Concrete　*90*	
	References　*91*	
6	**Chloride-Induced Corrosion**　*93*	
6.1	Pitting Corrosion　*94*	
6.2	Corrosion Initiation　*96*	
6.2.1	Chloride Threshold　*96*	
	Chloride Binding　*98*	
	Atmospherically Exposed Structures　*100*	
	Submerged Structures　*100*	
6.2.2	Chloride Penetration　*101*	
6.2.3	Surface Content (C_s)　*103*	
6.2.4	Apparent Diffusion Coefficient　*106*	
6.3	Corrosion Rate　*108*	
	Exceptions　*109*	
	References　*109*	
7	**Electrochemical Aspects**　*113*	
7.1	Electrochemical Mechanism of Corrosion　*113*	
	Polarization Curves　*115*	
7.2	Noncarbonated Concrete without Chlorides　*116*	
7.2.1	Anodic Polarization Curve　*116*	
7.2.2	Cathodic Polarization Curve　*118*	

7.2.3	Corrosion Conditions	*119*
7.3	Carbonated Concrete	*120*
7.4	Concrete Containing Chlorides	*122*
7.4.1	Corrosion Initiation and Pitting Potential	*122*
7.4.2	Propagation	*124*
7.4.3	Repassivation	*125*
7.5	Structures under Cathodic or Anodic Polarization	*126*
	References	*127*

8 Macrocells *129*

- 8.1 Structures Exposed to the Atmosphere *129*
  - Coated Reinforcement *130*
  - Protection Effect *130*
  - Presence of Different Metals *130*
  - Other Macrocell Effects *131*
- 8.2 Buried Structures and Immersed Structures *131*
  - Differential Aeration in Buried Structures *131*
  - Structures Immersed in Seawater *132*
  - Rebars Not Entirely Embedded in Concrete *133*
  - Buried Structures Connected with Ground Systems *133*
- 8.3 Electrochemical Aspects *134*
- 8.4 Modeling of Macrocells *137*
  - References *138*

9 Stray-Current-Induced Corrosion *141*

- 9.1 DC Stray Current *142*
- 9.1.1 Alkaline and Chloride-Free Concrete *142*
  - First Precondition *144*
  - Second Precondition *146*
- 9.1.2 Passive Steel in Chloride-Contaminated Concrete *147*
  - Interruptions in the Stray Current *148*
- 9.1.3 Corroding Steel *148*
- 9.2 AC Stray Current *149*
- 9.3 High-Strength Steel *150*
- 9.4 Fiber-Reinforced Concrete *151*
- 9.5 Inspection *151*
- 9.6 Protection from Stray Current *152*
  - References *153*

10 Hydrogen-Induced Stress Corrosion Cracking *155*

- 10.1 Stress Corrosion Cracking (*SCC*) *156*
  - Anodic Stress Corrosion Cracking *156*
  - Hydrogen-Induced Stress Corrosion Cracking (*HI-SCC*) *156*
- 10.2 Failure under Service of High-Strength Steel *157*
- 10.2.1 Crack Initiation *158*

10.2.2	Crack Propagation *158*	
	σ_s and K_{ISCC} *159*	
10.2.3	Fast Propagation *159*	
10.2.4	Critical Conditions *160*	
10.2.5	Fracture Surface *162*	
10.3	Metallurgical, Mechanical and Load Conditions *162*	
10.3.1	Susceptibility of Steel to HI-SCC *164*	
10.4	Environmental Conditions *165*	
	Critical Intervals of Potential and pH *166*	
10.5	Hydrogen Generated during Operation *166*	
	Noncarbonated and Chloride-Free Concrete *167*	
	Carbonated Concrete *167*	
	Concrete Containing Chlorides *167*	
	Cathodically Protected Structures *168*	
10.6	Hydrogen Generated before Ducts Are Filled *169*	
10.7	Protection of Prestressing Steel *169*	
	References *170*	

11 Design for Durability *171*

11.1	Factors Affecting Durability *172*
11.1.1	Conditions of Aggressiveness *172*
11.1.2	Concrete Quality *173*
11.1.3	Cracking *173*
11.1.4	Thickness of the Concrete Cover *175*
11.1.5	Inspection and Maintenance *176*
11.2	Approaches to Service-Life Modeling *177*
11.2.1	Prescriptive Approaches *178*
11.2.2	Performance-Based Approaches *179*
	Limit States and Design Equation *181*
	Variability *181*
11.3	The Approach of the European Standards *183*
11.4	The *fib* Model Code for Service-Life Design for Chloride-Induced Corrosion *189*
11.5	Other Methods *194*
11.6	Additional Protection Measures *197*
11.7	Costs *198*
	References *200*

12 Concrete Technology for Corrosion Prevention *203*

12.1	Constituents of Concrete *203*
12.1.1	Cement *203*
12.1.2	Aggregates *204*
12.1.3	Mixing Water *205*
12.1.4	Admixtures *205*
	Water Reducers and Superplasticizers *206*

12.2	Properties of Fresh and Hardened Concrete 206
12.2.1	Workability 207
	Measurement of Workability 207
12.2.2	Strength 208
	Compressive Strength and Strength Class 210
	Tensile Strength 210
12.2.3	Deformation 212
12.2.4	Shrinkage and Cracking 212
12.3	Requirements for Concrete and Mix Design 212
12.4	Concrete Production 215
12.4.1	Mixing, Handling, Placement and Compaction 215
12.4.2	Curing 217
12.5	Design Details 219
12.6	Concrete with Special Properties 219
12.6.1	Concrete with Mineral Additions 221
12.6.2	High-Performance Concrete (HPC) 223
12.6.3	Self-Compacting Concrete (SCC) 223
	References 225

13 Corrosion Inhibitors 227
- 13.1 Mechanism of Corrosion Inhibitors 228
- 13.2 Mode of Action of Corrosion Inhibitors 228
- 13.3 Corrosion Inhibitors to Prevent or Delay Corrosion Initiation 229
- 13.4 Corrosion Inhibitors to Reduce the Propagation Rate of Corrosion 234
- 13.5 Transport of the Inhibitor into Mortar or Concrete 236
- 13.6 Field Tests and Experience with Corrosion Inhibitors 238
- 13.7 Critical Evaluation of Corrosion Inhibitors 238
  - Concentration Dependence 239
  - Measurement and Control of Inhibitor Action 240
- 13.8 Effectiveness of Corrosion Inhibitors 240
  - References 240

14 Surface Protection Systems 243
- 14.1 General Remarks 243
- 14.2 Organic Coatings 245
- 14.2.1 Properties and Testing 248
- 14.2.2 Performance 250
- 14.3 Hydrophobic Treatment 251
- 14.3.1 Properties and Testing 253
- 14.3.2 Performance 255
- 14.4 Treatments That Block Pores 257
- 14.5 Cementitious Coatings and Layers 258
- 14.6 Concluding Remarks on Effectiveness and Durability of Surface Protection Systems 259
  - References 260

15 Corrosion-Resistant Reinforcement 263
15.1 Steel for Reinforced and Prestressed Concrete 263
15.1.1 Reinforcing Bars 263
15.1.2 Prestressing Steel 264
15.1.3 Corrosion Behavior 266
15.2 Stainless Steel Rebars 266
15.2.1 Properties of Stainless Steel Rebars 267
 Chemical Composition and Microstructure 267
 Mechanical Properties 268
 Weldability 268
 Other Properties 268
15.2.2 Corrosion Resistance 269
 Resistance to Pitting Corrosion 269
 Fields of Applicability 271
15.2.3 Coupling with Carbon Steel 273
15.2.4 Applications and Cost 275
15.2.5 High-Strength Stainless Steels 276
15.3 Galvanized Steel Rebars 276
15.3.1 Properties of Galvanized Steel Bars 277
15.3.2 Corrosion Resistance 279
15.3.3 Galvanized Steel Tendons 280
15.4 Epoxy-Coated Rebars 280
15.4.1 Properties of the Coating 280
15.4.2 Corrosion Resistance 281
15.4.3 Practical Aspects 282
15.4.4 Effectiveness 282
 References 283

16 Inspection and Condition Assessment 287
16.1 Visual Inspection and Cover Depth 288
16.2 Electrochemical Inspection Techniques 291
16.2.1 Half-Cell Potential Mapping 291
 Principle 292
 Procedure 292
 Data Collection and Representation 294
 Interpretation 295
16.2.2 Resistivity Measurements 298
 Measurements at the Concrete Surface 300
 Procedure 301
 Interpretation 301
16.2.3 Corrosion Rate 302
 Determination of the Polarization Resistance 304
 Execution of the Measurements 305
 Corrosion Rate Measurements Onsite 305
 Interpretation of the Results 306

- 16.3 Analysis of Concrete *307*
- 16.3.1 Carbonation Depth *307*
- 16.3.2 Chloride Determination *308*

  Chloride Profile Based on Cores or Powder Drilling *309*

  Dissolution of the Powder *309*

  Chemical Analysis *309*

  Interpretation *310*

  References *310*

17 Monitoring *315*
- 17.1 Introduction *315*
- 17.2 Monitoring with Nonelectrochemical Sensors *316*

  Sensors Based on Macrocell Measurements *317*

  Sensors Based on Indepth Resistivity Measurements *318*

  Macrocell Corrosion Monitoring *319*

  Relative Humidity Sensors *321*
- 17.3 Monitoring with Electrochemical Sensors *322*

  Corrosion Potential *322*

  Linear Polarization Resistance (*LPR*) *323*

  Chloride Content *323*

  pH Monitoring *323*

  Oxygen-Transport Monitoring *323*
- 17.4 Critical Factors *324*

  Objective of Monitoring *324*

  Monitoring Design *325*

  Choice of Sensors and Probes *325*
- 17.5 On the Way to "Smart Structures" *325*
- 17.6 Structural Health Monitoring *327*

  References *328*

18 Principles and Methods for Repair *333*
- 18.1 Approach to Repair *334*
- 18.1.1 Repair Options *334*
- 18.1.2 Basic Repair Principles *337*
- 18.2 Overview of Repair Methods for Carbonated Structures *339*
- 18.2.1 Methods Based on Repassivation *339*

  Conventional Repair *339*

  Repassivation with Alkaline Concrete or Mortar *340*

  Electrochemical Realkalization *341*

  Cathodic Protection *341*
- 18.2.2 Reduction of the Moisture Content of the Concrete *341*
- 18.2.3 Coating of the Reinforcement *342*
- 18.3 Overview of Repair Methods for Chloride-Contaminated Structures *342*
- 18.3.1 Methods Based on Repassivation *343*

	Repassivation with Alkaline Mortar or Concrete 344
	Electrochemical Chloride Removal (*ECR*) 345
18.3.2	Cathodic Protection 345
18.3.3	Other Methods 345
	Hydrophobic Treatment 345
	Coating of the Reinforcement 346
	Migrating Inhibitors 346
18.4	Design, Requirements, Execution and Control of Repair Works 346
	References 347

19 Conventional Repair 349

19.1	Assessment of the Condition of the Structure 349
19.2	Removal of Concrete 350
19.2.1	Definition of Concrete to be Removed 350
	Carbonation-Induced Corrosion 351
	Chloride-Induced Corrosion 352
	Variability 354
19.2.2	Techniques for Concrete Removal 355
19.2.3	Surface Preparation 356
19.3	Preparation of Reinforcement 356
19.4	Application of Repair Material 357
19.4.1	Requirements 357
	Alkalinity and Resistance to Carbonation and Chloride Penetration 357
	Cover Thickness 357
	Rheology and Application Method 357
	Bond to the Substrate and Dimensional Stability 358
	Mechanical Properties 358
19.4.2	Repair Materials 358
19.4.3	Specifications and Tests 359
19.5	Additional Protection 360
	Corrosion Inhibitors 360
	Surface Treatment of Concrete 360
	Coating of Rebars 361
19.6	Strengthening 361
	References 362

20 Electrochemical Techniques 365

20.1	Development of the Techniques 366
20.1.1	Cathodic Protection 366
20.1.2	Cathodic Prevention 368
20.1.3	Electrochemical Chloride Removal 368
20.1.4	Electrochemical Realkalization 369
20.2	Effects of the Circulation of Current 369
20.2.1	Beneficial Effects 369

Reactions on the Steel Surface 369
Migration 370
20.2.2 Side Effects 370
Hydrogen Embrittlement 371
Alkali Aggregate Reaction 371
Loss of Bond Strength 371
Anodic Acidification 371
20.2.3 How Various Techniques Work 373
20.3 Cathodic Protection and Cathodic Prevention 373
20.3.1 Cathodic Protection of Steel in Chloride-Contaminated Concrete 373
20.3.2 Cathodic Prevention 376
20.3.3 Cathodic Protection in Carbonated Concrete 377
20.3.4 Throwing Power 379
20.3.5 The Anode System 379
20.3.6 Practical Aspects 380
Design 380
Anode System 381
Power System 381
Electrical Connections 382
Zones 382
Repair Materials 382
The Monitoring System 383
Trials 384
Execution 384
Operation and Maintenance 384
20.3.7 Service Life 384
20.3.8 Numerical Modeling 386
20.4 Electrochemical Chloride Extraction and Realkalization 386
20.4.1 Electrochemical Chloride Extraction 387
Treatment Effectiveness 388
Durability after Chloride Extraction 390
Trials 392
Monitoring of the Process 392
Monitoring after Treatment 392
Side Effects 393
20.4.2 Electrochemical Realkalization 394
End-Point Determination and Treatment Effectiveness 396
Influence of the Cement Type 398
Durability 398
Side Effects 399
20.4.3 Practical Aspects 399
References 400

Index 407

Preface to the Second Edition

Since this book was first published, durability of reinforced concrete structures has continued to receive worldwide interest of materials scientists and designing engineers. Although some of the open questions raised in the preface of the first edition have found reasonable explanations in the past decade, others are still unanswered and new issues have arisen. For example, the need for sustainability has, on the one hand, increased the demand for durable structures and, on the other hand, promoted the development and use of new materials with lower environmental impact whose durability properties need to be verified. The increased demand for maintaining large numbers of existing structures and prolonging their service life poses technical and economical challenges of a larger scale. At the same time, increased experience with regard to repair techniques and materials must be incorporated in asset management on the scale of, for example, road networks. Challenges for the next decade are science-based models for the prediction of service life of new and existing structures and reliable accelerated tests that are able to provide durability-related design parameters both for concrete (e.g., with regard to resistance to carbonation or chloride penetration) and for steel (e.g., relating to the chloride threshold for corrosion initiation).

In the second edition of this book, all chapters have been revised and updated with recent findings and new perspectives. The structure of the book has been maintained, so that it may serve as a reference for students and materials scientists, who may learn from the explanation of corrosion and degradation mechanisms, as well as people involved in the design, execution, and management of reinforced concrete structures, who may concentrate on the parts of the book dealing with practical aspects of assessment, monitoring, prevention, and protection techniques.

With this second edition we also have a new co-author, Elena Redaelli, but, sadly, we lost Pietro Pedeferri, who passed away on 3 December 2008. Pietro strongly wanted the first edition of this book, and he was the driving force for its realization. He dedicated his life to the study of electrochemistry and corrosion science and technology, making important contributions to several aspects of corrosion and protection techniques, such as localized corrosion of stainless steels, cathodic protection, corrosion in the human body, corrosion in the oil industry, surface treatments of titanium, and corrosion and protection of steel in concrete. Corrosion of steel in concrete became a major field of interest for him in the 1980s.

Looking for a durable solution to prevent corrosion of reinforcing steel in motorway bridges exposed to chloride contamination due to de-icing salt application, he proposed the technique of *cathodic prevention*, and in explaining the advantages of cathodic prevention over cathodic protection, he developed the graph shown in Figure 20.4, now internationally acknowledged as *Pedeferri's diagram*. This graph definitely expresses Pietro's spirit as a researcher as well as a teacher, showing that he was able to transfer in a simple, although rigorous, manner a complex matter as the way of dealing with depassivation and repassivation of steel in chloride-contaminated concrete. Pietro Pedeferri contributed to the understanding of several mechanisms involved in the corrosion behavior of steel reinforcement in concrete, and his knowledge permeates the whole book, even in this new edition.

The front cover is a tribute to Pietro Pedeferri as an artist. In fact, he was able to conjugate his research studies on the anodic oxidation of titanium with his creativity, and he developed a unique technique for electrochemical painting of titanium. Mixing acids, electrical currents, flow of liquids, and his poetic inspiration, he generated beautiful and colorful drawings.

The Authors, January 2013

Preface to the First Edition

Over the millennia, concrete prepared by the Romans using lime, pozzolana and aggregates has survived the elements, giving proof of its durability. Prestigious concrete works have been handed down to us: buildings such as the Pantheon in Rome, whose current structure was completed in 125 A.D., and also structures in marine environments have survived for over two thousand years. This provides a clear demonstration that concrete can be as durable as natural stone, provided that specific causes of degradation, such as acids or sulphates, freeze–thaw cycles, or reactive aggregates, are not present.

Today, thanks to progress made over the past few decades in the chemistry of cement and in the technology of concrete, even these causes of deterioration can be fought effectively. With an appropriate choice of materials and careful, adequately controlled preparation and placement of the mixture, it is possible to obtain concrete structures which will last in time, under a wide variety of operative conditions.

The case of reinforced concrete is somewhat different. These structures are not eternal, or nearly eternal, as was generally supposed up until the 1970s. Instead, their service life is limited precisely because of the corrosion of reinforcement. Actually, concrete provides the ideal environment for protecting embedded steel because of its alkalinity. If the design of a structure, choice of materials, composition of the mixture, and placement, compaction and curing are carried out in compliance with current standards, then concrete is, under most environmental conditions, capable of providing protection beyond the 50 years typical of the required service life of many ordinary structures, at least in temperate regions. In fact, cases of corrosion that have been identified in numerous structures within periods much shorter than those just mentioned can almost always be traced to a failure to comply to current standards or to trivial errors in manufacturing of the concrete. However, under environmental conditions of high aggressiveness (generally related to the presence of chlorides), even concrete which has been properly prepared and placed may lose its protective properties and allow corrosion of reinforcement long before 50 years have elapsed, sometimes resulting in very serious consequences.

The problem of corrosion in reinforced concrete structures is thus a very real one and must be given special consideration. It is, in fact, only since the early

1980s that research has devoted much attention to this problem. From those years on main physiological aspects related to behaviour of steel in concrete, such as the nature of the aqueous pore liquid present in the hardened concrete, the electrochemistry of steel in this environment, the mechanism of protection of steel by an oxide film, etc. have been established. Passing to the pathological side, research has explained the phenomenology and mechanisms of corrosion, established the conditions which give rise to it and the laws governing its evolution, and developed techniques for diagnosing and controlling it. In particular, it has been shown that the only circumstances that can give rise to corrosion are those when both depassivation occurs (e.g., due to carbonation or chlorides) and oxygen and humidity are present.

Several points still need to be clarified. For example: the atlas of pathological anatomy has been defined clearly with regard to corroding reinforcement, but only sketchily in relation to the surrounding concrete; the body of diagnostics allows the state of corrosion in a structure to be evaluated for the more common forms of corrosion, but is still incomplete in the case of hydrogen embrittlement in high-strength steels of prestressed structures or corrosion caused by stray current; the handbook of anticorrosionistic pharmacology includes a long list of methodologies of prevention (from inhibitors to coatings, to corrosion resistant reinforcement, to electrochemical techniques); however, their long-term effects or their possible negative side-effects are not always clearly known. Probably, the greatest shortcomings have to do with the basic aspects of corrosion. For instance, in the area of physiopathology: the species around the passive reinforcement in concrete are known, but those around corroding reinforcement are not; the influence of species on the passivity or corrosion of steel is known in qualitative terms, but very little is known of the entity of their interaction with the constituents of cement paste, and thus of their mobility in electrical fields or in various concentration gradients, in relation to the type of cement or to the characteristics of the concrete, etc.

In the field of construction, notable progress has also been achieved: the problem of corrosion, and more in general of the durability of structures in reinforced concrete, is very seriously taken into consideration; new laws are in place and new technologies and products are available.

But the above must not lull us into a false sense of security. It is true that today there is greater sensitivity to this problem, often being the subject of conferences, seminars and publications. New rules and standards do exist, though they are perceived as compulsory, being the result of legislation. Finally, new technologies and sophisticated products are being adopted, for example in the field of repair of structures damaged by corrosion. All these aspects do not, however, in themselves eliminate the errors, often trivial, that are at the basis of most failures today, and even less are they able to solve those cases where structures operate under conditions of high aggressiveness. Substantial progress will be made, also with regard to durability, only when our current technology, based on empiricism and common sense, evolves into a technology based on a thorough knowledge of degradation processes and of methods for their control.

Education, and thus teaching in particular, has a very important role to play, not only by making professionals sensitive to the durability problem, but also by giving them the tools necessary to solve it.

We hope that this text may be useful for those who work in the field of civil and construction engineering, as well as for those involved in the area of maintenance and management of reinforced concrete structures. Its aim is to provide the knowledge, tools and methods to understand the phenomena of deterioration and to prevent or control them. In some sections of the text, because of our professional background, we have gone into details of some electrochemical aspects. These explanations go beyond what is strictly required in civil and construction engineering and are not essential to an understanding of the other sections.

Finally, we wish to thank the European commissions that, by promoting the cooperative actions COST 509, 521 and recently 534, gave to several European researchers the opportunity to meet, collaborate and exchange views in the field of corrosion of steel in concrete. This book was born from that cooperation. We gratefully acknowledge all friends and colleagues on COST Actions, RILEM technical committees and European Federation of Corrosion working groups for providing data, papers and, most of all, for stimulating discussion.

The Authors, November 2003

1
Cements and Cement Paste

The protection that concrete provides to the embedded steel and, more in general, its ability to withstand various types of degradation, depend on its microstructure and composition. Concrete is a composite material made of aggregates and hydrated cement paste, that is, the reaction product of the cement and the mixing water. This chapter illustrates the properties of the most utilized cements and the microstructure of hydrated cement pastes. The properties of concrete and its manufacturing are discussed in Chapter 12.

1.1
Portland Cement and Hydration Reactions

Cements are fine mineral powders that, when they are mixed with water, form a paste that sets and hardens due to hydration reactions. *Portland cement* is the basis for the most commonly used cements [1–5]. It is produced by grinding clinker, which is obtained by burning a suitable mixture of limestone and clay raw materials. Its main components are *tricalcium* and *dicalcium silicates* (C_3S and C_2S),[1] the *aluminate* and *ferroaluminate of calcium* (C_3A and C_4AF, respectively). *Gypsum* ($C\bar{S}$) is also added to clinker before grinding, to control the rate of hydration of aluminates. Table 1.1 shows the typical ranges of variation of the constituents of portland cement. Other components, such as sodium and potassium oxides, are present in small but variable amounts.

In the presence of water, the compounds of portland cement form colloidal hydrated products of very low solubility. Aluminates react first, and are mainly responsible for setting, that is, solidification of the cement paste. The hydration of C_3A and C_4AF, in the presence of gypsum, mainly gives rise to hydrated sulfoaluminates of calcium. Hardening of cement paste, that is, the development of strength that follows setting, is governed by the hydration of silicates. The hydration of C_3S and C_2S gives rise to *calcium silicate hydrates* forming a gel, indicated as *C–S–H*. It is composed of extremely small particles with a layer structure that

---

1) In the chemistry of cement, the following abbreviations are used: CaO = C; SiO_2 = S; Al_2O_3 = A; Fe_2O_3 = F; H_2O = H; SO_3 = $\bar{S}$.

Corrosion of Steel in Concrete: Prevention, Diagnosis, Repair, Second Edition. Luca Bertolini,
Bernhard Elsener, Pietro Pedeferri, Elena Redaelli, and Rob Polder.
© 2013 Wiley-VCH Verlag GmbH & Co. KGaA. Published 2013 by Wiley-VCH Verlag GmbH & Co. KGaA.

Table 1.1 Main components of portland cement and typical percentages by mass.

Tricalcium silicate	$3CaO \cdot SiO_2$	C_3S	45–60%
Dicalcium silicate	$2CaO \cdot SiO_2$	C_2S	5–30%
Tricalcium aluminate	$3CaO \cdot Al_2O_3$	C_3A	6–15%
Tetracalcium ferroaluminate	$4CaO \cdot Al_2O_3 \cdot Fe_2O_3$	C_4AF	6–8%
Gypsum	$CaSO_4 \cdot 2H_2O$	$C\bar{S}$	3–5%

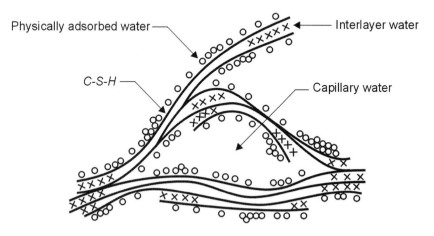

Figure 1.1 Feldman–Sereda model for C–S–H [2].

tend to aggregate in formations a few μm in dimension, characterized by interlayer spaces of small dimensions (<2 nm) and by a large surface area (100–700 m²/g). Figure 1.1 shows a model proposed to describe this structure. Due to the high surface area, C–S–H can give considerable strength to the cement paste. Its chemical composition is not well defined since the ratio between the oxides may vary as the degree of hydration, water/cement ratio, and temperature vary (for instance the C/S ratio may vary from 1.5 to 2). However, upon complete hydration, it tends to correspond to the formula $C_3S_2H_3$ usually used in stoichiometric calculations. C–S–H represents approximately 50–60% of the volume of the completely hydrated cement paste.

Hydration of calcium silicates also produces hexagonal crystals of calcium hydroxide (Ca(OH)₂, *portlandite*). These have dimensions of the order of a few μm and occupy 20 to 25% of the volume of solids. They do not contribute to the strength of cement paste. However, Ca(OH)₂, as well as NaOH and KOH, are very important with regard to protecting the reinforcement, because they cause an alkaline pH up to 13.5 in the pore liquid (Section 2.1.1).

The hydration reactions of tricalcium and dicalcium silicates can be illustrated as follows:

$$2C_3S + 6H = C_3S_2H_3 + 3Ca(OH)_2 \qquad (1.1)$$

$$2C_2S + 4H = C_3S_2H_3 + Ca(OH)_2 \qquad (1.2)$$

The reaction products are the same, but the proportions are different. The ratio between C–S–H and portlandite, passing from the hydration of C_3S to that of C_2S changes from 61/39 to 82/18, and the amount of water required for hydration from 23% to 21%. In principle, C_2S should lead to a higher ultimate strength of the cement paste by producing a higher amount of C–S–H. Nevertheless, the rate of hydration is much lower for C_2S compared with C_3S, and the strength of cement paste after 28 days of wet curing is mainly due to C_3S. Thus, the larger the amount of C_3S in a portland cement, the higher the rate of hydration and strength development of its cement paste. Increasing the fineness of cement particles can also increase the rate of hydration. The reactions leading to hydration of portland cement are exothermic; hence increasing the rate of hydration also increases the rate of generation of heat of hydration.

1.2
Porosity and Transport Processes

The cement paste formed by the hydration reactions contains interconnected pores of different sizes, as shown in Figure 1.2. Although the classification of concrete porosity is quite a complex matter [6], for the purposes of this book pores can be roughly divided into air voids, capillary pores and gel pores. The interlayer spacing within C–S–H (*gel pores*) have dimensions ranging from a few fractions of a nm to several nm. These essentially do not affect the durability of concrete and its

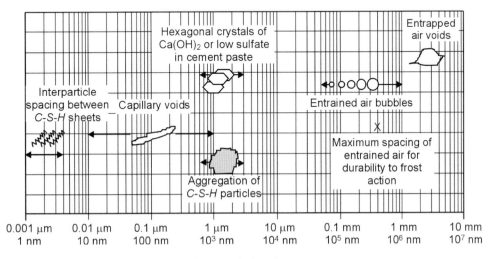

Figure 1.2 Dimensional range of solids and pores in hydrated cement paste [3].

protection of the reinforcement because they are too small to allow significant transport of aggressive species. The *capillary pores* are the voids not filled by the solid products of hydration within the hardened cement paste. They have dimensions of 10 to 50 nm if the cement paste is well hydrated and produced using low water/cement ratios (*w*/*c*), but can reach up to 3–5 μm if the concrete is made using high *w*/*c* ratios or it is not well hydrated. Larger pores of dimensions of up to a few mm are the result of the *air entrapped* during mixing and not removed by vibration of fresh concrete. Air bubbles with diameter of about 0.05–0.2 mm may also be introduced in the cement paste intentionally by means of air-entraining admixtures, so as to produce resistance to freeze–thaw cycles (Section 3.1.3). Both capillary pores and entrapped air are relevant to the durability of concrete and its protection of the rebars, since they determine the resistance to the penetration of aggressive species. The main factors affecting the capillary porosity, that is, water/cement ratio, curing, and type of binder, will be briefly analyzed in the following sections. Entrapped air can be reduced by providing adequate workability to the fresh concrete and proper compaction; this is dealt with in Chapter 12.

1.2.1
Water/Cement Ratio and Curing

The water/cement ratio, that is, the ratio between mass of mixing water and mass of cement, and curing have a major effect on the capillary porosity, which can be described by considering changes occurring in time. During the hydration of cement paste, the gross volume of the mixture practically does not change, so that the initial volume, equal to the sum of the volumes of mixed water (V_w) and cement (V_c) is equal to the volume of the hardening product. As indicated in Figure 1.3 from Neville and Brooks [7], the total volume consists in the sum of the volume of cement that has not yet reacted (V_{uc}), the hydrated cement ($V_p + V_{gw}$), the capillary pores that are filled by water (V_{cw}) or by air (V_{ec}). The volume of the products of hydration can be assumed to be roughly double that of the cement; hence during hydration these products fill the space previously occupied by the cement that has hydrated and part of the surrounding space initially occupied by water (Figure 1.4). Therefore, if the cement paste is kept moist (curing), the hydration proceeds and the volume of the capillary pores decreases and will reach a minimum when the hydration of cement has completed. Nevertheless, the porosity reached after complete hydration will be greater in proportion to the initial space between the cement particles and thus to the amount of mixing water. Figure 1.5 shows an example of the effect of *w*/*c* ratio and curing on the pore-size distribution, measured by mercury intrusion porosimetry. As the *w*/*c* ratio decreases, or as the curing time increases (and thus also the degree of hydration increases), the reduction of porosity is mainly due to the reduction in pores of larger dimensions that have been filled or have been connected only by C–S–H gel pores.

In conclusion, the volume of the capillary pores ($V_{cp} = V_{cw} + V_{ec}$) in the cement paste increases with the amount of water used in the paste and thus with the water/cement ratio (*w*/*c*) and decreases with the degree of hydration (*h*), that is,

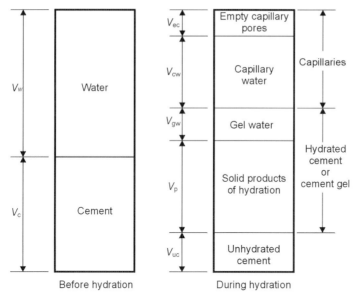

Figure 1.3 Schematic representation of the volumetric proportions in cement paste before and during hydration [7].

Figure 1.4 Example of microstructure of hydrated cement paste (scanning electron microscope).

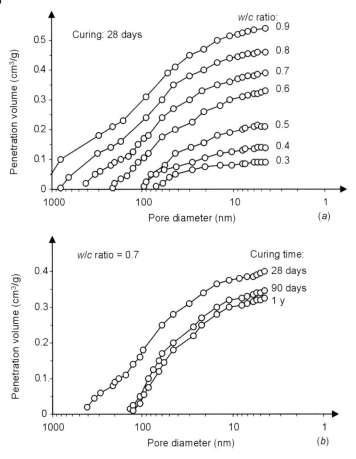

Figure 1.5 Influence of the water/cement ratio (a) and curing (b) on the distribution of pore size in hydrated cement pastes [3].

the fraction of hydrated cement. The effect of w/c and h on the volume of capillary pores (V_{cp} in liters per kg of cement) can be described by the following formula, proposed by Powers [8]:

$$V_{cp} = (w/c - 0.36h) \qquad (1.3)$$

When concrete is considered instead of cement paste, the w/c ratio and degree of hydration remain the main factors that determine the capillary porosity. Nevertheless, concrete is more complex because of the presence of the aggregates and the interfacial transition zone. The interfacial transition zone is a layer (usually several tens of micrometers thick) of hydrated cement paste in contact with aggregates. Especially for coarse aggregates, the hydrated cement paste in this area is typically heterogeneous and has a higher porosity compared to bulk cement paste [2–4].

1.2.2
Porosity, Permeability and Percolation

In determining the resistance to degradation of concrete and its role in protecting the embedded steel, not only the total capillary porosity (i.e., the percentage of volume occupied by capillaries) should be considered, but also the size and interconnection of capillary pores. Figure 1.6 shows the relation between the transport properties of cement paste (expressed as coefficient of water permeability) and the compressive strength as a function of the w/c ratio and degree of hydration. The decrease in capillary porosity increases the mechanical strength of cement paste and reduces the permeability of the hydrated cement paste. A distinction should

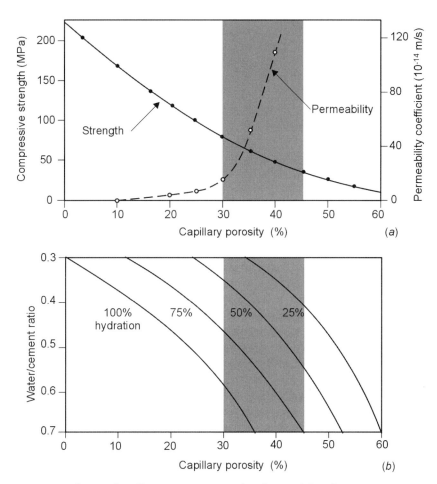

Figure 1.6 Influence of capillary porosity on strength and permeability of cement paste (a). Capillary porosity derives from a combination of water/cement ratio and degree of hydration (b) (Powers [7] from [3]).

be made between capillary pores of larger dimensions (e.g., >50 nm), or *macropores*, and pores of smaller dimensions, or *micropores* [3]. The reduction in porosity resulting of both the macro- and the micropores plays an essential role in increasing mechanical strength.

On the other hand, the influence of porosity on transport processes cannot be explained simply by the pore volume, the concept of connectivity or the degree of continuity of the pore system also has to be taken into account. At high porosities the interconnected capillary pore system (Figure 1.4) extends from the surface to the bulk of the cement paste. Permeability is high (Figure 1.6) and transport processes like for example, capillary suction of (chloride-containing) water can take place rapidly. With decreasing porosity the capillary pore system loses its connectivity, thus transport processes are controlled by the small gel pores. As a result, water and chlorides will penetrate only a short distance into cement paste. This influence of structure (geometry) on transport properties can be described with the percolation theory [9]: below a critical porosity, p_c, the percolation threshold, the capillary pore system is not interconnected (only finite clusters are present); above p_c the capillary pore system is continuous (infinite clusters). The percolation theory has been used to design numerical experiments and applied to transport processes in cement paste and mortars [10].

The steep increase of the water permeability above ca. 25% porosity (corresponding to a w/c ratio of 0.45 with a degree of hydration of 75%, Figure 1.6) is the background of the specified values in the codes of practice for high quality concrete. For instance, Table 1.2 shows the relationship between water/cement ratio and degree of hydration in order to achieve segmentation of the macropores in a paste of portland cement. This relationship was proposed by Powers for portland cement pastes in the 1950s.

1.3
Blended Cements

The use of portland cement clinker is continuously decreasing, being replaced by supplementary cementitious materials (SCM). These new types of cements are

Table 1.2 Curing times necessary to achieve a degree of hydration capable of segmenting the macropores in a portland cement paste (Powers from [2]).

w/c	Degree of hydration	Curing
0.40	50%	3 days
0.45	60%	7 days
0.50	70%	14 days
0.60	92%	6 months
0.70	100%	1 year
>0.70	100%	Impossible

obtained by intergrinding or blending portland cement with particular mineral substances.[2] Among these types of binders, those with the addition of pozzolanic materials or ground granulated blast furnace slag are of particular interest with regard to durability of reinforced concrete.

1.3.1
Pozzolanic Materials

Pozzolanic materials can be either natural, like pozzolana, or artificial, like *fly ash* and *silica fume* [4]. They are mainly glassy siliceous materials that may contain aluminous compounds but have a low lime (calcium hydroxide) content. In themselves they do not have binding properties, but acquire them in the presence of lime, giving rise to hydration products similar to those of portland cement.

The reaction between pozzolanic materials, lime and water is known as the *pozzolanic reaction*:

$$\text{pozzolana} + \text{water} + Ca(OH)_2 \rightarrow C-S-H \tag{1.4}$$

In cements containing pozzolanic additions, the lime needed to react with pozzolana is provided by the hydration of portland cement. The hardened cement paste (compared to that obtained with ordinary portland cement) has a lower lime content and higher content of C–S–H. The amount of pozzolanic addition to portland cement should be adjusted to the amount of lime produced in the hydration of portland cement. Any excess of the pozzolanic addition will not react and thus will behave as an inert addition ("filler").

Natural Pozzolana This is a sedimentary material, usually of pyroclastic origin, that is derived from the sediment of volcanic eruptions that has produced incoherent deposits or compact deposits that have been chemically transformed with time (such as Italian pozzolana that was already used by the Romans). Pozzolanic materials may also have other origins, such as diatomaceous earth composed of the siliceous skeletons of micro-organisms. The pozzolanic activity of these materials is related to their siliceous component in the vitreous state and to their fineness [11]. There are also pozzolana that are obtained by calcination of natural substances.

---

2) In the presence of blended cements the definition of the w/c ratio may be ambiguous and deserves some attention. Usually the mass of cement refers to the blended cement, and so it includes the addition. Unless otherwise indicated, this will be done in this book. It should be noted that the effect of w/c on durability related parameters is strictly related to the composition of the binder. So, the requirement for w/c should always be associated to a specific type of cement. If the addition is not interground with cement, but it is added at the concrete plant, the mass of cement in the w/c ratio will normally not include it and the concept of water/binder ratio (w/b) is introduced. Water/binder ratio will also be used as a general term when distinction between addition at the cement plant or at the concrete plant is not relevant.

Fly Ash Fly ash (also called pulverized fuel ash, PFA) is a byproduct of the combustion of coal powder in thermoelectric power plants [12]. It consists of fine and spherical particles (dimensions from 1 to 100 μm and specific surface area of 300 to 600 m^2/kg) that are collected from exhaust gases with electrostatic or mechanical filters. Its composition depends on the type of coal it derives from; the most common PFA is mainly siliceous. Because it is formed at high temperature and subsequently undergoes rapid cooling, its structure is mainly amorphous and thus reactive. Fly ash is normally added in amounts between 15% and 30%; binders with higher amounts are termed "high-volume fly ash" materials.

Silica Fume Silica fume (SF) is a waste product of manufacturing ferro-silicon alloys. It consists of an extremely fine powder of amorphous silica. Average particle diameter is about 100 times smaller than that of portland cement and the specific surface area is enormous: 13 000–30 000 m^2/kg compared to 300–400 m^2/kg for common portland cements. Silica fume shows an elevated pozzolanic activity and is also a very effective filler [13]. For these reasons, addition of silica fume to portland cement may lead to a very low porosity of the cement paste, increasing the strength and lowering the permeability. It is usually added in the proportion of 5 to 10% and it is combined with the use of a superplasticizer in order to maintain adequate workability of the fresh concrete.

1.3.2
Ground Granulated Blast Furnace Slag

Production of pig iron generates great quantities of liquid slag, which is composed like portland cement of lime, silica, and alumina, although in different proportions. The slag acquires hydraulic properties if it is quenched and transformed into porous granules with an amorphous structure. This is then ground to obtain a powder whose fineness is comparable to that of cement, which is called *ground granulated blast furnace slag* (GGBS) [2].

Unlike pozzolanic materials, which hydrate only in the presence of lime, slag has latent hydraulic characteristics and thus might be used as a hydraulic binder. The rate of hydration of this process is, however, too slow for practical purposes. This material shows good hardening properties when mixed with portland cement, because the hydration of portland cement creates an alkaline environment that activates the reaction of GGBS. Nevertheless, even when activated by portland cement, the hydration of GGBS is slower than that of portland cement. To achieve a high early strength, the slag content should be relatively low (35–50%). The hydration of GGBS, however, refines the pore structure of the cement paste; in order to achieve an optimal densification of the cement paste the GGBS content should be higher than 65%.

Blast furnace slag cement with improved properties with regard to both early strength and density has been introduced. It is a CEM III/A 52.5 containing 57% finely ground slag and rapid hardening portland cement. It has a better resistance to carbonation than CEM III/B, and a similarly good resistance to chloride penetration and alkali-silica reactions. Its relatively high early strength makes it suitable for use in the precast concrete industry.

1.3.3
Ground Limestone

Besides additions with pozzolanic or hydraulic properties, ground limestone may also be used [14]. In Europe the use of portland cement blended with about 15–20% of ground limestone is continuously increasing, mainly due to saving of CO_2 emissions, and in some countries this type of binder is by far the most widely used. Even though some reaction of fine carbonate particles with hydration products of portland cement may occur and replacement of small amounts (10–15% by mass) of portland cement with ground limestone may not significantly affect the strength of concrete (due to an accelerating effect on hydration of the C_3S phase), this addition is essentially inert.

1.3.4
Other Additions

A large number of new additions have been considered in order to study possible advantages of their incorporation in blended cements. This is often justified by environmental issues [15]. Materials that show pozzolanic or hydraulic properties may replace the clinker of portland cement, reducing the consumption of energy and raw materials as well as the pollution associated to its production (e.g., total CO_2 emissions for the production of clinker, coming both from the fuel and the decomposition of limestone, range from 0.84 to 1.15 kg per kg of clinker for kilns burning conventional fuels and raw materials [16]). Furthermore, most of the proposed materials, such as metakaolin, rice husk ash, recycled glass or even ashes from municipal solid waste incineration, are wastes, so they can be reused, avoiding their disposal [17].

Several mineral additions have also been shown to improve the resistance of concrete materials to the penetration of aggressive ions [18]. So the research in this field may produce interesting results also regarding the durability of reinforced concrete structures.

1.3.5
Properties of Blended Cements

Cement paste obtained with blended cements differs considerably from that obtained with portland cement. The hydration of pozzolanic materials or GGBS consumes lime and thus reduces its amount with respect to a cement paste obtained with portland cement. Figure 1.7 outlines the microstructures of hardened cement pastes made of portland cement and blended cements [19]. It can be observed that the addition of PFA or GGBS leads to the formation of very fine products of hydration that lead to a refinement of pores. Consequently, an increase

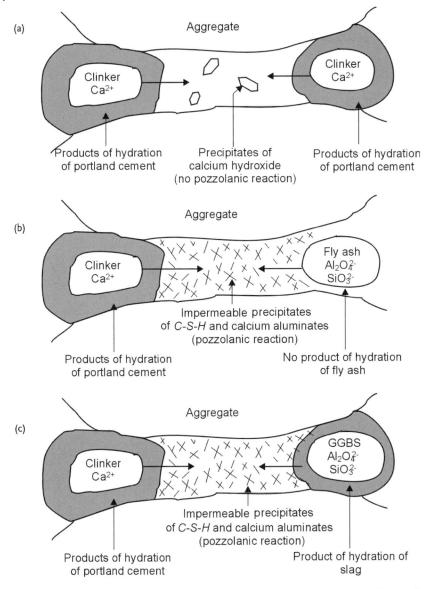

Figure 1.7 Microstructure of hydration of portland cement (a), and cements with addition of fly ash (b) and blast furnace slag (c) (Bakker from [19]).

in the resistance to penetration of aggressive agents can be obtained. However, the reactions of pozzolanic materials or GGBS are slower than hydration of portland cement; hence this effect will be achieved only if the wet curing of the concrete is long enough.

The rate of reaction of blast furnace slag and fly ash differs strongly. To show this, Figure 1.8 compares electrical resistivity measurements of wet cured con-

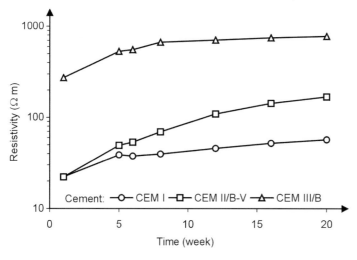

Figure 1.8 Electrical resistivity of concrete made with different cements from 7 days age while exposed in a fog room (w/c = 0.45, 2% chloride mixed in) [20].

crete with a water/cement ratio of 0.45 made with portland, portland fly ash and blast furnace slag cements. The development of resistivity of concrete at an early age shows the changes that occur in the microstructure of cement paste (Section 2.5.3). From the age of one week on, the resistivity of blast furnace slag cement (CEM III/B) was more than three times higher than for portland (CEM I) and portland fly ash (CEM II/B-V) cement [20]. The resistivity of PFA concrete became significantly higher than that of ordinary portland cement concrete from about eight weeks of age. Recent work has shown that CEM III/B mortar shows lower resistivity than CEM I mortar up to a few days age [21]; from three days on, the situation is reversed, as is shown in Figure 1.9. It appears, then, that the refinement of the pore structure by GGBS starts as early as a few days, while the effect of PFA takes several weeks to months to develop. Silica fume is also known to react quickly.

Regarding the addition of ground limestone, it was observed that the electrical resistivity measured on saturated mortar specimens made with limestone cement was similar to that measured on specimens made with portland cement, as is shown in Figure 1.10, which highlights the fact that unlike pozzolanic additions ground limestone does not improve the impermeability of hydrated cement paste [22].

1.4 Common Cements

According to the European standard EN 197-1 [23], cements can be classified on the basis of composition and standard strength at 28 days.

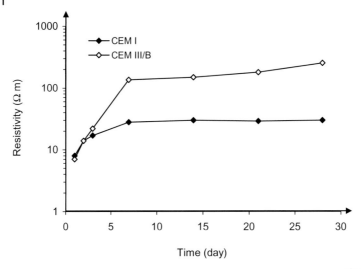

Figure 1.9 Electrical resistivity of mortar (w/c 0.50) made with portland and blast furnace slag cements at early age (based on data from [21]).

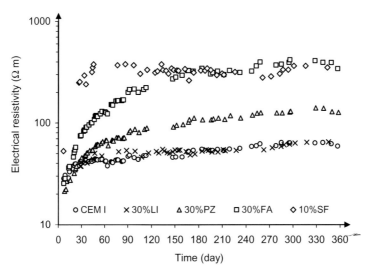

Figure 1.10 Electrical resistivity of water-saturated mortars with different additions as a function of time after casting ($w/b = 0.5$): CEM I = portland cement, 30%LI = 30% ground limestone, 30%PZ = 30% natural pozzolana, 30%FA = 30% coal fly ash, 10%SF = 10% silica fume (modified from [22]).

As far as composition is concerned, five main types of cement can be considered:

- CEM I (portland cement), with at least 95% of clinker (by total binder mass),
- CEM II (portland-composite cements), with addition of up to 35% of another single constituent,
- CEM III (blast furnace cement) with addition of 36–95% blast furnace slag,
- CEM IV (pozzolanic cement) with addition of 11–55% pozzolanic materials,
- CEM V (composite cement) with simultaneous addition of slag and pozzolana (18–49%).

Table 1.3 shows the composition and the designation of the types of cement provided by EN 197-1. For CEM II and CEM V, additions of substances that do not contribute to hydration, such as limestone, is also allowed. In each type of cement, minor additional constituents (such as fillers[3]) may be added up to 5% by mass.

To evaluate the performance of a cement at 28 days of moist curing, that is, when the strength of concrete is normally tested, the standard strength classes have been introduced. These are conventionally defined on the basis of the characteristic compressive strength measured on standardized mortar prisms with a *w/c* ratio equal to 0.5 and a sand/cement ratio of 3, cured for 28 days in moist conditions. For each type of cement, three classes of 28-day strength are potentially available; furthermore, depending on early strength each class is then divided into normal (*N*), high early strength (*R*) or low early strength (*L*, only applicable for CEM III cements), as shown in Table 1.4. The strength classes are essentially a measure of the rate of hydration of the cement: the higher the strength at a given time, the higher the rate of hydration.

Designation of cement indicates the cement type (Table 1.3) and the strength class (Table 1.4). For instance, CEM II/A-S 42.5N indicates portland-slag cement with addition of blast furnace slag in the range of 6–20% and strength class 42.5 with normal early strength.

EN 197-1 also provides other requirements of cement such as setting times and chemical properties. Methods of testing these properties are described in the European standards of the series EN 196.

1.5
Other Types of Cement

Other types of cement than those listed in Table 1.3 are available for special uses. These are for instance: *low heat cements* to be used when low heat of hydration is desired such as in massive structures, *sulfate-resisting cements* to be used to increase the resistance of concrete to sulfate attack, *expansive cements*, *quick setting cements*,

---

3) Fillers are finely ground natural or artificial inorganic materials, which may be added to improve the physical properties of cement such as its workability, or to achieve requirements of mechanical strength.

Table 1.3 Types of cement according to EN 197-1 and their composition (% by mass; values in the table refer to the nucleus of cement, excluding calcium sulfate) [23].

Type of cement	Designation	Clinker	Blast Furnace slag	Silica Fume	Pozzolana Natural	Pozzolana Natural calcined	Fly ash Siliceous	Calcareous	Burned Shale	Limestone	Minor ad-Additional constituents[a]
		K	S	D[b]	P	Q[c]	V	W	T	L[e]	
Portland cement	I	95–100	–	–	–	–	–	–	–	–	0–5
Portland-slag cement	II/A-S	80–94	6–20	–	–	–	–	–	–	–	0–5
	II/B-S	65–79	21–35	–	–	–	–	–	–	–	0–5
Portland-silica fume cement	II/A-D	90–94	–	6–10	–	–	–	–	–	–	0–5
Portland-pozzolana cement	II/A-P	80–94	–	–	6–20	–	–	–	–	–	0–5
	II/B-P	65–79	–	–	21–35	–	–	–	–	–	0–5
	II/A-Q	80–94	–	–	–	6–20	–	–	–	–	0–5
	II/B-Q	65–79	–	–	–	21–35	–	–	–	–	0–5
Portland-fly ash cement	II/A-V	80–94	–	–	–	–	6–20	–	–	–	0–5
	II/B-V	65–79	–	–	–	–	21–35	–	–	–	0–5

1.5 Other Types of Cement

Cement type	Code	Clinker	Blast furnace slag	Pozzolana/Fly ash	Silica fume	Pozzolana	Fly ash	Burnt shale	Limestone	Minor additional constituents
	II/B-W	65–79	–	–	–	–	–	21–35	–	–	0–5
Portland-burned shale cement	II/A-T	80–94	–	–	–	–	–	–	6–20	–	0–5
	II/B-T	65–79	–	–	–	–	–	–	21–35	–	0–5
Portland-limestone cement	II/A-L	80–94	–	–	–	–	–	–	–	6–20	0–5
	II/B-L	65–79	–	–	–	–	–	–	–	21–35	0–5
Portland-composite cement	II/A-M	80–88	<-------	----12	–20$^{d)}$--	–	<-------	-------	-------	------->	0–5
	II/B-M	65–79	<-------	----21	–35$^{d)}$--	–	<-------	-------	-------	------->	0–5
III Blast furnace cement	III/A	35–64	36–65	–	–	–	–	–	–	–	0–5
	III/B	20–34	66–80	–	–	–	–	–	–	–	0–5
	III/C	5–19	81–95	–	–	–	–	–	–	–	0–5
IV Pozzolanic cement	IV/A	65–89	–	11–35	–	<-------	------->	–	–	–	0–5
	IV/B	45–64	–	36–55	–	<-------	------->	–	–	–	0–5
V Composite cement	V/A	40–64	18–30	18–30	–	<-------	------->	–	–	–	0–5
	V/B	20–38	31–49	31–49	–	<-------	------->	–	–	–	0–5

a) Minor additional constituents may be fillers or one or more of the main constituents, if these have not been included as main constituents.
b) The proportion of silica fume is limited to 10%.
c) The proportion of nonferrous blast furnace slag is limited to 15%.
d) The proportion of filler is limited to 5%.
e) L indicates limestone with TOC (total organic carbon) not higher than 0.50%. LL indicates limestone with TOC not higher than 0.20%.

Table 1.4 Strength classes of cement according to EN 197-1 [23].

Class	Compressive strength (MPa)			
	Early strength		Standard strength	
	2 days	7 days	28 days	
32.5 L[a]	–	≥12	≥32.5	≤52.5
32.5 N	–	≥16	≥32.5	≤52.5
32.5 R	≥10	–	≥32.5	≤52.5
42.5 L[a]	–	≥16	≥42.5	≤62.5
42.5 N	≥10	–	≥42.5	≤62.5
42.5 R	≥20	–	≥42.5	≤62.5
52.5 L[a]	≥10	–	≥52.5	–
52.5 N	≥20	–	≥52.5	–
52.5 R	≥30	–	≥52.5	–

a) Strength class only defined for CEM III cements.

white or *colored cements*, etc. [2, 4]. New types of cement are under study in order to reduce the environmental impact and improve sustainability; care should be used in the selection of new types of cements, since the long-term durability performance is unknown.

High Alumina Cement (HAC) Mention should be made of high alumina cement (HAC). In fact, although nowadays it is generally not used for structural purposes, in the past its use has caused problems of durability (as well as from the structural point of view), especially in countries like Spain and the United Kingdom where it was extensively used. It is obtained by melting a suitable mixture of limestone and bauxite (mainly consisting of alumina) at about 1600 °C. Its primary constituent is $CaO \cdot Al_2O_3$ (CA). The hydration reaction (CA + 10H = CAH_{10}) mainly produces CAH_{10}, which is unstable. In humid environments whose temperature exceeds 25 °C a process of conversion takes place: hydration products transform into another compound (C_3AH_6). This transformation induces a considerable increase in porosity, and thus a drastic loss of strength and a decrease in the resistance to aggressive agents, especially to carbonation. The degree to which the porosity will increase and related consequences occur, depends on the w/c ratio used, and is much greater when this ratio exceeds 0.4.

In some countries, HAC has more recently been used to obtain acid-resistant (sewage) pipe or lining. During production, the w/c is kept below 0.4 and thermal treatment is applied during the manufacturing process, deliberately forcing the conversion to the stable compound (C_3AH_6). If the strength and density of the converted material are adequate, the durability is good. The stable product has a high resistance against attack by acids. HAC concrete also has good properties for high-temperature applications (e.g., furnaces).

Calcium Sulfoaluminate Cements (CSA) The use of raw materials such as limestone, bauxite, and calcium sulfate leads to 20–30% reduced CO_2 emission compared to equivalent ordinary portland cement and to a lower firing temperature of 1250 °C. Further, calcium sulfoaluminate (CSA) cement is easier to grind than portland cement clinker. The setting time of CSA is between 30 min and 4 h, most of the hydration heat evolution occurs between 2 and 24 h of hydration. Due to a denser pore structure concrete with CSA cements has a high resistance to freeze–thaw and against chemical attack by seawater, sulfates and magnesium and ammonium salts. However, the durability of reinforced concrete with CSA cement has to be considered lower: the pH of the pore solution is one unit lower compared to ordinary portland cement and the carbonation rate has been found to be higher [24].

References

1 Taylor, H.F.W. (1990) *Cement Chemistry*, Academic Press Inc., London.
2 Neville, A.M. (1995) *Properties of Concrete*, 4th edn, Longman Group Limited, Harlow.
3 Metha, P.K. and Monteiro, P.J.M. (2006) *Concrete: Microstructure, Properties, and Materials*, 3rd edn, McGrawHill.
4 Bensted, J. and Barnes, P. (eds) (2002) *Structure and Performance of Cements*, Spon Press, London.
5 Collepardi, M. (2006) *The New Concrete*, Tintoretto, Villorba.
6 Diamond, S. (2007) Physical and chemical characteristics of cement composites, in *Durability of Concrete and Cement Composites* (eds C.L. Page and M.M. Page), Woodhead Publishing Limited, pp. 10–44.
7 Neville, A.M. and Brooks, J.J. (1990) *Concrete Technology*, Longman Scientific and Technical, Harlow.
8 Powers, T.C. (1958) Structure and physical properties of hardened portland cement paste. *Journal of American Ceramic Society*, **41**, 1–6.
9 Stauffer, D. (1985) *Introduction to Percolation Theory*, Taylor & Francis, London.
10 Elsener, B., Flückiger, D., and Böhni, H. (2000) A percolation model for water sorption in porous cementitious materials, in *Materials for Buildings and Structures*, Euromat, vol. 6 (ed. F.H. Wittmann), Wiley-VCH Verlag GmbH, Weinheim, pp. 163–169.
11 Massazza, F. (1998) Pozzolana and pozzolanic cements, in *Lea's Chemistry of Cement and Concrete*, 4th edn (ed. P. Hewlet), Butterworth-Heinemann, Oxford, pp. 471–635.
12 ACI 232.2R-03 (2004) Use of fly ash in concrete, American Concrete Institute.
13 ACI 234R-06 (2006) Guide for the use of silica fume in concrete, American Concrete Institute.
14 Hawkins, P., Tennis, P., and Detwiler, R. (2003) *The Use of Limestone in Portland Cement: A State-of-the-Art Review*, Portland Cement Association, Skokie.
15 Mehta, P.K. (2001) Reducing the environmental impact of concrete. *Concrete International*, **23**(10), 61–66.
16 Damtoft, J.S., Lukasik, J., Herfort, D., Sorrentino, D., and Gartner, E.M. (2008) Sustainable development and climate change initiatives. *Cement and Concrete Research*, **38**, 115–127.
17 Siddique, R. (2008) *Waste Materials and By-Products in Concrete*, Springer-Verlag, Berlin.
18 Bertolini, L., Carsana, M., Cassago, D., Collepardi, M., and Quadrio Curzio, A. (2004) MSWI ashes as mineral additions in concrete. *Cement and Concrete Research*, **34** (10), 1899–1906.
19 Schiessl, P. (ed.) (1988) *Corrosion in Concrete*, RILEM Technical

Committee 60-CSC, Chapman and Hall, London.
20 Polder, R.B. (2000) Simulated de-icing salt exposure of blended cement concrete–chloride penetration. Proc. 2nd International RILEM Workshop Testing and modelling the chloride ingress into concrete (eds C. Andrade and J. Kropp), RILEM PRO 19, 189–202.
21 Caballero, J., Polder, R.B., Leegwater, G., and Fraaij, A. (2010) Chloride penetration into cementitious mortar at early age. Proc. 2nd International Conference on Service Life Design (eds Delft, K. van Breugel, G. Ye, and Y. Yuan), 65–72.
22 Bertolini, L., Carsana, M., Frassoni, M., and Gelli, M. (2011) Pozzolanic additions for durability of concrete structures. *Proceedings of ICE – Construction Materials*, **164** (6), 283–291.
23 EN 197-1 (2011) Cement – Part 1: Composition, Specifications and Conformity Criteria for Common Cements, European Committee for Standardization.
24 Juenger, M.C.G., Winnefeld, F., Provis, J.L., and Ideker, J.H. (2011) Advances in alternative cementitious binders. *Cement and Concrete Research*, **41** (12), 1232–1243.

2
Transport Processes in Concrete

Concrete can be penetrated, through its pores, by gases (e.g., nitrogen, oxygen, and carbon dioxide present in the atmosphere) and liquid substances (e.g., water, in which various ions are dissolved). The term *permeability* indicates in general the property of concrete to allow the ingress of these substances. The permeability of concrete is not only important for water-retaining structures and elements (pipes, canals or tanks), but is a decisive factor in the durability of reinforced concrete. Phenomena that lead to degradation of reinforced concrete depend on the processes that allow transport of water, carbon dioxide, chloride ions, oxygen, sulfate ions and electrical current within the concrete.

The movement of fluids and ions through concrete can take place according to four basic mechanisms: *capillary suction*, due to capillary action inside capillaries of cement paste, *permeation*,[1] due to pressure gradients, *diffusion*, due to concentration gradients, and *migration*, due to electrical potential gradients [1–3]. The kinetics of transport depend on the mechanism, on the properties of the concrete (e.g., its porosity and the presence of cracks), on the binding by the hydrated cement paste, of the substances being transported, as well as on the environmental conditions existing at the surface of the concrete (microclimate) and their variations in time (Figure 2.1).

In this chapter the mechanisms of transport operating in concrete and the parameters that define them are discussed. Since the liquid present in the pores has an important influence both on the transport of the various aggressive species and on the degradation phenomena that can take place in concrete, it is worthwhile looking first at the composition of the pore solution and the physical forms of water in concrete as a function of environmental conditions.

---

1) The term *permeation* is preferred to indicate the mechanism of transport by the action of a pressure difference, in order to avoid confusion with the word *permeability*, often used (even in this text, though not always with complete accuracy) to indicate in general the properties of concrete in relation to all transport mechanisms. Instead, since misunderstandings are not possible, the term coefficient of permeability will be used.

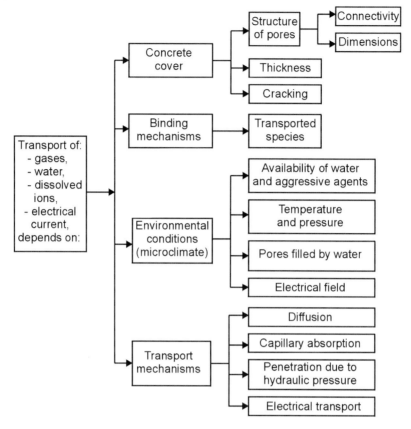

Figure 2.1 Principal factors involved in the transport processes in concrete, essential in the phenomenon of corrosion.

2.1
Composition of Pore Solution and Water Content

A certain amount of water is contained in the pores of the hydrated cement paste. The actual quantity of water in the pores of concrete, that is, the moisture content, depends on the humidity of the surrounding environment. Several ions produced by the hydration of cement are dissolved in the pore liquid, so that in reality it is a quite concentrated aqueous solution.

2.1.1
Composition of Pore Solution

The chemical composition of the solution in pores of hydrated cement paste depends on the type of binder used but also on the exposure conditions, for example, it changes due to carbonation or penetration of salts.

Hydration of cement produces a solution that consists mainly of NaOH and KOH. Depending on the composition of the cement, the pH of the pore solution may be between 13 and 14. When concrete undergoes carbonation (Chapter 5) the pH of the pore solution drops to values approaching neutrality (pH ≈ 9) as a consequence of a drastic reduction in the concentration of hydroxyl ions. Penetration of salts from the environment may also lead to a remarkable change in the composition of the pore solution.

Table 2.1 shows the ionic concentrations measured by different researchers in the pore solution of cement pastes, mortars and concretes, obtained both with portland cements (OPC) and blended cements [4–16]. Measurements were carried out by chemical analysis of the liquid extracted under pressure, using specific pore-extraction devices.

In noncarbonated and chloride-free concrete, the concentration of hydroxyl ions (OH^-) varies from 0.1 M to 0.9 M, due to the presence of both NaOH and KOH (the latter is predominant, especially in portland cement). Other ions, for example, Ca^{2+} and SO_4^{2-}, are present only in very low concentrations. Addition of blast furnace slag or fly ash to portland cement results in a moderate reduction of ionic concentration, and thus in pH. From hydroxyl ion concentrations in Table 2.1, values of pH 13.4–13.9 can be calculated for portland cement, and pH 13.0–13.5 for blended cements. Addition of silica fume in higher percentages may lead to a decrease in the pH to values lower than 13 [4, 10].

The penetration or addition of chloride bearing salts changes the chemical composition of pore solution, depending also on the type of salt (i.e., sodium chloride or calcium chloride). This is due to chemical or physical binding of chloride and hydroxyl ions; for instance, sodium chloride increases the OH^- concentration (and thus the pH), while calcium chloride decreases the pH [9].

Few results are available for carbonated concrete; in tests on portland cement pastes [7, 16] a very diluted solution was found after carbonation, with only small amounts of the alkali ions Na^+ and K^+ (Table 2.1).

2.1.2
Water in Concrete

Water may be present in the hydrated cement paste in many forms that may be classified on the basis of the degree of difficulty with which it can be removed [17].

Capillary Water The water contained in capillary pores accounts for the greatest part of water in concrete (and the most important part with regard to corrosion). The aqueous solution contained in pores larger than about 50 nm in diameter can be considered free of bonding forces with the solid surface (similarly as water present in larger voids due to entrapped or entrained air). As the relative humidity of the environment decreases below 100%, this water will evaporate without causing any significant shrinkage in the cement paste. This "free" solution has transport properties analogous to those of a bulk solution. For example, the coefficient of chloride diffusion, or the mobility of various ions, and thus the electrical

Table 2.1 Ionic concentration (in mmol/l) measured in the pore solution extracted from cement pastes, mortars, and concrete made with ordinary portland cement (OPC) and with additions of blast furnace slag (GGBS), fly ash (PFA) and silica fume (SF).

Cement	Chloride added (by mass of cement)	Water/binder	Age (days)	Sample	Source	[OH$^-$]	[Na$^+$]	[K$^+$]	[Ca^{++}]	[Cl$^-$]	[SO$_4^=$]
OPC	No	0.45	28	Paste	[10]	470	130	380	1	n.a.	n.a.
OPC	No	0.5	28	Mortar	[6]	391	90	288	<1	3	<0.3
OPC	No	0.5	28	Paste	[4]	834	271	629	1	n.a.	31
OPC	No	0.5	192	Mortar	[14]	251	38	241	<1	n.a.	8
OPC	No	0.5	–	Paste	[7]	288	85	228	n.a.	1	n.a.
OPC[a]	No	0.5	84	Paste	[5]	589	n.a.	n.a.	n.a.	2	n.a.
OPC[b]	No	0.5	84	Paste	[5]	479	n.a.	n.a.	n.a.	3	n.a.
80%GGBS	No	0.5	28	Mortar	[6]	170	61	66	<1	8	15
70%GGBS	No	<0.55	8 years	Concrete	[13]	95	89	42	n.a.	5	8
65%GGBS	No	0.5	84	Paste	[5]	355	n.a.	n.a.	n.a.	5	n.a.
25%PFA	No	0.5	28	Mortar	[6]	331	75	259	<1	<1	18
30%PFA	No	0.5	84	Paste	[5]	339	n.a.	n.a.	n.a.	2	n.a.
10%SF	No	0.5	28	Paste	[4]	266	110	209	1	n.a.	32
20%SF	No	0.5	28	Paste	[4]	91	59	109	1	n.a.	33
20%SF	No	0.45	28	Paste	[10]	98	36	53	1	n.a.	n.a.
30%SF	No	0.5	28	Paste	[4]	26	35	53	2	n.a.	35

2.1 Composition of Pore Solution and Water Content

Binder	Chloride	w/b	Age (days)	Type	Ref.						
OPC[a]	0.4% (NaCl)	0.5	84	Paste	[5]	741	n.a.	n.a.	n.a.	84	n.a.
OPC[b]	0.4% (NaCl)	0.5	84	Paste	[5]	661	n.a.	n.a.	n.a.	42	n.a.
OPC	0.4% ($CaCl_2$)	0.5	28	Mortar	[6]	62	90	161	<1	104	5
OPC	0.4% (NaCl)	0.5	35	Paste	[4]	835	546	630	2	146	41
OPC	1% (NaCl)	0.5	–	Paste	[7]	458	580	208	n.a.	227	n.a.
65%GGBS	0.4% (NaCl)	0.5	84	Paste	[5]	457	n.a.	n.a.	n.a.	28	n.a.
80%GGBS	0.4% ($CaCl_2$)	0.5	28	Mortar	[6]	138	41	299	<1	67	21
25%PFA	0.4% ($CaCl_2$)	0.5	28	Mortar	[6]	257	157	307	<1	85	28
30%PFA	1% (NaCl)	0.5	28	Mortar	[5]	457	n.a.	n.a.	n.a.	39	n.a.
10%SF	0.4% (NaCl)	0.5	35	Paste	[4]	192	264	256	1	216	37
10%SF	1% (NaCl)	0.5	35	Paste	[4]	158	614	345	2	615	50
OPC	No	0.5	–	Carbonated paste[c]	[7]	3×10^{-4}–2.8	10–23	4–20	31	8–16	n.a.
OPC	No	0.6	–	Carbonated paste[d]	[16]	3×10^{-4}–0.1	0.3–9.9	0.6–26.4	7.1–78.5	7.5–23.4	1.2–24.1
OPC	1% (NaCl)	0.5	–	Carbonated paste[c]	[7]	6×10^{-5}–0.3	330–369	20–31	21	533–651	n.a.

a) Low C_3A content (7.7%).
b) High C_3A content (14.3%).
c) Interval of values from pastes exposed to environment with different CO_2 concentrations.
d) Interval of values from pastes exposed to environment with different CO_2 concentrations and in the presence of different saturated salt solutions.
n.a. = concentration not available.

conductivity, can be considered analogous to those of an electrolytic solution of the same composition.

Water held by capillary tension in pores of diameters smaller than 50 nm will evaporate at lower values of the relative humidity as the diameter of the pores decreases. Indicatively, values of relative humidity from 95% to 60% are required when the diameter of capillary pores decreases from 50 nm to 5 nm [18]. In this case, evaporation can produce significant shrinkage of the cement paste. In addition, the mobility of ions (thus the electrical conductivity of the solution in these micropores) is affected by chemical and physical interactions between the liquid and the solid and is therefore lower than that of a solution of the same composition.

Adsorbed Water Even when water has evaporated from the capillary pores, some water will still remain adsorbed to the inner surface in the form of a very thin layer of adsorbed water. This water can be removed if the external humidity falls below 30%; it contributes little to transport phenomena, thus it is negligible with regard to corrosion of reinforcement. Its removal, however, causes shrinkage of the cement paste and influences creep behavior.

Interlayer Water Concrete loses the water retained between the C–S–H layers if the external humidity falls below 11%. This water does not matter as far as corrosion of reinforcement is concerned, since gel pores are too small to allow transport processes at any appreciable rate. It does influence shrinkage and creep.

Chemically Combined Water Chemically combined water (i.e., water that is an integral part of C–S–H or other hydration products) is not lost on drying, and it can only be released when the hydrates decompose on heating (>1000 °C). It does not contribute to any transport phenomena.

2.1.3
Water Content and Transport Processes

For concrete exposed to the atmosphere under equilibrium conditions and in the absence of wetting, the water content can be related to the relative humidity of the environment, as depicted in Figure 2.2. With regard to the capillary pores, water is first adsorbed on their surface and then, as the relative humidity increases, water condensates and fills up the pores, starting with the smallest and moving to those of larger dimensions.

In Figure 2.3 the capillary pores are graphically represented by spherical cavities connected by narrow capillary cylinders whose dimensions are considered to be statistically distributed (only accessible pores are represented, i.e., pores connected with each other and with the external environment). According to this diagram, inside concrete that is exposed to the atmosphere of a certain relative humidity, pores whose diameters are below a given value turn out to be filled with water, while those with diameters above this value are filled with air.

The presence of water-filled pores that are interconnected with each other has a marked influence on the kinetics of transport processes. It hinders those pro-

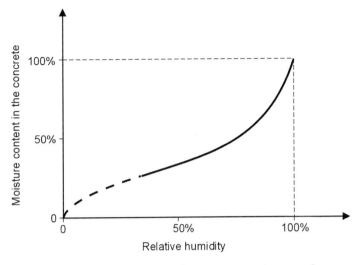

Figure 2.2 Schematic representation of water content in the pores of concrete as a function of the relative humidity of the environment, in conditions of equilibrium [1].

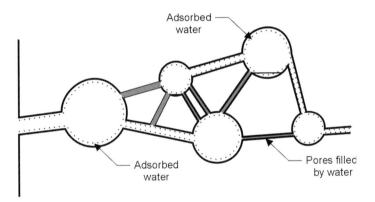

Figure 2.3 Representation of water present in capillary pores in concrete in equilibrium with a nonsaturated atmosphere [1].

cesses that take place easily in the gaseous phase, such as oxygen and carbon dioxide diffusion. On the other hand, it facilitates those processes that occur in aqueous solution, like the diffusion of chlorides, or ions in general.

2.2
Diffusion

The penetration of aggressive species within concrete often occurs by diffusion, that is, by the effect of a concentration gradient. O_2, CO_2, Cl^- or $SO_4^=$ move through pores from the surface, where they are present in higher concentrations, to internal zones where their concentration is lower. Gases diffuse much more rapidly through open pores than through water-saturated ones (diffusion of gases in water is 4–5 orders of magnitude slower than in air). On the other hand, chloride and sulfate ions diffuse only when dissolved in pore water; the diffusion is more effective in saturated than in partially saturated pores.

2.2.1
Stationary Diffusion

Under conditions of stationary (unidirectional and constant) mass transfer, Fick's first law describes the phenomenon of diffusion:

$$F = -D\frac{dC}{dx} \tag{2.1}$$

where: F is the flux (kg/m² s) and C is the concentration of the diffusing species (kg/m³) present at distance x from the surface. D is the diffusion coefficient, expressed in m²/s, which depends on the diffusing species, on the characteristics of the concrete and on the environmental conditions. This coefficient can change as a function of position and time, following variations in the pore structure (i.e., due to hydration of the cement paste), or of the external humidity (thus the degree of saturation of pores) or the temperature.

Laboratory tests of stationary diffusion of chloride ions in cementitious systems have been conducted since the 1970s for purposes of research and classification of different concrete compositions [5].

A thin sample of concrete, mortar or cement paste normally less than 10 mm thick, is placed inside suitable cells, separating chambers filled with two solutions, one concentrated (*upstream*) and the other diluted (*downstream*) with respect to the ion to be investigated. A sketch of the setup is given in Figure 2.4.

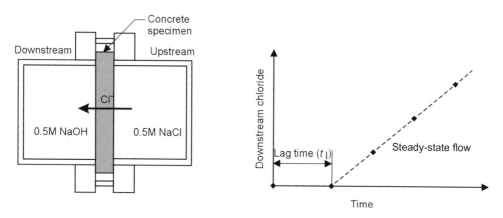

Figure 2.4 Setup for steady-state diffusion cell and example of typical results.

After a certain transition period (lag time, t_l), the flux of the species that diffuses through the sample becomes constant. If C_1 and C_2, respectively, stand for the concentrations in the upstream and downstream chambers ($C_2 \ll C_1$) measured at time t from the beginning of the test, L the thickness (m) of the sample, A the cross section (m²) and V the volume of the downstream chamber (m³), Fick's first law can be written as follows:

$$\frac{C_2 \cdot V}{A \cdot (t - t_1)} = \frac{D_{ss} \cdot C_1}{L} \tag{2.2}$$

From this relationship the steady-state diffusion coefficient D_{ss} can be calculated. This kind of test can be used to compare the characteristics of different concretes with regard to chloride diffusion. A practical complication may be that concrete with low porosity may require a very long time to reach a constant flux. Values found for D_{ss} may vary from 10^{-14} to 10^{-11} m²/s for concrete with various binders and w/c [19, 20].

2.2.2
Nonstationary Diffusion

As diffusion rarely reaches stationary conditions in concrete structures, the flux depends on time t and is governed by Fick's second law:

$$\frac{\partial C}{\partial t} = D \frac{\partial^2 C}{\partial x^2} \tag{2.3}$$

This equation is usually integrated under the assumptions that the concentration of the diffusing ion, measured on the surface of the concrete, is constant in time and is equal to C_s ($C = C_s$ for $x = 0$ and for any t), that the coefficient of diffusion D does not vary in time, that the concrete is homogeneous, so that D does not vary through the thickness of the concrete, and that it does not initially contain chlorides ($C = 0$ for $x > 0$ and $t = 0$). The solution thus obtained is:

$$\frac{C(x,t)}{C_s} = 1 - \mathrm{erf}\left(\frac{x}{2\sqrt{Dt}}\right) \tag{2.4}$$

$$\text{where: } \mathrm{erf}(z) = \frac{2}{\sqrt{\pi}} \int_0^z e^{-t^2} dt \tag{2.5}$$

is the error function.[2]

Equation (2.4) is often utilized to describe mathematically the experimental profiles of chloride concentration found in nonsteady-state laboratory tests or from structures exposed in the field (Chapter 6).

2) The values of the error function have been tabulated. Alternatively, they can be calculated by normal standard distribution (N, also tabulated), considering that:

$$\mathrm{erf}(z) = 2N(z\sqrt{2}) - 1 \quad N(z\sqrt{2}) = \frac{1}{\sqrt{2\pi}} \int_{-\infty}^{z\sqrt{2}} e^{\frac{t^2}{2}} dt$$

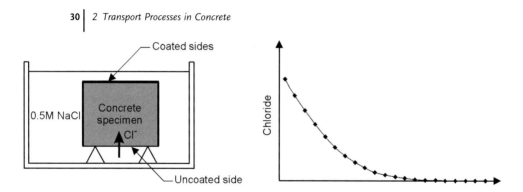

Figure 2.5 Setup for nonsteady-state diffusion and example of typical results.

In laboratory tests, cubic or cylindrical specimens are coated on all sides but one and submerged in a volume of chloride solution as shown in Figure 2.5. After some time (from several weeks to over a year), samples from incremental depths are taken by sawing slices or grinding off thin layers ("profile grinding") and analyzed for chloride [21]. The obtained chloride profile is analyzed by optimizing the fit of Eq. (2.4) to the experimental data with C_s and D as parameters, for instance using the least squares method. The result of such a procedure is the best fitting pair of the surface content and the diffusion coefficient. Values found for the nonsteady-state chloride diffusion coefficient may vary from 10^{-13} to $10^{-11}\,\mathrm{m^2/s}$ for concrete with various binders and w/c [2, 21]. A nonsteady-state test has been standardized in the Nordic countries [22].

Another application of Fick's second law of diffusion is the analysis of chloride penetration profiles in cores taken from structures that were actually exposed to chloride penetration from the outer surface. A profile of the chloride content is determined experimentally as a function of depth (Section 16.3.2). The values of C_s and D are then determined mathematically, by fitting Eq. (2.4) to the experimental data [20]. An example is given in Figure 2.6. The diffusion coefficient D (which may typically vary from 10^{-11} to $10^{-13}\,\mathrm{m^2/s}$ as a function of the concrete's characteristics) in this case is an effective diffusion coefficient D_{eff} as the transport properties in a structure may vary between pure diffusion in water-saturated concrete to capillary suction in the case of splash water. The effective diffusion coefficient D_{eff} can be used as the main parameter that describes the rate of chloride penetration. Together with the fitted surface content and using Eq. (2.4) the chloride penetration can be extrapolated to longer times. The limits of this approach will be discussed in Chapter 6.

2.2.3
Diffusion and Binding

Species that diffuse into concrete can bind to a certain degree with components of the cement matrix, for example, chlorides bind with aluminate phases or are

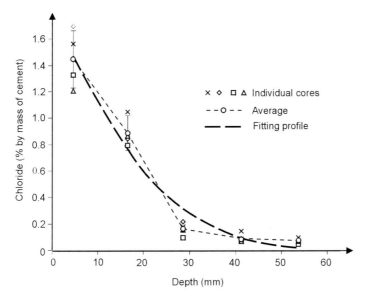

Figure 2.6 Chloride profiles of a quay wall on the Dutch North Sea coast after 8 years of splash zone exposure, blast furnace slag cement; average of measured profiles (thin line), individual results of four cores (only symbols) and best fitting profile according to Eq. (2.4), described by $C_s = 1.77\%$, $D_{app} = 0.89 \times 10^{-12}\,m^2/s$.

adsorbed on C–S–H; carbon dioxide reacts with alkaline components, in particular $Ca(OH)_2$. The gradual consumption of these compounds modifies the conditions of diffusion, which can no longer simply be described by Fick's second law but require a corrective term.

In particular for chloride penetration, often the total concentration of the diffusing species is taken into consideration and the effects of chemical reactions in concrete are disregarded. Indeed, it is difficult to estimate the corrective term. In fact, the binding capacity of a given cement paste is a function of various parameters, such as its porosity, the local concentration of a given substance and the temperature. It will also depend on the chemical composition of the concrete and thus on its variations (e.g., the binding capacity of chlorides is considerably reduced in carbonated concrete) [2]. The binding of diffusing species is important in experimentally determining the diffusion coefficient. In fact, as long as the binding capacity of the concrete has not been exhausted, the flux of material will be affected. Therefore, when a corrective term is not applied in evaluating the concentration of a diffusing compound, the diffusion coefficient derived in that way is an "apparent" one that is dependent on time. Anyway, it has been shown [23] that, even in the presence of binding, chlorides move into concrete as if diffusion determines the penetration rate. As a first approach, binding therefore can be neglected in the calculation and the diffusion coefficient is termed *apparent*. As chloride penetration in concrete structures is essentially a nonsteady-state process, tests should be carried out either under nonsteady-state conditions, or should take into account the effect of binding.

2.3
Capillary Suction

When water comes into contact with a porous material such as concrete, it is absorbed rapidly due to the underpressure in the pores caused by what is called capillary action. This action depends on the surface tension, viscosity, and density of the liquid, on the angle of contact between the liquid and the pore walls and on the radius of the pore. In concrete, the contact angle is small due to the presence of molecular attraction between the liquid and the substrate (i.e., between water and cement paste). Under these conditions, a drop will spread on a flat surface, while the meniscus of a capillary pore will rise above the level of the surrounding liquid and be concave towards the dry side; how this aspect can be changed by hydrophobic treatment will be seen in Chapter 14.

In theory, capillary action is stronger as the pore dimensions decrease. On the other hand, the smaller the pores become, the slower the transport will be due to increasing friction. In the usual range of pore dimensions, a more porous concrete absorbs more water and faster than a dense concrete. This general rule has become the basis for several test methods.

A sorptivity test is normally carried out to measure capillary suction in concrete. The bottom surface of a previously dried sample is placed in contact with water at atmospheric pressure. In general, capillary absorption is measured as mass (or volume) of liquid absorbed per unit of surface (i, g/m^2 for mass or m^3/m^2 for volume) in time t (a test normally lasts from four to 24 h). The development of i as a function of time is of the type:

$$i = S \cdot \sqrt{t} \tag{2.6}$$

This relationship, empirically derived from the observation of experimental data, is commonly used to define the parameter S. This is correct only for very porous materials or in the early stage of capillary action. In fact, in concrete with a low w/c ratio, the square-root law (Eq. 2.6) is changed to a power law with exponent lower than 0.5. This can be explained by the percolation theory and has been demonstrated experimentally [24].

The constant S is expressed in $g/(m^2 s^{0.5})$ (if mass change is determined) or $m/s^{0.5}$ (if absorbed volume is determined) and is adopted as a representative parameter of the characteristics of concrete with regard to capillary absorption. Typical values of S vary from $5 g/(m^2 s^{0.5})$ for normal-strength concrete (w/c 0.50, compressive strength 43 MPa) [25, 26], to $1.5 g/(m^2 s^{0.5})$ (w/c 0.40, compressive strength 67 MPa) and $0.7 g/(m^2 s^{0.5})$ for high-strength concrete (w/c 0.30, compressive strength 90 MPa). The actual values obtained with this type of test, however, depend on various factors; the main one is the degree of drying to which the samples have been previously subjected. For example, oven desiccation at 105 °C may modify the water content in hydration products and cause microcracks that favor capillary absorption. Other methods of drying give results that more closely reflect the real behavior of concrete, but that require much longer times. A procedure for preconditioning designed to resemble atmospheric exposure of concrete in practice as much as possible involves four weeks of drying in a climate room with air of 20 °C and 65% R.H. (relative atmospheric humidity) before capillary absorption tests on samples with dimensions between 50 and 100 mm [25, 26]. In EN 13057 a procedure for capillary absorption tests on repair materials is suggested, involving oven drying at 40 °C [27]. This method of drying is probably the best compromise between quickly establishing a well-defined and low-moisture condition and avoiding damage to the microstructure. Another practical aspect introduced is that if the absorption does not give a straight line on a i versus log(time) plot, one should calculate S from the value measured at 24 h.

A procedure for testing water absorption of concrete is given by BS 1881 part 122 [28]. It involves placing specimens (cubes, cores) for 72 h in a drying oven in which the temperature is controlled at 105 °C. After cooling down for 24 h in the air, specimens are weighed and then immersed for 30 min in a tank containing water with a temperature of 20 °C. Finally, the specimens are weighed after removing all free water from the surface. The water absorption is calculated as the increase in mass resulting from immersion as a percentage of the mass of the dry specimen; as an additional result, the wet density is calculated. Dense concrete has a water absorption determined according to this method of the order of 1 to 2% by mass. As discussed above, this method involves drying at high temperature, which may damage the microstructure and thereby render the result nonrepresentative for concrete under field conditions.

Various instruments have been developed for testing water absorption onsite, for example, the AutoClam [29]. It involves placing a circumferentially sealed cylinder with piston onto the concrete surface and allowing it to absorb water during for example, 15 min, while recording the volume absorbed by following the movement of the piston; a water absorption index is calculated. This instrument also allows measuring air permeability (applying a pressure of 500 mbar) and water permeability (applying a water pressure of 500 mbar) and is used both onsite and in the laboratory.

2.4 Permeation

When a liquid (assumed incompressible and entirely viscous) penetrates concrete that is previously saturated by the liquid by a pressure difference, the flow through the pores is defined by Darcy's law:

$$\frac{dq}{dt} = \frac{K \cdot \Delta P \cdot A}{L \cdot \mu} \tag{2.7}$$

where: dq/dt is the flow (m³/s), μ is the viscosity of the fluid (N s/m²), K is the intrinsic permeability of concrete (m²), ΔP is the pressure head applied (Pa), A is the surface of the cross section (m²), and L is the thickness (m) of the specimen.

2.4.1
Water Permeability Coefficient

Although the permeation tests may be carried out with any liquid, in general the coefficient of permeability is measured with water. One side of a concrete sample is placed in contact with water (with pressure up to 10 bar) and either the depth of penetration by water in a given time, the time necessary for water to penetrate the entire thickness of the sample or the flow through the sample are measured.

If the penetration involves the entire thickness of the sample, so that it is possible to measure the flow, Eq. (2.7) is written:

$$\frac{dq}{dt} = \frac{k \cdot H \cdot A}{L} \tag{2.8}$$

where: dq/dt is the flow (m³/s), H (m) represents the height of the column of water pressure differential across the sample ($\Delta P = H \cdot \delta \cdot g$, with δ = water density and g = acceleration of gravity). In this case the coefficient of permeability k is measured in m/s; it is connected to the intrinsic permeability coefficient by the equation $k = K \cdot \delta \cdot g / \mu$ and thus depends on the density and on the viscosity of the liquid (for water $K \cong 10^{-7} \cdot k$). The permeability of the cement paste depends on the capillary porosity, and the permeability coefficient decreases as the w/c ratio decreases and as hydration proceeds (as already observed in Figure 1.6). With a w/c ratio of 0.75, the permeability coefficient of water in concrete is very high, usually 10^{-10} m/s upon complete curing, while with a w/c ratio of 0.45 it can go below 10^{-12}–10^{-11} m/s.

For concrete of low porosity, high pressures may be required to obtain flow through the sample within reasonable time. In these cases, a permeability coefficient may be estimated (k_v, m/s) by measuring the average depth of the wet front x_p in time t (s), using Valenta's equation: $k_v = x_p^2 \cdot v_t / (2 \cdot H \cdot t)$ where H is the pressure head of water (m) and v_t is the volume of spaces filled during the test (m³/m³).

According to the European standard EN 12390-8 [30], water permeability should be measured on concrete specimens of specified geometry, under the application of a water pressure of 500 kPa for 72 h on a face of the specimen; the specimen is then split perpendicularly to the face on which the pressure was applied and the maximum depth of water penetration is measured. Another example of a practical laboratory test of water permeability is given by German standard DIN 1048 [31] that describes a test in which pressures of 1, 3, and 7 bar are applied consecutively for 48, 24, and 24 h, respectively; at the end the sample is split to measure the average depth to which the water front has advanced. Concrete is considered "impermeable" if this average depth does not exceed 30 mm.

It should be pointed out that this type of test, useful in evaluating suitability of concrete for use in hydraulic structures (canals, tanks, aqueducts, etc.), can be of no significant help in determining the durability of concrete in aggressive environments. It has been widely demonstrated that permeation tests do not necessarily measure the durability of a particular concrete, in that they are carried out with water under pressure, that is, they introduce the aggressive agent under conditions different from those found in practice (usually capillary absorption or diffusion).

2.4.2
Gas Permeability Coefficient

The gas permeability coefficient (usually permeability to air) is determined by measuring the flow of gas under pressure through a sample. In this case the fluid is compressible and the flow can be expressed by the equation:

$$\frac{dq}{dt} = \frac{k_{gas} \cdot A \cdot (P_1^2 - P_2^2)}{L \cdot \mu \cdot 2P_2} \tag{2.9}$$

where: dq/dt is the flow, P_1 is the upstream pressure, and P_2 the downstream pressure. Since the gas is compressible, it is important to measure the flow at the pressure P_2.

The gas permeability coefficient (k_{gas}, m^2) depends on the pressure chosen for the test [3]. For low pressures (up to 10 bar), the gas permeability coefficients for concrete are normally between 5×10^{-16} and 10^{-18} m^2. For concrete with low intrinsic permeability, the increase in pressure from 1 to 10 bar may lower the permeability coefficient by one order of magnitude. Consequently, it is essential to indicate the test pressure and to compare only those data that have been normalized with respect to a single pressure. The humidity contained in the concrete will also influence the permeability coefficient.

A test for determining oxygen permeability was described by a RILEM committee under the name *Cembureau* method [32]. It involves applying oxygen gas under pressures from 1.5 to 3.5 bar to concrete cylinders, either cast or cored, of 150 mm diameter and 50 mm thickness, and measuring the flow of oxygen permeating. Preconditioning may be carried out by either storage in air of 20 °C and 65% R.H. for 28 days, or 7 days drying at 105 °C. It is made clear that the two methods for preconditioning produce different results.

2.5
Migration

2.5.1
Ion Transport in Solution

The transport of ions in solution under an electric field is called *migration*. The velocity of ion movement is proportional to the strength of the electric field and the charge, and (hydrated) size of the ion. A comparison of different ions is

Table 2.2 Values of mobility of various ions at infinite dilution at 25 °C (in $10^{-4}\,m^2\,V^{-1}\,s^{-1}$).

Ion	H$^+$	Na$^+$	K$^+$	Ca^{2+}	OH$^-$	Cl$^-$	½SO$_4^{2-}$	½CO$_3^{2-}$
u_i	349	50.1	73.9	59.5	198	75.2	79.8	69.5

possible based on their mobility u (Table 2.2). The hydrogen and hydroxyl ions show the highest ion mobilities due to their interaction with the solvent water.[3]

The ion mobility u_i (describing the ion movement under an electric field) is directly related to the diffusion coefficient D_i (describing the movement under a concentration gradient) by [33, 34]:

$$D_i = RTu_i/|z_i|F \tag{2.10}$$

where: R is the gas constant (J/[K mol]), T the temperature (K), F Faraday's constant (96 490 C/mol) and z_i the valence of ion i.

The contribution of a certain ion (i) to the total current flowing I_{tot} is called the transference number or transport number t_i, which increases with the concentration c_i and the mobility u_i of the ion (other ions being constant):

$$t_i = I_i/I_{tot} = c_i \cdot u_i \cdot |z_i|/\Sigma(c_i \cdot u_i \cdot |z_i|) \tag{2.11}$$

It has to be noted that the temperature has a marked effect on ion migration (and thus on current flow and electrical resistivity).

2.5.2
Ion Transport in Concrete

The principles that apply to aqueous solutions are, basically, also valid for concrete, because the transport of electrical current is due to ion movement in the water-filled pore system (Section 2.1). Positive ions (Na$^+$, K$^+$, Ca^{++}) migrate in the direction of the current; negative ones (OH$^-$, SO$_4^=$, Cl$^-$) in the opposite direction.

Applying the concept of transport numbers (Eq. 2.11) to concrete it can be shown that for chloride-free concrete, assuming that the pore solution contains 0.5 mole/liter of NaOH, the transport numbers (equal to the fraction of current transported by the given ion) for OH$^-$ and Na$^+$ are 0.8 and 0.2, respectively. For concrete contaminated by chloride salts, assuming that the pore solution contains 0.5 mole/liter of NaOH and 0.5 mole/liter of NaCl, the transport numbers of OH$^-$, Na$^+$, and Cl$^-$ are 0.52, 0.20, and 0.28, respectively. In a general sense, these estimated ion transport numbers have been confirmed by experimental studies [35, 36].

---

[3] It can be seen that the conductivities of ions H$^+$ and OH$^-$ are much greater than those of other ions because of the special bonds that they have with water, the transport of these ions in reality does not require their actual displacement, but a series of breaks and reformations of bonds with water molecules.

Unlike in bulk solutions, the ions are not able to move by the shortest route, but they have to find their way along the narrow and tortuous capillary pores. So the actual distance (e.g., from the concrete surface to the reinforcement) is much longer than the geometrical one. Moreover, the ions can be transported only in water-filled and interconnected pores. Therefore, the velocity of migration (and diffusion) in concrete is governed by the pore volume and the pore geometry and distribution. For this reason ion migration (and diffusion) even in water-saturated cement-based materials (cement paste, mortar, concrete) is 2–3 orders of magnitude lower compared to bulk solutions [36]. This drastic influence of the pore volume can be modeled using the percolation theory [24, 37].

Electrical current flow by ion migration in concrete is important for electrochemical rehabilitation techniques such as chloride removal (Chapter 20), but also for (macrocell) corrosion processes (Chapter 8).

Ion migration of concrete can be considered a measure of its resistance to chloride penetration using migration tests, of which various versions exist. Some of these tests measure nonsteady-state migration, expressed as the depth of penetration of chloride ions into a specimen in an electrical field [38, 39]. Other methods apply an electrical field across a specimen until steady-state flow of chloride ions is detected in the downstream cell [40]. Because of their specific nature, a detailed description of these methods will be omitted.

2.5.3
Resistivity of Concrete

The electrical resistivity of concrete is an important parameter used to describe for example, the degree of water saturation, the resistance to chloride penetration or the corrosion rate. The resistivity of concrete may have values from a few tens to many thousands of Ω m (Table 2.3) as a function of the water content in the concrete, the type of cement used (portland or blended cements), the w/c, the presence of chloride ions or whether the concrete is carbonated or not [41, 42]. At early ages, the resistivity of concrete is low and considerable increases occur due to hydration of the cement, as was shown in Figure 1.9. For a constant relative

Table 2.3 Electrical resistivity (Ω m) of concrete made with portland cement (OPC), blast furnace slag cement (GGBS), and portland cement with addition of 5% silica fume (SF); w/c = 0.45. Values were determined after 1.5–2.5 years exposure in a fog room or 20 °C 80% R.H. climate [41, 42].

	Immersed and splash zone (fog room)	Atmospheric zone (20 °C, 80% R.H.)
Portland cement	135	300–700
70% blast furnace slag	800	2200
5% silica fume	250	300–2000

Table 2.4 Global reference values at 20 °C for the electrical resistivity in Ω m of dense-aggregate concrete of mature structures (age >10 years); conditions in square brackets are corresponding laboratory climates [43–45].

Environment	Concrete resistivity (Ω m)	
	Ordinary portland cement (CEM I)	Blast furnace slag cement CEM III/B (>65% slag) or fly ash cement CEM II/B-V (>25%) or with silica fume (>5%)
Very wet, submerged, splash zone (fog room)	50–200	300–1000
Outside, exposed	100–400	500–2000
Outside, sheltered, coated, hydrophobized (not carbonated) (20 °C/80% R.H.)	200–500	1000–4000
Ditto, carbonated	1000 and higher	2000–6000 and higher
Indoor climate (carbonated) (20 °C/50% R.H.)	3000 and higher	4000–10 000 and higher

humidity and in stationary conditions, resistivity is increased by a lower water to cement ratio (w/c), longer curing times (hydration) or by the addition of reactive minerals such as blast furnace slag, fly ash and/or silica fume. The resistivity of concrete increases when the concrete is drying out and when the concrete carbonates (in particular in portland cement concrete). Carbonation reduces the number of ions available for carrying the current and may densify the concrete.

All of these factors can be rationalized on the basis of ion migration in the porous and tortuous concrete microstructure: a high relative humidity increases the amount of water-filled pores (decrease of resistivity), the w/c ratio and type of cement determine the pore volume and pore-size distribution (fewer but coarser pores with pure portland cement; more but finer pores and less interconnectivity of pores with blast furnace slag or fly ash); chloride ions increase the conductivity of the pore solution and carbonation decreases it. Table 2.4 shows resistivities determined for mature concrete in various climates [43–45].

Temperature Dependence Temperature changes have important effects on concrete resistivity. A higher temperature causes the resistivity to decrease and vice versa (for a constant relative humidity). This is caused by changes in the ion mobility in the pore solution and by changes in the ion–solid interaction in the cement paste. As a first approach, an Arrhenius equation can be written to describe the effect of temperature on conductivity:

$$\sigma(T_i) = \sigma(T_0) \cdot \exp\left(b \cdot \left(\frac{1}{T_0} - \frac{1}{T_i} \right) \right) \quad (2.12)$$

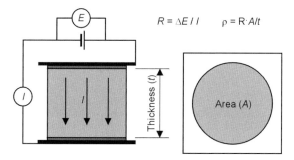

Figure 2.7 Measurement of electrical resistivity of concrete by a two-electrode method.

with σ the conductivity ($1/\Omega$ m), T_0 the reference temperature (K), T_i the actual temperature (K), and b an empirical factor (K). For steady-state conditions, b was found to be in the range of 1500 to 4500 [11, 14, 15, 24]. The coefficient b was found to increase with decreasing relative humidity for a given cement paste, mortar or concrete and to decrease with w/c ratio of the mix for a given relative humidity.

Concrete Resistivity and Corrosion Rate The corrosion rate of steel in concrete has often been found to decrease with increasing concrete resistivity [46–49]. Further work has shown indeed that this relationship is not universal but depends on concrete composition [15, 50].

Measuring Concrete Resistivity The resistivity of concrete can be measured in a variety of ways. The simplest laboratory method is pressing steel plates to two parallel faces of a cube or cylinder, via wetted cloth for good electrical contact, as shown in Figure 2.7. The resistivity is calculated from the resistance measured between the plates by:

$$\rho = R \cdot C \tag{2.13}$$

where: ρ is resistivity (Ω m), R resistance between plates (Ω), and C is a geometrical cell constant, calculated by $C = A/t$, with A the area of the parallel faces and t the thickness of the specimen [43].

Measuring the resistivity on a very local scale inside concrete using small embedded electrodes has been shown to be a good method for measuring and monitoring the effects of water uptake and drying, as a function of depth, binder type and environment [51, 52].

Measuring concrete resistivity onsite may reveal local variations of concrete quality (porosity) and local variations of concrete humidity. A detailed description of measuring and interpreting concrete resistance on structures is given in a RILEM TC 154 recommendation [45]. Frequently, resistance is measured with a probe according to Wenner consisting of four equally spaced point electrodes that are pressed onto the concrete surface (4-point method). This method has long been known and used for determining soil resistivities [53], and was studied for the application to concrete structures by Stratfull [54] and Naish and coworkers [55]. The resistance R measured by the four-point probe can be converted to concrete resistivity using a cell constant based on theoretical considerations by:

$$\rho = 2\pi \cdot a \cdot R \qquad (2.14)$$

with a the electrode spacing. Several commercial instruments based on the four-point method are available. Further details and precautions regarding onsite measurements are given in [45]. The use of resistance measurements in condition assessment of reinforced concrete structures is discussed in Chapter 16.

2.6
Mechanisms and Significant Parameters

Each of the transport processes that lead to corrosion of the reinforcement and then govern its kinetics can be characterized by a parameter (D, S, K, ρ) which depends on the concrete properties and can be determined experimentally. Table 2.5 shows the parameters that are relevant to different situations. At least theoretically, these parameters can be used in the design of concrete structures to calculate the evolution in time of corrosion (initiation or propagation) or any other type of degradation as a function of concrete properties and environmental conditions.

Correlations The kinetics of the different transport processes that can take place in concrete, and thus the parameters that define them, are correlated among themselves because they depend on the porous structure of concrete. These correlations, however, are not of a general nature, but vary in relation to the composition or other properties of concrete. For example, for concrete made with a certain type of cement, correlations between the water permeability and the diffusion coefficient of chlorides, that is between K and D, can be established. Nevertheless, when variations are made in the type of cement used, for example, changing from portland cement to blended cement, the relationships are no longer valid (D may vary even two orders of magnitude with no corresponding variation in K [3]).

Analogously, correlations exist between the coefficient of water permeability k and that of capillary absorption S, but these lose their validity if the surface of the concrete is subjected to a hydrophobic treatment, which will reduce considerably capillary absorption but not permeation. For concrete obtained with portland cement, correlations between the coefficient of water permeability (k) and conductivity ($\sigma = 1/\rho$) measured for a given value of relative humidity are available. On the other hand, conductivity varies greatly in concrete made with blended cements or carbonated concrete, while there is no significant change in water permeability.

It is therefore necessary, in general, to determine the specific property of concrete concerned. Analogous considerations can be made for correlations between the transport properties of concrete and its compressive strength. However, these

Table 2.5 Properties of concrete relevant with regard to different durability requirements; S = capillary absorption, K = permeation, D = diffusion, ρ = electrical resistivity (modified from [3]).

Function requirements	S	K		D		ρ	Primary influencing factors
		H_2O	Gas	Ion	Gas		
Protection of rebar							
Chloride ingress				✓		✓	Type of cement
Carbonatation			✓		✓		Moisture content
Reduction in corrosion rate	✓			✓	✓	✓	Moisture content affects S, K, D, and ρ
Resistance to concrete degradation (sulfate attack, ASR, etc.)							For ASR and sulfates chemical effects dominate
Low pressure	✓			✓			
High pressure		✓		✓			
Containment of fluids							
Low pressure	✓						
High pressure		✓	✓				Time, moisture content

correlations, although valid in some cases, cannot be generalized. In fact, unlike compressive strength, transport properties are not a simple function of the volume of the capillary pores, but also depend on the dimensions, distribution, form, tortuosity, and continuity of these pores and thus may vary even if the pore volume remains constant (Section 1.2.2).

Nevertheless, these correlations may be useful for the design of structures. As will be seen later, they can be used to fix the requirements for durability in terms of the requirements of strength. For example, Parrott [56] proposed the following correlation between the coefficient of permeability for air k_{60} (measured at 60% R.H.) and the compressive strength measured on cube specimens (R_c):

$$\log(k_{60}) = (30 - R_c)/10 \quad (2.15)$$

Figure 2.8 shows the correlations found between the diffusion coefficient for chlorides and the compressive strength on different types of concrete [57]. It can be seen that the relationship between the two parameters may be determined only for a given concrete, while no generalized correlation for all types of concrete can be determined.

From theoretical and experimental work there appears to be a relationship between resistivity and chloride diffusion in a particular water-saturated concrete [58–60]. For example, concrete with a high percentage of blast furnace slag has a

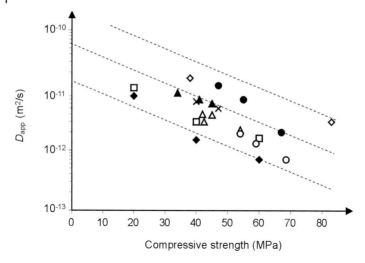

Figure 2.8 Correlations between the diffusion coefficient for chlorides and the compressive strength of different types of concrete (each symbol represents a particular type of concrete) [57].

high resistivity and a low chloride diffusion coefficient. Complete quantitative evaluation may require knowing the pore-water conductivity [61] and information on chloride binding (and other ion–solid interactions). The relationship takes the following form:

$$\rho \cdot D = A \tag{2.16}$$

where: A is a constant with a magnitude of about $1000\,\Omega\,\text{m}^3/\text{s}$, with resistivity ($\rho$) in Ω m and diffusion coefficient (D) in m^2/s.

It has been confirmed in a general sense that this relationship holds true for nonsaturated concrete as well, both over a range of relative humidities [62] as for a range of binders including blast furnace slag and fly ash in different climates [63]. In both series of tests it was seen that the diffusion coefficient decreased and the resistivity increased with lower humidity and drying out of the pore system.

Various laboratory test methods described above (or similar) have been applied to a range of concrete compositions used for making blocks that were exposed in the marine splash zone on the South East coast of the UK [20]. For three of the investigated mixes, Table 2.6 reports binder type, cement content, w/c and data of compressive strength, sorptivity, and oxygen and chloride diffusion obtained at 28 days. Other tests were carried out on the exposure site at 6 years age [20, 64]. In this environment, penetration of chloride is the main factor determining the durability. It is interesting to see that at 28 days age, some physical tests indicate that the mixes perform similarly (sorptivity, oxygen diffusion) and that the slag mix has a higher water permeability than the other two (portland and 30% fly ash), while the steady-state chloride diffusion is much lower for the fly ash and slag

Table 2.6 Test results of various methods on three concrete compositions, at 28 days age and after 6 years marine splash zone exposure [20, 64]; ranges given are in most cases related to the level of curing, where poor curing produced results indicating more permeable concrete.

Type of binder	OPC	30% fly ash	70% blast furnace slag
Binder content (kg/m³)	288	325	365
w/c	0.66	0.54	0.48
Tests at 28 days age			
Compressive cube strength (MPa)	39	50	38
Sorptivity (mm/min$^{0.5}$)	0.08–0.20[a]	0.04–0.17	0.04–0.30
Water permeability (10^{-10} m/s)	2.0	0.5–1.1	0.3–8.0
Oxygen diffusion (10^{-8} m²/s)	2.8–6.5	1.6–6.6	1.3–5.0
Steady-state chloride diffusion (10^{-12} m²/s)	1.5–2.4	0.03–0.04	0.02
After 6 years South East UK coast, splash zone (Folkestone)			
D_{app} (10^{-12} m²/s)	6.4–14.5	0.34–0.47	0.40–1.4
Autoclam[b] air permeability index (–)	1.2–3.1	0.03–0.2	0.15–0.5
Autoclam[b] water permeability index (–)	9–14	0.01–0.8	1–5
Resistivity (Ωm)	60	575	400–600

a) Sorptivity was generally higher for the outer 15 mm than for deeper layers.
b) AutoClam: portable instrument for onsite determination of water absorption, water and air permeability [29].

mixes than for the portland mix. After 6 years, the apparent chloride diffusion coefficient, the resistivity and the water and air permeability indicate less permeability for the fly ash and the slag mixes. It appears that purely physical tests at early age have a poor correlation to the permeability for chloride, while in particular the electrical resistivity and to a lesser extent the physical tests at later age correlate well to chloride permeability.

Presence of More Than One Transport Mechanism In reality, the transport of aggressive species through concrete may take place by a combination of mechanisms, as shown in Figure 2.9 and consequently a range of concrete properties will be important. For example, the case of a structure subjected to wetting–drying cycles, which is reached by spray or periodically wetted by rainfall, may be considered. Water will penetrate the concrete by capillary absorption or permeation when the external surface is wetted. This water penetration will be as much as 10–20 mm in a few hours if the concrete is of average quality and considerably less if the concrete is of good quality and well cured. Subsequent evaporation, which always occurs much more slowly than absorption, facilitates the ingress of oxygen and deposits the salts that were absorbed with the water. Depending on the type of salt, precipitation may occur, causing efflorescence (e.g., sodium sulfate) or simultaneous enrichment in the pore solution and accumulation in the solid (chloride). A succession of wet and dry conditions, or more generally, changes in humidity, will therefore favor (even if in subsequent periods) both the penetration of water

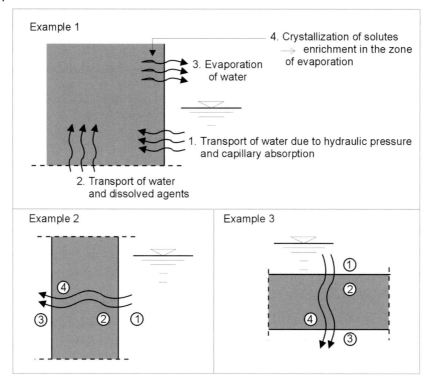

Figure 2.9 Examples of some transport mechanisms in concrete [1].

soluble species (e.g., chlorides) and the transport of gases (e.g., oxygen). It will thus favor, in the final analysis, the corrosion process itself.

In a quantitative investigation of the various processes occurring under wetting and drying conditions, McCarter and coworkers [51, 52] have carried out series of tests on concrete specimens with embedded arrays of small stainless steel electrodes for monitoring resistivity. For cycles of 48 h of wetting (under a 400-mm water head) and 49 days of drying, they have found that at young ages (6 weeks) the wetting/drying front penetrates 25–30 mm; however, after about 1.5 years a reduction of this depth was found for fly ash and slag cement concrete (down to 15 mm), while it remained 20–30 mm in portland cement concrete [52]. It appears that the response of concrete under variations of the external humidity is a complex function of the length of wet and dry cycles, the concrete composition and the hydration of the cement. Understanding corrosion promoting processes like accumulation of chloride, carbonation and penetration of oxygen to the reinforcement under varying conditions of wetting and drying, which is typical for most real structures, will be even more complex.

References

1. CEB (1992) Durable Concrete Structures, Bulletin d'information N.183, Lausanne.
2. Frederiksen, J.M. (ed.) (1996) HETEK, Chloride penetration into concrete. State of the art. Transport processes, corrosion initiation, test methods and prediction models, The Road Directorate, Report No. 53, Copenhagen.
3. Bamforth, P.B. (1995) Specifications and tests to determine the permeability of concrete, CEMCO-95, XXIII curso de estudios mayores de la construcciòn, Instituto E. Torroja, CSIC, Madrid, 24–26 April 1995.
4. Page, C.L. and Vennesland, Ø. (1983) Pore solution composition and chloride binding capacity of silica fume cement pastes. *Materials and Structures*, **16**, 91.
5. Holden, W.R., Page, C.L., and Short, N.R. (1983) The influence of chlorides and sulphates on durability, in *Corrosion of Reinforcement in Concrete Construction* (ed. A.P. Crane), Ellis Horwood Limited, pp. 143–150.
6. Polder, R.B. (1986) *Chloride in cement-sand mortar – A. Expression experiments*, TNO report BI-86-21, Delft (NL), April 1986.
7. Sergi, G. (1986) *Corrosion of steel in concrete: cement matrix variables*, PhD Thesis, University of Aston, Birmingham.
8. Tritthart, J. (1989) Chloride binding in cement–I. Investigations to determine the composition of porewater in hardened cement. *Cement and Concrete Research*, **19**, 586–594.
9. Tritthart, J. (1989) Chloride binding in cement–II. The influence of the hydroxide concentration in the pore solution of hardened cement paste on chloride binding. *Cement and Concrete Research*, **19**, 586–594.
10. Larbi, J.A., Fraay, A.L.A., and Bijen, J.M. (1990) The chemistry of the pore solution of silica fume-blended cement systems. *Cement and Concrete Research*, **20**, 506–516.
11. Elkey, W. and Sellevold, E.J. (February 1995) Electrical resistivity of concrete, in *Corrosion of Reinforcement – Field and Laboratory Studies for Modelling and Service Life* (ed. K. Tuutti), University of Lund, Report TVBM-3064, pp. 303–334.
12. Sandberg, P. (February 1995) Pore solution chemistry in concrete, in *Corrosion of Reinforcement – Field and Laboratory Studies for Modelling and Service Life* (ed. K. Tuutti), University of Lund, Report TVBM-3064, pp. 161–170.
13. Polder, R.B., Walker, R.J., and Page, C.L. (1995) Electrochemical desalination of cores from a reinforced concrete coastal structure. *Magazine of Concrete Research*, **47** (173), 321–327.
14. Bürchler, D., Elsener, B., and Böhni, H. (1996) Electrochemical resistivity and dielectrical properties of hardened cement paste and mortar, in *Proc. Fourth Int. Symp. on Corrosion of Reinforcement in Concrete Structures* (eds C.L. Page, P.B. Bamforth, and J. Figg), Society of Chemical Industry, London, pp. 283–293.
15. Bertolini, L. and Polder, R.B. (1997) Concrete resistivity and reinforcement corrosion rate as a function of temperature and humidity of the environment, TNO report 97-BT-R0574.
16. Anstice, D.J., Page, C.L., and Page, M.M. (2005) The pore solution phase of carbonated cement pastes. *Cement and Concrete Research*, **35**, 377–383.
17. Metha, P.K. and Monteiro, P.J.M. (2006) *Concrete: Microstructure, Properties, and Materials*, 3rd edn, McGrawHill.
18. Parrott, L.J. (1991/92) Carbonation, moisture and empty pores. *Advances in Cement Research*, **4** (15), 111–118.
19. Page, C.L., Short, N.R., and Tarras, A.E. (1981) Diffusion of chloride ions in hardened cement paste. *Cement and Concrete Research*, **11**, 395–406.
20. Bamforth, P.B. and Chapman-Andrews, J. (24–29 July 1994) Long term performance of RC elements under UK coastal conditions, in *Proc. Int. Conf. on Corrosion and Corrosion Protection of Steel in Concrete* (ed. R.N. Swamy), Sheffield Academic Press, pp. 139–156.
21. Polder, R.B. (1996) Laboratory testing of five concrete types for durability in a

marine environment, in *Proc. Fourth Int. Symp. on Corrosion of Reinforcement in Concrete Construction* (eds C.L. Page, P.B. Bamforth, and J.W. Figg), Society of Chemical Industry, London, pp. 115–123.

22 NT (1995) BUILD 443, Concrete, Hardened: Accelerated Chloride Penetration, Nordtest.

23 Yu, S.W., Sergi, G., and Page, C.L. (1993) Ionic diffusion across an interface between chloride-free and chloride-containing cementitious materials. *Magazine of Concrete Research*, **45** (165), 257–261.

24 Elsner, B., Flückiger, D., and Böhni, H. (2000) A percolation model for water sorption on porous cementitious materials, in *Materials for Buildings and Structures* (ed. F.H. Wittmann), Wiley-VCH Verlag GmbH, Weinheim, pp. 163–169.

25 Polder, R.B., Borsje, H., and de Vries, J. (1996) Hydrophobic treatment of concrete against chloride penetration, in *Proc. Fourth Int. Symp. on Corrosion of Reinforcement in Concrete Construction* (eds C.L. Page, P.B. Bamforth, and J.W. Figg), Society of Chemical Industry, London, pp. 546–565.

26 de Vries, J., Polder, R.B., and Borsje, H. (21–24 June 1998) Durability of hydrophobic treatment of concrete, in *Proc. 2nd Int. Conf. on Concrete under Severe Conditions*, CONSEC'98, vol. 2 (eds O.E. Gjørv, K. Sakai, and N. Banthia), E & FN Spon, London, pp. 1341–1350.

27 EN 13057 (2002) Products and Systems for the Protection and Repair of Concrete Structures. Test Methods. Determination of Resistance of Capillary Absorption, European Committee for Standardization.

28 British Standards Institution (1983) BS 1881 part 122, *Testing concrete – method for determination of water absorption*.

29 Basheer, P.A.M., Long, A.E., and Montgomery, F.R. The Autoclam for measuring the surface absorption and permeability of concrete on site. Int. Symp. on Advances in Concrete Technology, Athens, 1992, 107–132.

30 EN 12390-8 (2009) Testing Hardened Concrete. Part 8 – Depth of Penetration of Water under Pressure, European Committee for Standardization.

31 DIN (1978) 1048:1978 part 4.7, *Prüfverfahren für Beton, Wasserundurchlässigkeit*.

32 Kollek, J.J. (1989) The determination of the permeability of concrete to oxygen by the Cembureau method – a recommendation. *Materials and Structures*, **22**, 225–230.

33 Moore, W. (1972) *Physical Chemistry*, 4th edn, Longman, London.

34 Ackermann, G., Jugelt, W., Möbius, H.-H., Suschke, H.D., and Werner, G. (1974) *Elektrolytgleichgewichte Und Elektrochemie*, Verlag Chemie.

35 Castellote, M., Andrade, C., and Alonso, C. (1999) Modelling of the processes during steady-state migration tests: quantification of transference numbers. *Materials and Structures*, **32**, 180–186.

36 Elsener, B. (1990) *Ionenmigration und elektrische Leitfähigkeit im Beton*, SIA Doc. D065 Schweiz. Ingenieur- und Architektenverein, Zürich, 51–59.

37 Bürchler, D., Elsener, B., and Böhni, H. (1996) Electrical resistivity and dielectrical properties of hardened cement paste and mortar. MRS Proceedings "Electrically based microstructural characterization", 1996, **411**, 407.

38 Tang, L. and Nilsson, L.O. (1992) Rapid determination of chloride diffusivity of concrete by applying an electric field. *ACI Materials Journal*, **49** (1), 49–53.

39 Tang, L. (1996) Electrically accelerated methods for determining chloride diffusivity in concrete. *Magazine of Concrete Research*, **48**, 173–179.

40 Andrade, C. (1993) Calculation of chloride diffusion coefficients in concrete from ionic migration measurements. *Cement and Concrete Research*, **23**, 724–742.

41 Polder, R.B. (1996) *Durability of new types of concrete for marine environments*, CUR report 96-3, Gouda.

42 Polder, R.B. (1996) The influence of blast furnace slag, fly ash and silica fume on corrosion of reinforced concrete in marine environment. *Heron*, **41** (4), 287–300.

43 COST 509 (1997) *Corrosion and protection of metals in contact with concrete*, Final

report, R.N. Cox, R. Cigna, O. Vennesland, T. Valente (eds), European Commission, Directorate General Science, Research and Development, Brussels, EUR 17608 EN.

44 Polder, R.B., Andrade, C., Elsener, B., Vennesland, Ø., Gulikers, J., Weidert, R., and Raupach, M. (2000) Draft RILEM technical recommendation test methods for on site measurement of resistivity of concrete. *Materials and Structures*, **33**, 603–611.

45 Polder, R.B. (2001) Test methods for on site measurement of resistivity of concrete–a RILEM TC 154 recommendation. *Construction and Building Materials*, **15**, 125–132.

46 Polder, R.B., Peelen, W.H.A., Bertolini, L., and Guerrieri, M. (2002) Corrosion rate of rebars from linear polarization resistance and destructive analysis in blended cement concrete after chloride loading. 15th International Corrosion Congress, Granada, 22–27 September 2002 (CD-ROM).

47 Andrade, C., Alonso, M.C., Gonzalez, J.A., and Feliu, S. (1990) Similarity between atmospheric/underground corrosion and reinforced concrete corrosion, in *Corrosion of Reinforcement in Concrete* (eds C.L. Page, K.W.J. Treadaway, and P.F. Bamforth), Elsevier, pp. 39–48.

48 Alonso, M.C., Andrade, C., and Gonzalez, J.A. (1988) Relation between resistivity and corrosion rate of reinforcement in carbonated mortar made with several cement types. *Cement and Concrete Research*, **8**, 687–698.

49 Glass, G.K., Page, C.L., and Short, N.R. (1991) Factors affecting the corrosion of steel in carbonated mortars. *Corrosion Science*, **32** (12), 1283–1294.

50 Fiore, S., Polder, R.B., and Cigna, R. (1–4 July 1996) Evaluation of the concrete corrosivity by means of resistivity measurements, in *Proc. Fourth Int. Symp. on Corrosion of Reinforcement in Concrete Construction* (eds C.L. Page, P.B. Bamforth, and J.W. Figg), Society of Chemical Industry, Cambridge, UK, pp. 273–282.

51 McCarter, W.J., Emerson, M., and Ezirim, H. (1995) Properties of concrete in the cover zone: developments in monitoring techniques. *Magazine of Concrete Research*, **47**, 243–251.

52 Chrisp, T.M., McCarter, W.J., Starrs, G., Basheer, P.A.M., and Blewett, J. (2002) Depth-related variation in conductivity to study cover-zone concrete during wetting and drying. *Cement and Concrete Composites*, **24**, 415–426.

53 Wenner, F. (1915) A method for measuring earth resistivity. *Bulletin of the Bureau of Standards*, **12**, 469–478.

54 Stratfull, R.F. (1968) *Materials Protection*, **7** (3), 29–34.

55 Naish, C.C., Harker, A., and Carney, R.F.A. (1990) Concrete inspection: interpretation of potential and resistivity measurements, in *Corrosion of Reinforcement in Concrete* (eds C.L. Page, K.W.J. Treadaway, and P.F. Bamforth), Elsevier, pp. 314–332.

56 Parrot, L.J. (1994) Design for avoiding damage due to carbonation-induced corrosion. Proc. of Canmet/ACI Int. Conf. on Durability of concrete, Nice, 283.

57 Bamforth, P.B. (1994) Prediction of the onset of reinforcement corrosion due to chloride ingress. Proc. of Concrete across borders, Odense, Denmark, 22–25 June 1994.

58 Andrade, C., Sanjuán, M.A., Recuero, A., and Río, O. (1994) Calculation of chloride diffusivity in concrete from migration experiments, in non steady-state conditions. *Cement and Concrete Research*, **24**, 1214–1228.

59 Polder, R.B. (1997) Chloride diffusion and resistivity testing of five concrete mixes for marine environment. Proc. RILEM. Int. workshop on Chloride penetration into concrete, St-Remy-les-Chevreuses, 15–18 October 1995 (eds L.O. Nilsson and P. Ollivier), RILEM, 225–233.

60 Polder, R.B. (2000) Simulated de-icing salt exposure of blended cement concrete–chloride penetration. Proc. 2nd Int. RILEM Workshop Testing and modelling the chloride ingress into concrete, (eds C. Andrade and J. Kropp), RILEM PRO 19, 189–202.

61 Arup, H., Sørensen, B., Frederiksen, J., and Thaulow, N. (1993) The rapid

chloride permeation test – an assessment, paper 334, NACE, Corrosion/'93.

62 de Vera, G., Climent, M.A., López, J.F., Viqueira, E., and Andrade, C. (2002) Transport and binding of chlorides through non-saturated concrete after an initial limited chloride supply. Proc. 3rd International RILEM Workshop Testing and modelling chloride ingress into concrete (eds C. Andrade and J. Kropp), RILEM PRO 38, 205–218.

63 Polder, R.B. and Visser, J. (2002) Redistribution of chloride in blended cement concrete during storage in various climates. Proc. 3rd International RILEM Workshop Testing and modelling chloride ingress into concrete (eds C. Andrade and J. Kropp), RILEM PRO 38, 347–359.

64 Polder, R.B., Bamforth, P.B., Basheer, M., Chapman-Andrews, J., Cigna, R., Jafar, M.I., Mazzoni, A., Nolan, E., and Wojtas, H. (24–29 July 1994) Reinforcement corrosion and concrete resistivity – state of the art, laboratory and field results, in *Proc. Int. Conf. on Corrosion and Corrosion Protection of Steel in Concrete* (ed. R.N. Swamy), Sheffield Academic Press, pp. 571–580.

3
Degradation of Concrete

Degradation processes of concrete can be classified (Figure 3.1) as: *physical* (caused by natural thermal variations such as freeze–thaw cycles, or artificial ones, such as those produced by fire), *mechanical* (abrasion, erosion, impact, explosion), *chemical* (attack by acids, sulfates, ammonium and magnesium ions, pure water, or alkali aggregate reactions), *biological* (fouling, biogenic attack), and *structural* (overloading, settlement, cyclic loading). In practice, these processes may occur simultaneously, frequently giving rise to synergistic action.

Alterations that occur in concrete before the structure has been completed, that is within the first hours to months after casting (among these are cracking due to plastic settlement, plastic or drying shrinkage, creep, thermal shrinkage), are traditionally not considered among the phenomena of deterioration, although they should be prevented since they are important to the durability of the structure (see Section 12.2.4).

The processes of deterioration of concrete and corrosion of reinforcement are closely connected (Figure 3.1). The former provoke destruction of the concrete cover or cause microcracking that compromises its protective characteristics. On the other hand, corrosion attack can produce cracking or delamination of the concrete if voluminous corrosion products are formed that exert expansive action.

In this chapter, only a few of the most common forms of physical and chemical deterioration of concrete will be mentioned (effects of freeze–thaw cycles, acid solutions, pure water, sulfates, and alkali aggregate reactions). Other forms of deterioration, such as the action of certain aggressive liquids, while important in specific cases, will not be dealt with. For more details, the reader is referred to the classic literature on degradation of concrete, which has formed the source of useful information for many decades [1, 2]. Modern reference texts and standards rely for a large part on these classic sources [3, 4].

Deterioration due to structural causes such as accidental mechanical stresses (often due to improper use of the structure) or fatigue is also neglected here, since it lies in the area of structural engineering.

Corrosion of Steel in Concrete: Prevention, Diagnosis, Repair, Second Edition. Luca Bertolini,
Bernhard Elsener, Pietro Pedeferri, Elena Redaelli, and Rob Polder.
© 2013 Wiley-VCH Verlag GmbH & Co. KGaA. Published 2013 by Wiley-VCH Verlag GmbH & Co. KGaA.

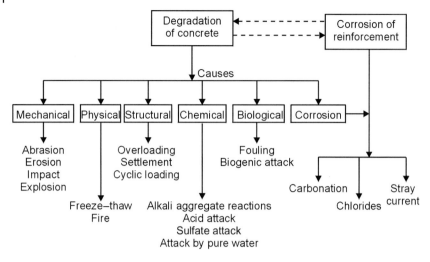

Figure 3.1 Causes of deterioration of reinforced concrete structures.

3.1
Freeze–Thaw Attack

When the temperature reaches values below 0 °C, water contained in the pore system of concrete can freeze, causing an increase in volume by about 9%. The tensile stresses generated may result in scaling, cracking or spalling of the concrete and, eventually, to its complete disintegration.

The freezing of pore water is usually a gradual process for three reasons: heat transfer towards the inside of the concrete is hindered by the relatively low thermal conductivity of concrete; the freezing point is lowered by the gradual increase in concentration of the ions dissolved in the part of the pore water that is not yet frozen; the freezing point also decreases on decreasing the diameter of the pores. Therefore, freezing begins in the outer layers and in the largest pores and extends to the inner parts and to smaller pores only if the temperature drops further. Indicatively and for saturated portland cement paste, free water in pores larger than 0.1 mm freezes between 0 and −10 °C; water in pores between 0.1 and 0.01 mm freezes between −20 and −30 °C and gelwater (pores smaller than 10 nm) freezes below −35 °C [5, 6]. Besides depending on the microstructure of concrete, the consequences of this type of deterioration also depend on environmental conditions, in particular on the degree of water saturation, on the number of freeze–thaw cycles, on the rate of freezing and on the minimum temperature reached. The presence of deicing salts like calcium and sodium chloride in contact with concrete is a further detrimental factor. The outer layers where these salts are present are more strongly affected by frost despite the lowering of the freezing point, probably due to increased water saturation caused by their hygroscopic effect, causing scaling and detachment of cement paste that covers the aggregate.

3.1.1
Mechanism

The action of freezing occurs mainly in capillary pore water. In fact, the gel pores are so small that they do not allow freezing, unless the temperature falls below −35 °C [3]. On the other hand, spaces of larger dimensions (above all, those introduced intentionally using air-entraining admixtures) are generally filled by air and, at least initially, are not affected by frost action.

Different theories have been put forward to explain the mechanism of frost damage. The most important are the theory of *hydraulic pressure* and the theory of *ice overpressure*.

According to the *hydraulic pressure* theory proposed by Powers [7], water freezes inside the capillary pores and the expanding ice pressurizes the remaining liquid. The pressure can only be released if there are (partially) empty pores in the vicinity. This pressure release can be calculated with Darcy's law (2.7). Summarizing: the pressure increases with decreasing pore diameter, with increasing distance over which the water must travel to empty pores or to the outside, with increasing flow of pressurized water, proportional to the rate at which ice forms inside the pores themselves.

Consequently, rapid cooling will lead to greater damage than cooling, which occurs more slowly. This is important because often in laboratory tests, for obvious reasons of time, concrete is cooled and heated much more rapidly than it would be in reality. The CDF-test [8], for example, provides for a cooling rate of 10 °C/h, while in reality typical values of 0.7 to 1 °C/h have been documented [9]. It appears that the conditions during testing are much more severe than in reality. Concrete compositions that on the basis of laboratory tests appear to be susceptible to freezing action may therefore perform well in the field.

It will be clear that the hydraulic pressure mechanism will have more severe consequences in a system of fully saturated pores, because in that case the pressure can only be released if the microstructure expands, which may quickly result in cracking.

An additional factor to the hydraulic pressure mechanism is the occurrence of *osmotic pressure*: as freezing proceeds in the capillary pores, ions become concentrated in the remaining solution and a difference in concentration is generated between the solution in the capillaries and that in the gel pores. Under these conditions, water tends to move from the gel pores to dilute the solution in the capillary pores, consequently increasing the internal pressure.

The *ice overpressure* mechanism is based on transport of liquid water (or vapor) from smaller pores to ice already formed in larger pores, where it freezes and increases the ice volume and consequently the pressure. Because this type of water transport is relatively slow, the ice overpressure mechanism is more important for longer freezing periods.

3.1.2
Factors Influencing Frost Resistance

Frost resistance is determined by the number of freeze–thaw cycles that a particular concrete can withstand before reaching a given level of degradation. In general, the loss of mass or the decrease of dynamic elastic modulus are applied as indexes of degradation.

Frost resistance is heavily influenced by the degree of saturation of the pores; in general there is a critical value, characteristic of each type of concrete (indicatively when about 80–90% of the total pore volume is water filled), below which the concrete is capable of withstanding a high number of freeze–thaw cycles, while above this value only a few cycles might be sufficient to damage the concrete.

Another important parameter is the water/cement ratio, on which the porosity of the cement matrix depends. Air-filled pores exert a beneficial action because they collect water that is forced out of the capillaries when ice crystals form, and thus lower the pressure. Nevertheless, space available to collect this water must be easily accessible and near the location of ice formation. Very porous concrete will facilitate the movement of water and can give more space to the growth of ice crystals. However, porous concrete rapidly becomes saturated with water and is thus sensitive to frost action. In practice, dense concrete of low w/c is the most resistant to freezing. In fact, for w/c values of practical interest (i.e., lower than 0.85), frost resistance increases rapidly with a decrease of the w/c ratio, both for ordinary concrete and for air-entrained concrete (Figure 3.2) [10, 11].

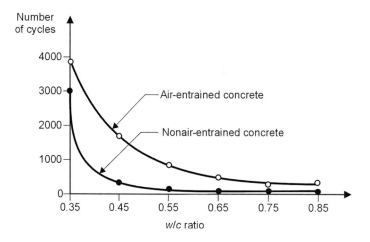

Figure 3.2 Number of freeze–thaw cycles that lead to a mass loss of 25% as a function of the w/c ratio of concrete after wet curing for 28 days, with and without entrained air [10, 11].

Prolonged curing of concrete before it undergoes freezing is beneficial, because it improves the mechanical resistance of the concrete and reduces its content of free water.

Finally, in order to obtain frost-resistant concrete, aggregates must also be able to withstand this type of attack. Although cement paste is usually more susceptible compared to aggregate, certain types of aggregate can also be damaged by frost action. The main parameter influencing the aggregate resistance to freezing and thawing is porosity, both in terms of pore-size distribution and total porosity. For convenience, a threshold value of water absorption is often considered to classify aggregates: when water absorption is lower than 1–2%, the aggregate can usually be considered durable. However, aggregates with very high porosity are usually frost resistant, due to their high permeability and low pressure generated after freezing, and mineralogy can also play a role (chert is the most susceptible aggregate). If frost-susceptible aggregate cannot be avoided, other means should be taken into account to ensure durability, as, for instance, reducing aggregate size (below a certain threshold the hydraulic pressure generated by frost is low) and/or reducing the permeability of concrete (therefore reducing the overall amount of water) [12].

3.1.3
Air-Entrained Concrete

Air that is accidentally entrapped in concrete does not improve the frost resistance of concrete, since it is distributed in voids that are relatively large, few in number, and unevenly distributed. Conversely, by introducing air-entraining admixtures to the concrete, it is possible to incorporate a system of very tiny and uniformly distributed bubbles inside the cement paste.

These air bubbles are of the order of 0.05 to 1 mm in size. Air bubbles can avoid generation of stresses in the capillaries when water freezes, if they are not too far apart [13]. Experience has shown that good freeze–thaw resistance of concrete requires distances between bubbles less than 0.1–0.2 mm. The volume of air entrained in concrete is of the order of 4–7% with respect to the volume of the concrete; however, for every mixture there is a minimum content of entrained air below which the presence of air bubbles is not effective.

It is important to realize that air-entrained concrete will have a lower compressive strength than nonair-entrained concrete with similar composition. As a rule, an increase of 1% in the content of air corresponds to a reduction in compressive strength of 5%. Therefore, to guarantee the same strength, the w/c ratio needs to be reduced.

The European standard EN 206-1 distinguishes four classes of severity of freeze–thaw attack: with or without deicing agents and with moderate or high water saturation (XF1 through XF4, Table 3.1). It further recommends a maximum w/c, a minimum cement content and minimum strength for each of these classes (Table 3.2). For XF2 and higher, at least 4% air is recommended.

Table 3.1 Exposure classes for freeze–thaw and chemical attack, according to EN 206-1 [4].

Class designation	Description of the environment	Informative examples where exposure classes may occur
\multicolumn{3}{l}{5 – Freeze–thaw attack with or without deicing agents}

5 – Freeze–thaw attack with or without deicing agents
Where concrete is exposed to significant attack by freeze–thaw cycles while wet, the exposure shall be classified as follows:

Class designation	Description of the environment	Informative examples where exposure classes may occur
XF1	Moderate water saturation, without deicing agent	Vertical concrete surfaces exposed to rain and freezing.
XF2	Moderate water saturation, with deicing agent	Vertical concrete surfaces of road structures exposed to freezing and airborne deicing agents.
XF3	High water saturation, without deicing agent	Horizontal concrete surfaces exposed to rain and freezing.
XF4	High water saturation, with deicing agent or seawater	Road and bridge decks exposed to deicing agents. Concrete surfaces exposed to direct spray containing deicing agents and freezing. Splash zone of marine structures exposed to freezing.

6 – Chemical attack
Where concrete is exposed to chemical attack from natural soils and ground water as given in Table 3.3, the exposure shall be classified as given below. The classification of seawater depends on the geographical location; therefore, the classification valid in the place of use of the concrete applies.[a]

Class designation	Description of the environment	Informative examples where exposure classes may occur
XA1	Slightly aggressive chemical environment according to Table 3.3	
XA2	Moderately aggressive chemical environment according to Table 3.3	
XA3	Highly aggressive chemical environment according to Table 3.3	

a) A special study may be needed to establish the relevant exposure condition where there is:
 – limits outside of Table 3.3;
 – other aggressive chemicals;
 – chemically polluted ground or water;
 – high water velocity in combination with the chemicals in Table 3.3.

3.2
Attack by Acids and Pure Water

3.2.1
Acid Attack

The hydrated components (C–S–H, portlandite, sulfoaluminates) in the cement matrix of concrete are in equilibrium with the pore liquid that is characterized by a high pH, due to the presence of OH^- (balanced by Na^+ and K^+). When concrete comes into contact with acid solutions, these compounds may dissolve with a rate that depends on the permeability of the concrete, the concentration and the type of acid. In soil with acidic ground water, the rate of refreshing is important.

Table 3.2 Recommendations (informative) for the choice of the limiting values of concrete composition and properties in relation to exposure classes for the exposure classes shown in Table 3.1 [4].

Exposure class		Maximum w/c	Minimum strength class	Minimum cement content (kg/m³)	Minimum air content (%)	Other requirements
Freeze–thaw attack	XF1	0.55	C30/37	300	–	Aggregates in accordance with EN 12620
	XF2	0.55	C25/30	300	4.0[a]	
	XF3	0.50	C30/37	320	4.0[a]	
	XF4	0.45	C30/37	340	4.0[a]	
Aggressive chemical environments	XA1	0.55	C30/37	300	–	Sulfate-resisting cement[b]
	XA2	0.50	C30/37	320	–	
	XA3	0.45	C35/45	360	–	

The values in this table refer to the use of cement type CEM I conforming to EN 197-1 and aggregates with nominal maximum size in the range of 20 to 32 mm. The minimum strength classes were determined from the relationship between the water/cement ratio and the strength class of concrete made with cement of strength class 32.5. The limiting values for the maximum w/c ratio and the minimum cement content apply in all cases, while the requirements for concrete strength class may be additionally specified.

a) Where the concrete is not air entrained, the performance of concrete should be tested according to an appropriate test method in comparison with a concrete for which freeze–thaw resistance for the relevant exposure class is proven.

b) When SO_4^{2-} leads to exposure class XA2 and XA3 it is essential to use sulfate–resisting cement. Where cement is classified with respect to sulfate resistance, moderate or high sulfate-resisting cement should be used for exposure class XA2 (and in exposure class XA1 when applicable) and high sulfate resisting cement should be used in exposure class XA3.

Acids that can attack concrete are: sulfuric acid, hydrochloric acid, nitric acid, organic acids such as acetic acid and humic acids, and solutions of CO_2. The rate of attack on the cement matrix depends on the solubility of the salts that are formed and thus on the nature of the anions involved. For more-soluble reaction products, higher rates of attack result than for insoluble products. With hydrochloric acid, soluble calcium chloride is formed, while with sulfuric acid, much less soluble calcium sulfate (gypsum) is formed. Water containing dissolved CO_2 may be acidic and trigger an ion-exchange reaction between carbonic acid and the constituents of the hydrated cement paste, in particular calcium hydroxide (portlandite). At lower concentrations, CO_2 and calcium hydroxide form calcium carbonate, which is sparsely soluble. In the presence of high concentrations of CO_2, calcium carbonate is transformed into the more-soluble calcium bicarbonate.

EN 206-1 [4] distinguishes three classes of acid aggressiveness, depending on the pH or the aggressive CO_2 content, XA1, XA2, and XA3 (see Table 3.3). It recommends maximum w/c ratios, minimum cement contents and strengths for each of these classes as indicated in Table 3.2.

Table 3.3 Limiting values for exposure classes for chemical attack from natural soil and ground water, according to EN 206-1 [4].

The aggressive chemical environments classified below are based on natural soil and ground water at water/soil temperatures between 5–25 °C and a water velocity sufficiently slow to approximate to static conditions.

The most onerous value for any single chemical characteristic determines the class.

Where two or more aggressive characteristics lead to the same class, the environment shall be classified into the next higher class, unless a special study for this specific case proves that it is not necessary.

Chemical characteristic	Reference test method	XA1	XA2	XA3
Ground water				
SO_4^{2-} mg/l	EN 196-2	≥ 200 and ≤ 600	>600 and ≤ 3000	>3000 and ≤ 6000
pH	ISO 4316	≤ 6.5 and ≥ 5.5	<5.5 and ≥ 4.5	<4.5 and ≥ 4.0
CO_2 mg/l aggressive	EN 13577	≥ 15 and ≤ 40	>40 and ≤ 100	>100 up to saturation
NH_4^+ mg/l	ISO 7150-1 or ISO 7150-2	≥ 15 and ≤ 30	>30 and ≤ 60	>60 and ≤ 100
Mg^{2+} mg/l	ISO 7980	≥ 300 and ≤ 1000	>1000 and ≤ 3000	>3000 up to saturation
Soil				
SO_4^{2-} mg/kg[a)] total	EN 196-2[b)]	≥ 2000 and ≤ 3000[c)]	>3000[c)] and $\leq 12\,000$	$>12\,000$ and $\leq 24\,000$
Acidity ml/kg	DIN 4030-2	>200 Baumann Gully	Not encountered in practice	

a) Clay soils with a permeability below 10^{-5} m/s may be moved into a lower class.
b) The test method prescribes the extraction of SO_4^{2-} by hydrochloric acid; alternatively, water extraction may be used, if experience is available in the place of use of the concrete.
c) The 3000 mg/kg limit is reduced to 2000 mg/kg where there is a risk of accumulation of sulfate ions in the concrete due to drying and wetting cycles or capillary suction.

3.2.2
Biogenic Sulfuric Acid Attack

A special form of acid attack of concrete may be caused by a particular mechanism in a specific environment. In structures containing waste water (sewer pipes, pumping stations, treatment plants) anaerobic conditions may occur in the liquid, where dissolved sulfate is reduced to sulfide by anaerobic bacteria. As hydrogen sulfide gas it can escape to the atmosphere inside the structure, where it is oxidized to sulfuric acid by aerobic bacteria that live on the concrete surface above the water line. The acid thus formed can severely attack concrete; hence *biogenic sulfuric acid attack* (BSA) [14]. The role of sulfate in this type of attack is of minor importance. The principle is sketched in Figure 3.3. BSA was discovered in California in the

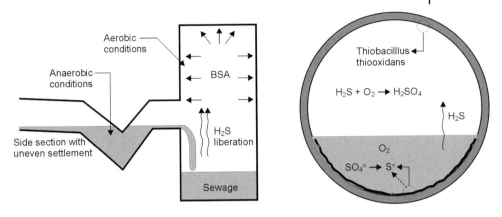

Figure 3.3 Biogenic sulfuric acid attack: (a) conditions in sewer system that promote BSA; (b) main reactions involved.

1940s and later identified and studied in other hot regions like Australia and South Africa [15]. In the 1970s, BSA also became manifest in temperate climates in Europe, for example, in Hamburg, where large sewers were extensively damaged by BSA. This stimulated new research into the biological origins [16], the identification [17], the influence of concrete composition [18], and possible countermeasures. Some aspects will be treated here briefly.

Anaerobic conditions can occur in sewers due to long retention times of waste water, for example, unexpectedly due to uneven settlement, as illustrated in Figure 3.3; waste water in (completely filled) pressure mains becomes anaerobic after being transported for a few hours. Liberation of the H_2S formed is subsequently promoted by turbulent overflow of anaerobic sewage into aerobic parts of the system. Various types of thiobacilli develop colonies on the concrete surface, which have increasing tolerance for acidic conditions. The final type in this series is thiobacillus thiooxidans (also called concretivorus), which is able to produce (and survive) sulfuric acid with concentrations up to 10% by mass with a pH below 1. The cement matrix is converted by the reaction with the acid to mainly gypsum and eventually, the converted layer of concrete falls off. Exposure testing for three years in sewers at Rotterdam showed that the rate of attack can be as high as 3 mm per year, with insignificant differences between (both very dense) portland and blast furnace slag cement concrete [17]. However, high alumina cement showed superior behavior [19]. A particular sewer system can be tested for BSA by measuring the oxygen and sulfide contents of the waste water and the pH of the concrete surface (using color indicator solutions). The presence of turbulent overflows must be checked and the sewage temperature taken into account [17]. Avoiding long retention times is the best preventative design strategy. Adding oxygen, hydrogen peroxide or nitrate to sewage in order to counteract anaerobic conditions has been successful. In some cases, increasing the flow by connecting rain drainages solved the problem. In large sewer pipe elements, protection of concrete by polymeric sheeting placed in the mold prior to concrete casting is used as a preventative measure.

3.2.3
Attack by Pure Water

Pure water, that is water with a low amount of dissolved solids, in particular calcium ions, acts aggressively towards concrete because it tends to dissolve calcium compounds. If the water flow rate is high, hydrolysis of hydration products continues, because the solution in contact with the concrete is continually being refreshed. Initially, calcium hydroxide, the most soluble component of cement paste, is removed. Then other components are attacked, producing a more open matrix, making the concrete more penetrable to further attack by aggressive solutions. Eventually this will have a deleterious effect on its strength. In the presence of cracks or construction defects, water can more easily percolate through the concrete, aggravating the aforementioned processes.

The degree of the attack by pure water depends to a large extent on the permeability of the concrete, but its $Ca(OH)_2$ content also plays an important role. Concrete types with a low level of $Ca(OH)_2$, like blast furnace slag cement concrete, have improved resistance with regard to this type of degradation. In addition to a lower $Ca(OH)_2$ content, there is probably an effect of the lower permeability of concrete made with slag cement.

Usually concrete withstands normal potable water because the free lime reacts with CO_2 forming $CaCO_3$, which is not readily soluble and forms a protective skin on the concrete surface. Only in soft water with a high concentration of free CO_2 is this protective layer dissolved by the formation of soluble $Ca(HCO_3)_2$. A special case is the attack of concrete by very soft waters (e.g., condensation), that dissolve the free lime (solubility 1.7 g/l). When free lime is dissolved or leached out, the other constituents of the cement paste are attacked and the resistance of the concrete diminishes. The aggressiveness of waters to concrete can be classified for example, according to EN 206-1 [4], where pH, free CO_2 content, ammonium, magnesium, and sulfate ions are taken as criteria (Table 3.3).

The calcium content in normal rain is not low enough to render it aggressive to concrete. "Acid" rain (with pH 3.5–4.5) in principle is aggressive, but the amount of precipitation is generally so low that the rate of attack is negligible to rather low.

3.2.4
Ammonium Attack

Ammonium ions, NH_4^+, act upon concrete as acids. This is due to their ability to react with OH^-, producing NH_3 and water. Because the NH_3 formed escapes as a gas, their action is comparable to acids with highly soluble reaction products. Aggressiveness due to NH_4^+ is classified by EN 206-1 [4] as shown in Table 3.3.

3.3
Sulfate Attack

Concrete often comes into contact with water or soil containing sulfates. These ions can penetrate the concrete and react with components of the cement matrix to cause expansive chemical reactions. This is referred to as *external sulfate attack*. Swelling may occur that, starting from the corners or edges of a concrete element, gives rise to cracking and disintegration. Sulfate attack can also manifest itself as a progressive loss of strength of the cement paste due to loss of cohesion between the hydration products. Concentrations of soluble sulfates in ground water of over several hundreds of mg/l should be considered potentially aggressive [1, 4].

Sulfate attack can also occur in the absence of exposure to a sulfate-bearing environment. In this case, it is referred to as *internal sulfate attack*. Aside from internal attack due to the presence of excessive amounts of sulfate in cement or aggregate, there also exists a more subtle form of attack that mainly occurs in steam-cured concrete and that is indicated as *delayed ettringite formation*.

3.3.1
External Sulfate Attack

External sulfate attack starts with penetration of sulfate ions into the concrete. Then mainly two detrimental reactions can occur[1]: sulfate can react with calcium hydroxide to form gypsum:

$$Ca(OH)_2 + SO_4^{2-} + 2H_2O \rightarrow CaSO_4 \cdot 2H_2O + 2OH^- \qquad (3.1)$$

and with calcium aluminates to form ettringite:

$$C_3A + 3CaSO_4 + 32H_2O \rightarrow 3CaO \cdot Al_2O_3 \cdot 3CaSO_4 \cdot 32H_2O \qquad (3.2)$$

The most important reaction is that connected to the formation of ettringite, in that it gives rise to greater expansive effects than gypsum. The mechanism by which formation of ettringite causes expansion is not well understood; however, it is generally attributed to the pressure exerted by the growth of ettringite crystals and to the swelling due to the absorption of water by poorly crystalline ettringite (colloidal ettringite, composed of tiny needle-shaped crystals). In addition, part of the ettringite is commonly located in the interface between paste and aggregate, resulting in loss of bond.

The severity of the attack depends mainly on the concentration of the sulfates in the soil or in the water in contact with concrete; and to a degree on the counterion.

---

1) Under very particular conditions, which are the presence of CO_2, a high humidity (>95% R.H.) and possibly low temperatures, sulfates can react with calcium hydroxide and calcium silicate hydrate to give thaumasite: $Ca(OH)_2 + CO_2 + CaSO_4 \cdot 2H_2O + CaO \cdot SiO_2 \cdot H_2O \rightarrow$ $CaCO_3 \cdot CaSO_4 \cdot CaSiO_3 \cdot 15H_2O$ and lead to the complete loss of strength of the concrete [20]. This mechanism has been identified as important in shotcrete-lined tunnels in particular rock soils in Norway, presumably related to the ground-water composition [21].

For instance, magnesium sulfate has a more marked effect than other sulfates because it also leads to decomposition of the hydrated calcium silicates. Ammonium sulfate is especially aggressive, because the ammonia produced escapes as a gas.

Protection With regard to protection against sulfate attack, the quality of the concrete is a crucial factor: a low permeability is the best defense against this type of attack, since it reduces sulfate penetration. This can be obtained by decreasing the w/c ratio and using blended cement (i.e., pozzolanic or blast furnace slag cement that reduce the calcium hydroxide content and refine the pore structure of the matrix). Finally, the severity of the attack depends on the content of C_3A and, to a lesser extent, of C_4AF in the cement. Standards in different countries provide for sulfate-resistant cements with a C_3A content below 3–5%.

The European standard EN 206-1 distinguishes three classes of aggressiveness (XA1, XA2, XA3, Table 3.1) with regard to chemical attack based on the concentrations of aggressive substances in (ground) water for sulfate, H^+ (pH), aggressive CO_2, ammonium and magnesium ions; and for sulfate also in soil (Table 3.3). It further recommends a maximum w/c, a minimum cement content, and minimum strength for each of these classes (Table 3.2). When sulfate is the aggressive substance, a sulfate-resistant cement is recommended. When the concentration is outside the limits given, a special study may be needed. American Concrete Institute recommendation ACI 201 fixes analogous limits for the w/c ratio and the type of cement and, furthermore, requires the use of blended cements for highly aggressive environments [22].

3.3.2
Internal Sulfate Attack

Internal sulfate attack is also called delayed ettringite formation (*DEF*). *DEF* is a form of sulfate attack connected with the decomposition of initially formed ettringite and its subsequent formation in completely hardened concrete, after months or even years [23, 24]. Unlike external sulfate attack, the occurrence of *DEF* does not require exposure to an environment containing sulfates, since the source of these ions is in the concrete itself. Moreover, although an increased presence of sulfates, due for instance to contamination of aggregates or an excessive content in the cement, promotes the attack, *DEF* may occur even in concretes with sulfate content within ordinary limits. *DEF* is a relatively young type of deterioration, as it began to be observed and recognized in the 1980s; it has been argued that this is due to the higher amounts of gypsum that are present in recent cements [25].

DEF mainly concerns precast and steam-cured concrete elements [26], although it has been reported that even other causes of temperature increase (e.g., heat of hydration of cement) may promote it [27]. This is why it is also referred to as "heat-induced internal sulfate attack". Ettringite decomposition occurs when concrete reaches temperatures higher than 70–80 °C within a few hours after casting.

In the massive scientific literature dedicated to the topic there is no agreement on the mechanisms proposed for the occurrence of *DEF*, which are quite complex and controversial [28, 29].

DEF requires two necessary conditions: concrete temperature higher than 70 °C in the period following placement, and continuous or intermittent exposure to a wet or water-saturated environment in operating conditions. However, these conditions are not sufficient, and this may be due to the fact that many other factors can either promote or hinder *DEF*.

DEF damage is an expansion-related phenomenon that manifests itself with extensive cracking of cement paste. Cracks are localized both in the paste and along the paste/aggregate interface. The crack opening in the paste/aggregate interface usually increases with aggregate size. The presence of large ettringite crystals in cracks that may be observed on a microscopic scale does not necessarily indicate distress, and is often a consequence of the cracks rather than a cause.

DEF attack is enhanced by any form of pre-existing weakening of concrete, as cracking due to other types of deterioration (freeze–thaw, alkali silica reaction, *ASR*) or to thermal variations. Moreover, there exist specific interactions with other phenomena: for instance, *ASR* has a strong chemical interaction with *DEF*, in that the local reduction of alkalinity brought about by *ASR* at paste/aggregate interface can trigger ettringite formation. Similarly, surface leaching may promote *DEF*, reducing the alkalinity of the pore solution [30].

Prevention Attention during production, such as keeping curing temperature within limits (below 65–70 °C) and delaying steam-curing treatment, is believed to avoid the occurrence of *DEF*. Other preventative measures can be considered in specific cases. *DEF* is often associated with cements that give high early strength (as those mainly used in precasting), so the use of cements with lower strength class, if applicable, or cements with mineral additions (fly ash and slag) may be of help in preventing *DEF*. The use of lower w/b ratios is beneficial since it counteracts cracking of cement paste [24], however, it has also been suggested that it may promote expansion [26]. Limestone aggregates have been reported to be beneficial in preventing expansion, compared to quartz aggregates, maybe due to their intrinsic higher bond with cement paste. Possibly, attention should also be dedicated to exposure conditions, avoiding wetting, since no attack will occur in low-humidity environments.

3.4
Alkali Silica Reaction

Some types of aggregate can react with Na^+, K^+, and OH^- ions in the pore solution, giving rise to detrimental expansion. The principal reactions can take place with aggregate containing certain forms of amorphous or poorly crystalline silica (alkali silica reaction, *ASR*) and with dolomitic limestone aggregate (alkali carbonate reaction).

Alkali silica and alkali aggregate reactions are a significant cause of damage to concrete structures in many countries around the world, for instance in the UK, Denmark, Norway, France, Belgium, The Netherlands, Canada, and the USA. As concrete is usually made with aggregate from local sources, the experience in each of these countries is based on a particular range of aggregates, see for example, [31]. However, a more or less general view has been formed during the last decades [32]. Within the scope of this book, only *ASR* will be discussed here, as it is by far the most important.

The significance of *ASR* for concrete structures is that deleterious expansion causes cracking of the concrete and reduces the tensile strength [33], which may have consequences for the structural capacity [34]; moreover, repair is difficult and specific repair techniques are still in the development stage. In some cases, repair of cracked areas with shotcrete has led to reappearance of cracking due to the introduction of new alkalis and needed to be repeated. An effective repair approach may be removal of cracked concrete and repair with low-alkali mortar or, even better, with slag or fly ash cement mortar; followed by installing a waterproof membrane and/or hydrophobic treatment to prevent further water ingress. A repair project of this type on two bridges in The Netherlands was accompanied with an intensive testing program, including monitoring of further expansion and drying out of the concrete after corrective measures had been taken [35]. Finally, if new sources of aggregate are to be developed, experience from the past cannot be relied on and tests are needed for which methods have to be developed [36].

A series of test methods for the evaluation of alkali reactivity of aggregates have been developed within the framework of RILEM Technical Committee 106-AAR, considering various procedures and degree of acceleration.

3.4.1
Alkali Content in Cement and Pore Solution

Cement contains small amounts of alkali oxides (Na_2O and K_2O), deriving from the minerals from which it is produced. Following hydration these pass into solution as hydroxides:

$$Na_2O + H_2O \rightarrow 2NaOH \tag{3.3}$$

$$K_2O + H_2O \rightarrow 2KOH \tag{3.4}$$

The alkali content in cement is generally expressed as an equivalent percentage of Na_2O by mass of cement. Since the molecular weights of Na_2O and K_2O are, respectively, 62 and 94, the equivalent percentage of Na_2O is calculated with the formula:

$$\%Na_2O_{eq} = \%Na_2O + 0.658 \cdot \%K_2O \tag{3.5}$$

A cement with an effective sodium equivalent content (Na_2O_{eq}) less than or equal to 0.6% by mass is considered to have a low alkali content. It has already been

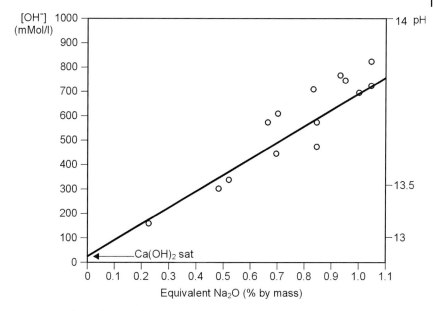

Figure 3.4 Relation between the equivalent percent of alkali in portland cement and OH⁻ concentration in the pore solution of cement paste [32].

seen that the liquid phase present in pores of a portland cement paste or concrete is, for the most part, composed of a solution of NaOH and KOH (Section 2.1). The concentration of these hydroxides increases as Na_2O_{eq} increases, as illustrated in Figure 3.4. Blended cements containing pozzolana, fly ash or blast furnace slag give a resulting alkaline solution of slightly lower pH. Addition of silica fume leads to the lowest pH.

Besides the cement, the other components of concrete (mixing water, aggregates, admixtures and additives) can contain alkalis, although generally in more modest levels. Of these alkalis, only those dissolved in the pore solution can react with aggregates.

The alkali content in concrete (expressed as kg/m³) is determined by multiplying the percentage of Na_2O_{eq} of each component in the mixture by its relative proportion (in kg/m³) and dividing by 100. In general, concrete that contains more than 3 kg/m³ of alkali can be considered to have a high alkali content [37].

3.4.2
Alkali Silica Reaction (ASR)

The reaction mechanism between alkali and reactive siliceous aggregate is complex. It requires the presence of hydroxyl, alkali metal and calcium ions, and water. Simplifying, the hydroxyl ions provoke the destruction of atomic bonds of the siliceous compounds, the alkali ions react with $Si(OH)_4$ complexes to form a fluid

(Na,K)–Si–OH gel, which then exchanges Na and K for Ca, upon which the gel solidifies. The solidified gel absorbs water and swells.

This swelling can induce tensile stresses within the concrete and lead to the appearance of cracking (with a pattern that depends on the geometry of the structure, the layout of the reinforcement and the level of tensile stress). For example, in floors or in foundations, elements that are usually only lightly reinforced and do not have significant tensile stress, the typical pattern of cracking is the so-called "map cracking" with randomly distributed cracks, from which whitish gel may leak. Another typical phenomenon are pop-outs, that is, the expulsion of small portions of concrete.

Development of *ASR* may be very slow and its effects may show even after long periods (up to several decades) [31]. The principal factors influencing this reaction are analyzed below.

Presence and Quantity of Reactive Aggregate The reactivity of silica minerals depends on their crystal structure and composition. The porosity, permeability, and specific surface of aggregates and the presence of Fe- and Al-rich coatings may influence the kinetics of the reaction. Opal, which has a very disordered structure, is the most reactive form of silica; pure, undeformed quartz, having an ordered structure, is generally not reactive. Cristobalite, moganite, quartzine, chalcedony, and microcrystalline quartz have intermediate reactivities.

The extent of reaction depends on the amount of reactive silica present in the aggregate mix. As a rule, there is a range of concentrations of reactive components over which expansion occurs, with a maximum that is called *pessimum*. As the quantity of reactive silica increases beyond the value of the pessimum, expansion decreases gradually until it becomes negligible. For opal, for example, the "reactive" range is 2–10% by volume, with the pessimum at about 5%, while for other, less-reactive, types of silica it may lie at contents of about 30%. For mixes of aggregates from various sources, there may be no pessimum at all.

Alkali Content in the Pore Liquid of Concrete Use of concrete with a low effective alkali content will prevent deleterious *ASR* even in the presence of potentially reactive aggregates. Figure 3.5 shows how the expansive effect induced by the reaction between alkalis and aggregate with chalcedony will be negligible if the equivalent content of Na_2O in concrete is less than $3\,kg/m^3$.

Theoretically, the alkali content in concrete may derive also from external sources, as in marine structures or in bridges where deicing salts are employed. However, the penetration of alkalis into concrete is relatively slow (slower than diffusion of chloride ions) and in most practical cases, external alkalis do not contribute significantly to the total alkali content and thus do not increase the risk of *ASR* [39].

Type and Quantity of Cement Use of pozzolanic or blast furnace slag cement, or addition of a pozzolana or GGBS, can prevent damage caused by *ASR* [31, 37]. As a matter of fact, besides having a diluting effect, the mineral addition reduces the

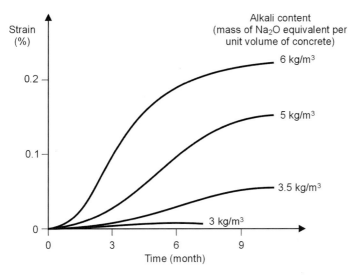

Figure 3.5 Behavior of concrete with reactive aggregate containing chalcedony as a function of alkali content [38].

concentration of OH⁻ ions in the pore solution of cement paste. This is because hydroxyl ions are consumed by the pozzolanic reaction occurring during hydration. Furthermore, the alkali transport is slowed down because of the lower permeability of pozzolanic and blast furnace cement paste. Finally, the hydration products of the mineral additions bind alkali ions to a certain extent, preventing them taking part in the reaction with the silica. Addition of silica fume has a similar effect.

Environment ASR can occur only in moist environments: it has in fact been observed that in environments with a relative humidity below 80–90%, alkalis and reactive aggregate can coexist without causing any damage. The temperature influences the reaction, by favoring it as the temperature increases.

Prevention In conclusion, to prevent ASR it is preferable to use nonreactive aggregate, low-alkali portland cement or blended cements with sufficient amounts of fly ash or slag, such as CEM II/B-V (>25% fly ash), CEM III/B or CEM III/A with >50% slag [37]. However, it should be noted that it is not always possible to avoid using reactive aggregates, because of the limits of local availability. In addition, methods for the evaluation of aggregate reactivity are sometimes difficult because tests can be laborious and sampling may not be easy. Often in a large amount of aggregate, reactive components are localized in a few granules, and the tested sample may not be sufficiently representative. If the presence of reactive aggregate cannot be ruled out with certainty, then the alkali content of the concrete should be limited. This can be achieved, above all, by controlling the percentage of alkali in the cement and the cement content of the concrete. An alternative strategy is using blended cement or adding a pozzolana or GGBS to portland cement. Additions rich in silica, such as silica fume and siliceous fly ashes, are more effective in counteracting expansion compared to additions rich in calcium, such as slag or calcareous fly ashes, and so can be added in lower amounts [40, 41]. Obviously, the above applies to structures exposed to moisture; ASR does not take place in concrete exposed to dry environments.

The use of lithium-based compounds, and in particular of lithium nitrate ($LiNO_3$), is a possible means for preventing ASR in concrete. Lithium compounds can be added to the concrete mix (usually as aqueous solutions, e.g., 30% $LiNO_3$). They can be effective provided they are added in amounts higher than a certain threshold: for instance, a lithium to sodium-plus-potassium molar ratio ([Li]/[Na+K]) of the order of 0.74 is often considered. However, the amount of lithium required depends on the nature of the compound, as well as on the concrete alkali content and aggregate reactivity. The simplest mechanism that explains the protection effect is that lithium interacts with reactive silica similarly to sodium and potassium, but the product is insoluble and does not absorb water nor swell. It seems that the effectiveness is higher for very reactive aggregates (such as opal, chert and volcanic glass), and lower for aggregates that react slowly (such as those containing microcrystalline quartz) [41].

The importance of preventing ASR is evident, since few remedies are available to stop it. Among these, the application of treatments on the surface (that seal cracks and keep concrete dry) and the impregnation of concrete with lithium compounds (either by spraying the concrete surface, by vacuum impregnation or by electrochemical techniques) have been proposed. Their effectiveness depends on the existing level of concrete cracking: the complete suppression of ASR would be difficult to achieve in the presence of extensive cracking.

3.5
Attack by Seawater

Attack of concrete by seawater can be of various types: superficial erosion caused by waves or tides, swelling caused by crystallization of salts, chemical attack by salts dissolved in the water (sulfates, chlorides). Freeze–thaw cycles can aggravate seawater action. The most critical parts of a structure are the tidal and splash zones. Cyclic drying and capillary suction occur in the concrete just above sea level, and water carries the dissolved salts into the concrete. Subsequent evaporation causes these salts to crystallize in the pores, producing stresses that can cause microcracking. The motion of waves and alternating tides also contribute to deterioration of the concrete, favoring removal of the concrete damaged by crystallization of new solid phases. A similar process may occur on concrete structures occasionally reached by waves or sea spray, or on structures on which the wind deposits salt aerosols. The rate of deterioration depends on the quantity of seawater absorbed by the concrete, so that all factors that contribute to obtaining a lower permeability will improve its resistance, even to the point of rendering these processes harmless. Permanent submersion is less severe than tidal or splash zone exposure. For example, it was found that a rather poorly graded plain portland cement concrete with a w/c of 0.54 and cement content of $300\,kg/m^3$ performed well during 16 years of submersion in the North Sea. Examination by polarizing and fluorescence microscopy showed a good and unaltered microstructure, apart from chemical reactions in the outer few millimeters [42]. Some of these superficial changes relate to the formation of brucite, $Mg(OH)_2$, the product of magnesium ions from the seawater and hydroxyl ions from the concrete pore solution. Brucite formation is sometimes regarded as protective; however, in the study mentioned above, the brucite observed did not form continuous layers.

The sensitivity of concrete towards the action of seawater is above all due to the presence of calcium hydroxide and hydrated calcium aluminates, which are both susceptible to reactions with sulfate ions from the seawater. Expansive effects decrease, therefore, as the percentage of C_3A in the cement or the content of $Ca(OH)_2$ in the concrete diminishes. The traditional approach to avoid these reactions has been to reduce the C_3A content of portland cement. For instance, reference [43] recommends a cement content not below $400\,kg/m^3$, while keeping in mind the effects of such high cement contents on cracking due to shrinkage or thermal stresses; a content of tricalcium aluminate below 12% and preferably between 6% and 10%; and the use of good-quality aggregate.

Limiting C_3A may, on the other hand, have an adverse effect on the chloride penetration resistance of concrete. The use of cement with low C_3A (such as ASTM type V) in many marine structures, in particular in the Middle East built in the 1970s and 1980s, has caused many of them to suffer from extensive reinforcement corrosion. Here, it was wrongly assumed that a better resistance against sulfate attack would also mean a better resistance to all adverse effects of seawater [43].

The modern view is that the cements best adapted to seawater are blast furnace slag, fly ash, and pozzolanic cement, because a much lower amount of calcium hydroxide is present in the hydration products and the finer pore structure strongly reduces the transport rate of both sulfate and chloride ions. More detailed recommendations may vary on local experience.

References

1 Biczok, I. (1972) *Concrete Corrosion, Concrete Protection*, 8th edn, Akademiai Kiado, Budapest, Wiesbaden.
2 Lea, F.M. and Hewlett, P.C. (1997) *Lea's Chemistry of Cement and Concrete*, 4th edn, Butterworth-Heinemann.
3 Metha, P.K. and Monteiro, P.J.M. (2006) *Concrete: Microstructure, Properties, and Materials*, 3rd edn, McGrawHill.
4 EN 206-1 (2006) Concrete – Part 1: Specification, Performance, Production and Conformity, European Committee for Standardization.
5 Beddoe, R. and Setzer, M.J. (1988) A low temperature DSC investigation of hardened cement paste subjected to chloride action. *Cement and Concrete Research*, **18**, 249–256.
6 Beddoe, R. and Setzer, M.J. (1990) Phase transformations of water in hardened cement paste. A low temperature DSC investigation. *Cement and Concrete Research*, **20**, 236–242.

7 Powers, T.C. (1975) Freezing effects in concrete, in *Durability of Concrete*, ACI publication SP-47, American Concrete Institute, Detroit, pp. 1–11.

8 RILEM Technical Committee 117-FDC, Setzer, M.J. (ed.) (1995) Draft Recommendation for test method for the freeze–thaw resistance of concrete – tests with water (CF) or with sodium chloride solution (CDF). *Materials and Structures*, **28** (3), 175–182.

9 Visser, J.H.M. and Peelen, W.H.A. (2000) Freeze–thaw damage of concrete: climatic and de-icing salt data in the Netherlands, TNO Building and Construction Research report 2000-BT-MK-R0130 (in Dutch).

10 Massazza, F. (1987) La durabilità del calcestruzzo, in *La Corrosione Delle Strutture Metalliche Ed in Cemento Armato Negli Ambienti Naturali* (ed. P. Pedeferri), Clup, Milan, pp. 365–418.

11 U.S. Bureau of Reclamation (1956) Concrete Laboratory, Report No. C-824, Denver.

12 Thomas, M.D.A. and Folliard, K.J. (2007) Concrete aggregates and the durability of concrete, in *Durability of Concrete and Cement Composites* (eds C.L. Page and M.M. Page), Woodhead Publishing Limited, pp. 247–281.

13 Richardson, M. (2007) Degradation of concrete in cold weather conditions, in *Durability of Concrete and Cement Composites* (eds C.L. Page and M.M. Page), Woodhead Publishing Limited, pp. 282–315.

14 Bieliecki, R. and Schremmer, H. (1987) *Biogenic Sulphuric Acid Attack in Gravity Sewers*, Heft 94 der Mitteilungen des Leichtweiss-Institut für Wasserbau der Technische Universität Braunschweig (in German).

15 Thistlethwayte, D.K.B. (1972) *Control of Sulphides in Sewage Systems*, Butterworths, Sydney–Melbourne–Brisbane; also *Sulfide in Abswasseranlagen: Ursachen, Auswirkungen, Gegenmassnahmen* (ed. N. Klose), Beton-Verlag, Düsseldorf, 1979.

16 Sand, W., Bock, E., and White, D.C. (1987) Biotest system for rapid evaluation of concrete resistance to sulphur-oxidising bacteria. *Materials Performance*, **26**, 14–17.

17 van Mechelen, T. and Polder, R.B. (1992) Biogenic Sulphuric Acid attack on concrete in sewer environment, in *Proc. Int. Conf. Implications of Ground Chemistry and Microbiology for Construction* (ed. A.B. Hawkins), University of Bristol.

18 Dumas, T. (1990) Calcium aluminate binders: an answer to bacterial corrosion. Proc. Symp. Corrosion and Deterioration in Building, CSTB/CEFRACOR, Paris.

19 Heijnen, W.M.M., Borsje, H., and Polder, R.B. (1994) Aluminous cement in sewers. *Cement*, **11**, 68–74 (in Dutch).

20 Page, C.L. (2012) Degradation of reinforced concrete: some lessons from research and practice. *Materials and Corrosion*, **63** (12), 1052–1058.

21 Hagelia, P. and Sibbick, R.G. (2009) Thaumasite Sulfate Attack, Popcorn Calcite Deposition and acid attack in concrete stored at the «Blindtarmen» test site Oslo, from 1952 to 1982. *Materials Characterization*, **60** (7), 686–699.

22 ACI 201.2R-01 (2001) *Guide to durable concrete*, American Concrete Institute.

23 Bensted, J., Brough, A.R., and Page, M.M. (2007) Chemical degradation of concrete, in *Durability of Concrete and Cement Composites* (eds C.L. Page and M.M. Page), Woodhead Publishing Limited, pp. 86–135.

24 Skalny, J., Marchand, J., and Odler, I. (2002) *Sulfate Attack on Concrete*, Spon Press.

25 Hime, W.G. (1996) Delayed ettringite formation – a concern for precast concrete? *PCI Journal*, **41**(4), 26–30.

26 Taylor, H.F.W., Famy, C., and Scrivener, K.L. (2001) Delayed ettringite formation. *Cement and Concrete Research*, **31**, 683–693.

27 Diamond, S. (1996) Delayed ettringite formation – processes and problems. *Cement and Concrete Composites*, **18**, 205–215.

28 Collepardi, M. (2003) A state-of-the-art review on delayed ettringite attack on concrete. *Cement and Concrete Composites*, **25**, 401–407.

29 Scrivener, K. and Skalny, J. (2005) Conclusions of the International RILEM TC 186-ISA workshop on internal sulfate attack and delayed ettringite formation

(4–6 September 2002, Villars, Switzerland). *Materials and Structures*, **38**, 659–663.

30. Diamond, S. (2000) The relevance of laboratory studies on delayed ettringite formation to DEF in field concretes. *Cement and Concrete Research*, **30**, 1987–1991.

31. Nijland, T.G. and Siemes, A.J.M. (2002) Alkali-silica reaction in the Netherlands: experiences and current research. *Heron*, **47** (2), 81–86.

32. Nixon, P. and Page, C.L. (1987) Pore solution chemistry and alkali aggregate reaction, in *ACI SP-100 Concrete Durability*, vol. 2 (ed. J.M. Scanlon), American Concrete Institute, pp. 1833–1862.

33. Siemes, A.J.M., Han, N., and Visser, J.H.M. (2002) Unexpectedly low tensile strength in concrete structures. *Heron*, **47** (2), 111–124.

34. den Uijl, J.A. and Kaptijn, N. (2002) Structural consequences of ASR: an example on shear capacity. *Heron*, **47** (2), 125–140.

35. Borsje, H., Peelen, W.H.A., Postema, F.J., and Bakker, J.D. (2002) Monitoring alkali-silica reaction in structures. *Heron*, **47** (2), 95–110.

36. Larbi, J.A. and Visser, J.H.M. (2002) A study of the ASR of an aggregate with high chert content by means of ultra-accelerated mortar bar test and pore fluid analysis. *Heron*, **47** (2), 141–159.

37. Nijland, T.G. and de Bruijn, W.A. (2002) New Dutch guideline on ASR-prevention. *Heron*, **47** (2), 87–94.

38. Sibbick, R.G. and Page, C.L. (1992) Threshold alkali contents for expansion of concretes containing British aggregates. *Cement and Concrete Research*, **22**, 990–994.

39. Visser, J.H.M. and Polder, R.B. (2000) Alkali-silica reaction in concrete: penetration of alkalis, TNO Building and Construction Research report 2000-BT-MK-R0207 (in Dutch).

40. Hobbs, D.W. (2002) Alkali-silica reaction in concrete, in *Structure and Performance of Cements*, 2nd edn (eds J. Bensted and P. Barnes), Spon Press, pp. 265–281.

41. Thomas, M.D.A., Fournier, B., Folliard, K.J., Ideker, J.H., and Resendez, Y. (March 2007) *The Use of Lithium to Prevent or Mitigate Alkali-Silica Reaction in Concrete Pavements and Structures*, FHWA-HRT-06-133, Federal Highway Administration, McLean, VA.

42. Polder, R.B. and Larbi, J.A. (1995) Investigation of concrete exposed to North Sea water submersion for 16 years. *Heron*, **40** (1), 31–56.

43. Mehta, P.K. (1988) Durability of concrete exposed to marine environment: a fresh look, in *Concrete in Marine Environment* (ed. V.M. Malhotra), ACI SP-109, American Concrete Institute, pp. 1–30.

4
General Aspects

During hydration of cement a highly alkaline pore solution (pH between 13 and 13.8), principally of sodium and potassium hydroxide, is obtained (Section 2.1.1). In this environment the thermodynamically stable compounds of iron are iron oxides and oxyhydroxides. Thus, on ordinary reinforcing steel embedded in alkaline concrete a thin protective oxide film (the *passive film*) is formed spontaneously [1–3]. This passive film is only a few nanometers thick and is composed of more or less hydrated iron oxides with varying ratios between Fe^{2+} and Fe^{3+} [4]. The protective action of the passive film is immune to mechanical damage of the steel surface. It can, however, be destroyed by carbonation of concrete or by the presence of chloride ions, and the reinforcing steel is then depassivated [5].

4.1
Initiation and Propagation of Corrosion

The service life of reinforced concrete structures can be divided in two distinct phases (Figure 4.1). The first phase is the *initiation phase*, in which the reinforcement is passive but phenomena that can lead to loss of passivity, for example, carbonation or chloride penetration in the concrete cover, take place; this phase ends with the initiation of corrosion (the way of defining the initiation of corrosion will be discussed later). The second phase is the *propagation phase*, which begins when the steel is depassivated (i.e., corrosion initiates) and finishes when a limiting state is reached beyond which consequences of corrosion cannot be further tolerated [6, 7].

4.1.1
Initiation Phase

During the initiation phase aggressive substances (CO_2, chlorides) that can depassivate the steel penetrate from the surface into the bulk of the concrete:

Corrosion of Steel in Concrete: Prevention, Diagnosis, Repair, Second Edition. Luca Bertolini,
Bernhard Elsener, Pietro Pedeferri, Elena Redaelli, and Rob Polder.
© 2013 Wiley-VCH Verlag GmbH & Co. KGaA. Published 2013 by Wiley-VCH Verlag GmbH & Co. KGaA.

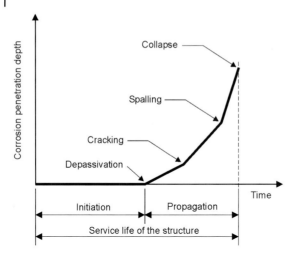

Figure 4.1 Initiation and propagation periods for corrosion in a reinforced concrete structure (from Tuutti's model) [7].

- *carbonation:* beginning at the surface of concrete and moving gradually towards the inner zones, the alkalinity of concrete may be neutralized by carbon dioxide from the atmosphere, so that the pH of the pore liquid of the concrete decreases to a value around 9, where the passive film is no more stable (Chapter 5);
- *chloride ions* from the environment can penetrate into the concrete and reach the reinforcement; if their concentration at the surface of the reinforcement reaches a critical level, the protective layer may be locally destroyed (Chapter 6).

The duration of the initiation phase depends on the cover depth and the penetration rate of the aggressive agents as well as on the concentration necessary to depassivate the steel. The influence of concrete cover is obvious and design codes define cover depths according to the expected environmental class (Chapter 11). The rate of ingress of the aggressive agents depends on the quality of the concrete cover (porosity, permeability) and on the microclimatic conditions (wetting, drying) at the concrete surface. Additional protective measures can be used to prolong the initiation phase (Chapters 13–15, 20) [8].

The initiation time can also be affected by any form of polarization of the reinforcement: for instance, a small amount of cathodic polarization makes the occurrence of pitting less probable and increases the initiation time, while an anodic polarization, due for instance to an electric connection with reinforcement with higher potential or external electric fields, may promote the opposite. In structures affected by electric fields, DC *stray current* in the concrete can enter the reinforcement in some areas (i.e., it passes from the concrete to the steel) and return to the concrete in a remote site. This may promote corrosion initiation (Chapter 9).

4.1.2
Propagation Phase

Breakdown of the protective layer is the necessary prerequisite for the initiation of corrosion. Once this layer is destroyed, corrosion will occur only if water and oxygen are present on the surface of the reinforcement. The corrosion rate, which varies considerably depending on temperature and humidity, determines the time it takes to reach any one of several undesired events in the life of the structure, such as severe loss of cross section of the reinforcement, cracking of the concrete cover, spalling and delamination of the concrete cover, and, eventually, collapse (Figure 4.1).

Carbonation of concrete leads to complete dissolution of the protective layer. Chlorides, instead cause localized breakdown, unless they are present in very large amounts. Therefore:

- corrosion induced by carbonation takes place on the whole surface of steel in contact with carbonated concrete (*general corrosion*);

- corrosion induced by chlorides is localized (*pitting corrosion*), with penetrating attack of limited area (*pits*) surrounded by noncorroded areas. Only when very high levels of chlorides are present (or the pH decreases) may the passive film be destroyed over wide areas of the reinforcement and the corrosion will be of a general nature.

If depassivation due to carbonation or chlorides occurs only on a part of the reinforcement, a *macrocell* can develop between corroding bars and those bars that are still passive (and connected electrically to the former). This may increase the rate of attack of the reinforcement that is already corroding (Chapter 8).

On high-strength steel used in prestressed concrete (but not with common reinforcing steel) under very specific environmental, mechanical loading, metallurgical and electrochemical conditions, *hydrogen embrittlement* can occur, which may lead to brittle fracture of the material (Chapter 10).

4.2
Corrosion Rate

The corrosion rate is usually expressed as the penetration rate and is measured in µm/year. Often, especially in laboratory tests, it is expressed in electrochemical units, that is, in mA/m^2 or in $\mu A/cm^2$. In the case of steel, $1\,mA/m^2$ or $0.1\,\mu A/cm^2$ corresponds to a loss of mass equal to approximately $9\,g/m^2$ year and a penetration rate of about $1.17\,\mu m/year$. As a general figure, the corrosion rate can be considered negligible if it is below $2\,\mu m/year$, low between 2 and $5\,\mu m/year$, moderate between 5 and $10\,\mu m/year$, intermediate between 10 and $50\,\mu m/year$, high between 50 and $100\,\mu m/year$ and very high for values above $100\,\mu m/year$.

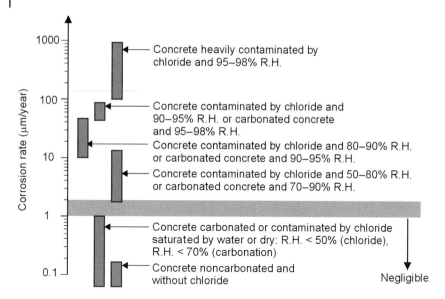

Figure 4.2 Schematic representation of corrosion rate of steel in different concretes and exposure conditions (after [9], modified).

Figure 4.2 shows the typical ranges of variation of the corrosion rate in carbonated or chloride-contaminated concrete as a function of relative humidity (R.H.) of the environment.

4.3
Consequences

The consequences of corrosion of steel reinforcement involve several aspects connected with the condition of the structure, such as its esthetic appearance, serviceability, safety, and structural performance. Corrosion of reinforcement often (but by no means always) manifests itself through cracking of the concrete cover much earlier than structural consequences occur; however, even spalling and/or delamination of concrete cover may affect the serviceability and safety of a structure.

The main consequences of corrosive attack are shown in Figure 4.3. Corrosion is often indicated by rust spots that appear on the external surface of the concrete, or by cracking of the concrete cover produced by the expansion of the corrosion products. These products in fact occupy a much greater volume than the original steel bar. The volume of the corrosion products can be from 2 to 6 times greater than that of iron they are derived from, depending on their composition and the degree of hydration. For example, the volumes of the oxides Fe_3O_4, $Fe(OH)_2$, $Fe(OH)_3$, $Fe(OH)_3 \cdot 3H_2O$ are respectively 2, 3, 4, and 6 times greater than that of iron. In general, the volume of the products of corrosion, a mixture of these oxides,

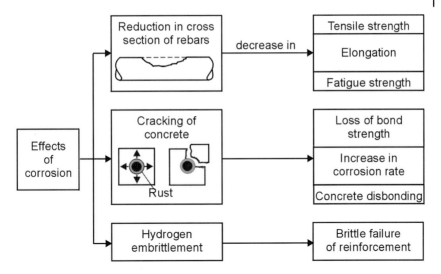

Figure 4.3 Structural consequences of corrosion in reinforced concrete structures [10].

can be considered 3–4 times the volume of iron. Consequently, the products of corrosion can produce tensile stresses that generate cracks in the concrete cover (Figure 4.4a), spalling in a localized area, or complete delamination (Figure 4.4b). Reduction of bonding of the reinforcement to the concrete may also occur.

In cases of chloride-induced localized corrosion, the cross section of the reinforcement and thus its loading capacity and its fatigue strength can be significantly reduced long before any sign of corrosion becomes visible at the concrete surface (Figure 4.4c). This type of corrosion attack is typical for wet, chloride-containing concrete as it may occur under defective membranes on bridge decks. The corrosion products formed are either soluble (chloro-hydroxy complexes) or in the form of "black rust" (magnetite) that occupies a low volume [11].

Finally, under very special conditions that lead to hydrogen embrittlement of high-strength steel, brittle failure of some types of prestressing steel can take place (Figure 4.4d).

4.4
Behavior of Other Metals

The behavior of metallic materials other than ordinary carbon steel used as reinforcement should be briefly mentioned. Zinc, galvanized steel, and stainless steels will be dealt with in Chapter 15.

Aluminum is passive in environments with a pH between 4 and 9, but not in alkaline media such as the pore solution of noncarbonated concrete. When it comes in contact with fresh cement paste, aluminum reacts by developing hydrogen,

Figure 4.4 Examples of consequences of corrosion of steel in concrete: (a) cracking of column and cross beam; (b) spalling and delamination of the concrete cover; (c) reduction of cross section of the rebar due to pitting corrosion; (d) brittle failure of prestressing tendons due to hydrogen embrittlement.

sometimes vigorously. Once the concrete has hardened, aluminum and aluminum alloys suffer general corrosion; the higher the moisture content of concrete, the higher the corrosion rate. In dry concrete the corrosion rate is negligible. Considerable loss of mass can take place in wet concrete, leading to the production of voluminous corrosion products that can spall off the concrete cover. The corrosion rate tends to decrease over time, since the corrosion products hinder the transport of alkalis to the corroding surface. The mass loss is lower in concrete with lower pH, and it becomes negligible in carbonated concrete [12]. The reactivity of aluminum in alkaline environments is used to make cellular concrete or mortar, or expansive grouts: aluminum powder is added to portland cement paste and, when the mix is plastic, hydrogen gas develops, producing air entrainment and volume increase. *Lead* too corrodes in alkaline environments. Its initial reactivity is lower than that of aluminum. However, the attack in humid concrete continues, because the environment prevents the formation of a layer of basic lead carbonate, which is usually present on the surface of lead exposed to neutral environments (e.g., the atmosphere or fresh water). *Copper* and its alloys do not corrode, unless in the presence of ammonia compounds, for instance coming from some admixtures; in this case, phenomena of stress corrosion cracking can occur. *Titanium* is perfectly passive in all types of concrete (with or without chlorides, carbonated or uncarbonated) even at very oxidizing potentials. That is why it is used as a base metal for anodes in cathodic protection of reinforcement. *Nickel* and its alloys behave optimally, even in the presence of chlorides and in carbonated concrete.

In principle, galvanic coupling may develop in concrete if steel reinforcement is electrically connected to reinforcement or inserts made of other metals. According to the potential difference between the two and to humidity conditions, possible galvanic coupling may take place and promote corrosion of either metal.

References

1. Pourbaix, M. (1973) *Lectures on Electrochemical Corrosion*, Plenum Press, New York.
2. Gouda, V.K. (1970) Corrosion and corrosion inhibition of reinforcing steel–I. Immersed in alkaline solutions. *British Corrosion Journal*, **5**, 198–203.
3. Arup, H. (1983) The mechanisms of the protection of steel by concrete, in *Corrosion of Reinforcement in Concrete Construction* (ed. A.P. Crane), Hellis Horwood Ltd, Chichester, pp. 151–157.
4. Rossi, A., Puddu, G., and Elsener, B. (2007) The surface of iron and Fe10Cr alloys in alkaline media, in *Corrosion of Reinforcement in Concrete. Mechanisms, Monitoring, Inhibitors and Rehabilitation Techniques* (eds M. Raupach, B. Elsener, R. Polder, and J. Mietz), European Federation of Corrosion Publication number 38, Woodhead Publishing Limited, Cambridge, pp. 44–61.
5. Pedeferri, P. and Bertolini, L. (2000) *Durability of Reinforced Concrete*, McGrawHill Italia, Milan (in Italian).
6. Bažant, Z.P. (1979) Physical model for steel corrosion in concrete sea structures. *Journal of the Structural Division*, ASCE, **105** (ST6), 1137–1166.
7. Tuutti, K. (1982) *Corrosion of Steel in Concrete*, Swedish Foundation for Concrete Research, Stockholm.
8. Page, C.L. (6–8 April 1998) Corrosion and its control in reinforced concrete, The sixth Sir F. Lea Memorial Lecture, 26th Annual convention of the Institute of Concrete Technology, Bosworth (UK).
9. Andrade, C., Alonso, M.C., and Gonzales, J.A. (1990) An initial effort to use the corrosion rate measurements for estimating rebar durability, in *Symposium on Corrosion Rates of Reinforcement in Concrete* (eds N.S. Berke, V. Chaker, and D. Whiting), American Society for Testing Materials, Philadelphia, ASTM STP 1065, pp. 29–37.
10. CEB (1992) Durable Concrete Structures, Bulletin d'information N.183, Lausanne.
11. Angst, U., Elsener, B., Jamali, A., and Adey, B. (2012) Concrete cover cracking owing to reinforcement corrosion–theoretical considerations and practical experience. *Materials and Corrosion*, **63** (12), 1069–1077.
12. Nürnberger, U. (2007) Corrosion of metals in contact with mineral building materials, in *Corrosion of Reinforcement in Concrete. Mechanisms, Monitoring, Inhibitors and Rehabilitation Techniques* (eds M. Raupach, B. Elsener, R. Polder, and J. Mietz), European Federation of Corrosion Publication number 38, Woodhead Publishing Limited, Cambridge, pp. 1–9.

5
Carbonation-Induced Corrosion

Carbonation is the reaction of carbon dioxide from the air with alkaline constituents of concrete [1]. It has important effects with regard to corrosion of embedded steel. The first consequence is that the pH of the pore solution drops from its normal values of pH 13 to 14, to values approaching neutrality. If chlorides are not initially present in the concrete, the pore solution following carbonation is composed of almost pure water (Section 2.1.1). This means that the steel in humid carbonated concrete corrodes as if it was in contact with water [2, 3]. A second consequence of carbonation is that chlorides bound in the form of calcium chloroaluminate hydrates and otherwise bound to hydrated phases may be liberated, making the pore solution even more aggressive [2–4].

5.1
Carbonation of Concrete

In moist environments, carbon dioxide present in the air forms an acid aqueous solution that can react with the hydrated cement paste, which tends to neutralize the alkalinity of concrete (this process is known as *carbonation*). Also, other acid gases present in the atmosphere, such as SO_2, can neutralize the alkalinity of concrete, but their effect is normally limited to the surface and the outer few millimeters.

The alkaline constituents of concrete are present in the pore liquid (mainly as sodium and potassium hydroxides, Section 2.1.1) but also in the solid hydration products, for example, $Ca(OH)_2$ or C–S–H. Calcium hydroxide is the hydrate in the cement paste that reacts most readily with CO_2. The reaction, that takes place in aqueous solution, can be schematically written as:

$$CO_2 + Ca(OH)_2 \xrightarrow{H_2O, \; NaOH} CaCO_3 + H_2O \quad (5.1)$$

This is the reaction of main interest, especially for concrete made of portland cement, even though carbonation of C–S–H also occurs when $Ca(OH)_2$ becomes depleted, for instance by pozzolanic reaction in concrete made with blended cement [1]. Analysis of the reactions of major cement paste constituents and experimental work have been used to model carbonation several decades ago [5].

Corrosion of Steel in Concrete: Prevention, Diagnosis, Repair, Second Edition. Luca Bertolini,
Bernhard Elsener, Pietro Pedeferri, Elena Redaelli, and Rob Polder.
© 2013 Wiley-VCH Verlag GmbH & Co. KGaA. Published 2013 by Wiley-VCH Verlag GmbH & Co. KGaA.

This work includes accelerated carbonation and provides a basis for the square-root model for the progress of carbonation (see below). Recent studies using instrumental techniques such as thermogravimetric analysis (TGA) and gammadensitometry of accelerated carbonation have supported views that C–S–H plays a role in carbonation of concrete [6]. One of the results is that the carbonation front may not be sharp, depending on kinetic effects, such as deposition of calcium carbonate on calcium hydroxide particles, slowing down their further reaction. Modeling studies have supported the view that C–S–H participates and that the effect of unhydrated cement minerals C_3S and C_2S, although they react too, is minor [7]. Another study using neutron diffraction applied in real time to accelerated carbonation and TGA has shown that ettringite is also carbonated [8]. In contrast to other studies the authors suggest that C–S–H is carbonated faster than $Ca(OH)_2$. Further work using advanced instrumentation is needed to improve understanding of the details of the processes involved. In any case, volumetric data suggest that carbonation of $Ca(OH)_2$ causes expansion and thus a decrease in porosity, while carbonation of C–S–H causes shrinkage, thus an increase in porosity. This is particularly important for blended cements, with lower amounts of $Ca(OH)_2$. A study using various microstructural methods (environmental scanning electron microscopy, mercury intrusion porosimetry, and nanoindentation) has shown that in cement with a high blast furnace slag percentage the microstructure coarsens and the porosity increases upon carbonation, in particular in the interfacial transition zone [9]. Consequently, the mechanical properties on the microscale in slag cement concrete deteriorate due to carbonation, which is relevant for its relatively lower resistance against frost salt scaling. On the other hand, in the case of concrete made with portland cement, carbonation reduces the porosity and leads to an increased strength. It should be kept in mind that these studies have involved accelerated carbonation, using CO_2 concentrations of a few percent up to 100% by volume. Most authors point out that under natural carbonation, phenomena may evolve differently.

5.1.1
Penetration of Carbonation

The carbonation reaction starts at the external surface and penetrates into the concrete producing a low pH front. The measurement of the depth of carbonation is normally carried out by spraying an alcoholic solution of phenolphthalein on a freshly broken face. The areas where pH is greater than 9 take on a pinkish color typical of phenolphthalein in a basic environment, while the color of carbonated areas remains unchanged (Section 16.3.1).

The rate of carbonation decreases with time, as CO_2 has to diffuse through the pores of the already carbonated outer layer. The penetration with time of carbonation can be described by:

$$d = K \cdot t^{1/n} \tag{5.2}$$

where d is the depth of carbonation (mm) and t is time (years). Often the exponent n is approximately equal to two and, therefore, a square-root trend can be considered: $d = K \cdot \sqrt{t}$. The carbonation coefficient K (mm/year$^{1/2}$) includes the diffusion coefficient D, the CO_2 concentration at the surface and the amount of alkaline components that have to be consumed by the CO_2, assuming that they are constant over time. Under these conditions K can be assumed as a measure of the rate of penetration of carbonation for given concrete and environmental conditions. In dense and/or wet concrete, however, the reduction of the carbonation rate with time is stronger than that described by the parabolic formula, so that $n > 2$; in very dense concrete the carbonation rate even tends to become negligible after a certain time.

5.1.2
Factors That Influence the Carbonation Rate

The rate of carbonation depends on both environmental factors (humidity, temperature, concentration of carbon dioxide) and factors related to the concrete (mainly its alkalinity and permeability).

Humidity The rate of carbonation varies with humidity of concrete due to two causes. First, as already seen in Chapter 2, diffusion of carbon dioxide within concrete is facilitated through the air-filled pores, but it is very slow through those filled with water (the diffusion of CO_2 in water is four orders of magnitude slower than in air).

The rate of diffusion of CO_2 consequently decreases with an increase in humidity of the concrete until it becomes nil in water-saturated concrete. This means that when the concrete is wet, CO_2 does not penetrate it. On the other hand, the carbonation reaction occurs only in the presence of water so that it becomes negligible in dry concrete.

The carbonation rate, and thus the value of K, will change on passing from a wet or humid climate to a dry one. Under conditions of equilibrium with an environment of constant relative humidity, the carbonation rate may be correlated to the humidity of the environment as shown in Figure 5.1 [4]. The interval of relative humidity most critical for promoting carbonation is from 60% to 70%, that is, a value similar to or slightly lower than the annual averages of the relative humidity of the atmosphere in North-Western Europe. Nevertheless, the carbonation rate may be lower if the structure is subjected to periodic wetting [10, 11]. For example, in the case of the external walls of a building, the rate of carbonation will be much higher in an area sheltered from rain than in an unsheltered area. Furthermore, on the outside of a building located in an area exposed to wetting, the rate of carbonation is lower than that on the inside, where the concrete is drier (Figure 5.2) [10]. Important parameters are the wetting time, as well as the frequency and duration of wetting–drying cycles. Since wetting of concrete is faster than drying, more frequent although shorter periods of wetting are more effective in reducing the penetration of carbonation than less frequent and longer periods, as can be seen in Figures 5.3 and 5.4 [10, 12].

It should be stressed that the microclimate plays an essential role in carbonation of real structures. The carbonation rate may vary from one part of a structure to

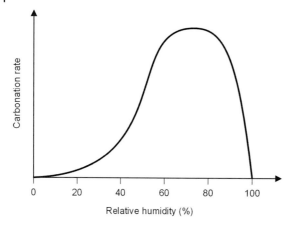

Figure 5.1 Schematic representation of the rate of carbonation of concrete as a function of the relative humidity of the environment, under equilibrium conditions [4].

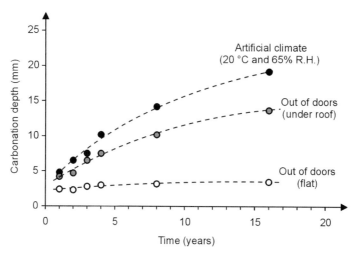

Figure 5.2 Example of penetration of carbonation in concrete in relation to the microclimate (average of 27 concretes) [10].

another (e.g., if one part of the structure is permanently sheltered, the rate of penetration will be considerably higher than in other parts exposed to rain), or passing from the outer layers of the concrete to the inner ones (the outer layers will be drier during the phase of drying, while they will be wetter during periods of wetting). The carbonation of concrete can thus be very variable even in different parts of a single structure.

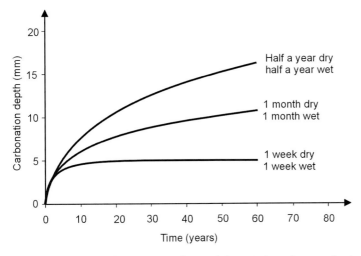

Figure 5.3 Influence of the duration of wet and dry periods on the rate of carbonation for a portland cement concrete with w/c ratio of 0.6 and during the dry cycle exposed to 65% R.H. [10].

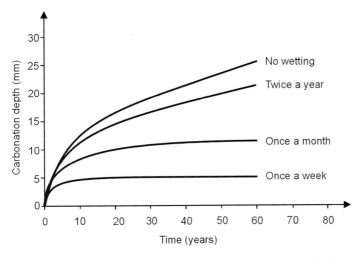

Figure 5.4 Influence of the cycles of wetting and drying on the rate of carbonation for a portland cement concrete with w/c ratio of 0.6 and exposed to 65% R.H. [10].

CO_2 Concentration The concentration of carbon dioxide in the atmosphere may vary from 0.03% (300 ppm by volume, or ppm[V]) in rural environments to more than 0.1% in urban environments. An increase of the CO_2 concentration in the atmosphere will accelerate carbonation. Comparatively high concentrations can be reached under specific exposure conditions, such as inside motor vehicle tunnels.

Monitoring of CO_2 in two tunnels in the Netherlands has shown that while mean values inside and outside the tunnels were in the range of 400 to 600 ppm, near the exits maximum values were measured of 1000 to 1200 ppm [13]. As the CO_2 content in the air increases, the carbonation rate increases. Accelerated tests carried out in the laboratory to compare the resistance to carbonation in different types of concrete show that, indicatively, one week of exposure to an atmosphere containing 4% CO_2 will cause the same penetration of carbonation as a year of exposure to a normal atmosphere [14]. Some researchers suggest that with a high concentration of CO_2 the porosity of carbonated concrete is higher than that obtained by exposure to a natural atmosphere, particularly if the concrete has been made with blended cement or has a high cement content. However, this is controversial, since it was shown that even 100% CO_2 under increased pressure, produced the same microstructure as natural carbonation [15].

Temperature All other conditions being equal, especially that of humidity, which is in general the most important single parameter, an increase in temperature will raise the rate of carbonation.

Concrete Composition The permeability of concrete has a remarkable influence on the diffusion of carbon dioxide and thus on the carbonation rate. A decrease in the w/c ratio, by decreasing the capillary porosity of the hydrated cement paste, slows down the penetration of carbonation as shown in the example of Figure 5.5 [16, 17]. Nevertheless, the advantages of a lower w/c ratio can only be achieved if concrete is properly cured, since poor curing hinders the hydration of the cement paste and leads to a more porous cement matrix. It should be stressed that poor curing will mainly affect the concrete cover, that is, the part that is aimed at pro-

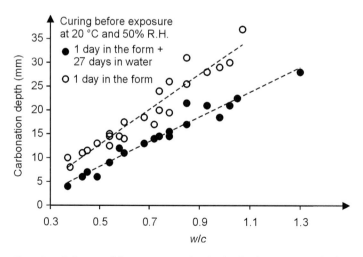

Figure 5.5 Influence of the w/c ratio on the depth of carbonation (portland cement concrete, 6 years, 20 °C, 50% R.H.) [16, 17].

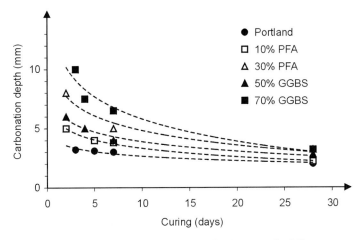

Figure 5.6 Influence of curing on the depth of carbonation for different cement pastes (6 months, 0.03% CO_2) of portland cement and blended cements with fly ash (PFA) and ground granulated blast furnace slag (GGBS) [17, 18].

tecting the reinforcement. In fact, the outer layer of concrete is the part most susceptible to evaporation of water (resulting in poor curing).

The type of cement also influences the carbonation rate. In fact, the capacity of concrete to fix CO_2 is proportional to the alkalinity in its cement paste. In portland cement, about 64% of the mass of the original cement is composed of CaO (mainly converted to solid portlandite) and about 0.5–1.5% of Na_2O and K_2O (mainly in solution as NaOH and KOH). Considering calcium oxide only, the quantity of CO_2 (molecular weight 44) that can react with a concrete produced with $300 \, kg/m^3$ portland cement that we can suppose is composed by 64% of CaO (molecular weight 56) is: $300 \times 0.64 \times 44/56 \approx 150 \, kg/m^3$. In the case of blast furnace slag cement with 70% of GGBS, the percentage of CaO is only 44%. For other blended cements, the quantity of CaO is somewhere between these two values [3]. For blended cement, hydration of pozzolanic materials or GGBS also leads to a lower $Ca(OH)_2$ content in the hardened cement paste that may increase the carbonation rate. Nevertheless, the w/c ratio and curing are still the most important factors. In fact, the denser structure produced by hydration of blended cements (Section 1.3) may slow down the diffusion of CO_2. In Figure 5.6 it can be seen that the lower alkalinity of cements with addition of fly ash or blast furnace slag can be compensated by the lower permeability of their cement pastes, if cured properly [17, 18].

5.2
Initiation Time

The time for initiation of carbonation-induced corrosion is the time required for the carbonation front to reach a depth equal to the thickness of the concrete cover.

It depends on all the factors mentioned above (that influence the carbonation rate) and on the thickness of the concrete cover. If the evolution of carbonation in the course of time and the thickness of the concrete cover are known, the initiation time can be evaluated. It should, however, be taken into consideration that the carbonation front may not be uniform across the concrete surface.

5.2.1
Parabolic Formula

Usually the square-root formula is used to describe the penetration of carbonation:

$$d = K \cdot \sqrt{t} \tag{5.3}$$

This gives fairly accurate predictions, though it tends to overestimate, at least for longer times, the penetration in the case of low-porosity portland cement concrete. Values of K depend on all the factors discussed above and thus change as a function of concrete properties and environmental conditions. In practice, an accurate prediction of K, and thus the time of depassivation of the steel, is complex, above all because this parameter may change in time or in different parts of a single structure.

For existing structures the carbonation depth of any part of the structure can be measured after a given period of time and thus the value of K and its spatial variation can be calculated. By assuming that the average exposure conditions will not change in the future, these values of K may be utilized to extrapolate the carbonation depth to a later point in time.

Values of K found for real structures exposed to the atmosphere, but protected from rain, vary from 2 to 15 mm/year^{½}. Indicatively: $2 < K < 6$ for concrete of low porosity (that is well compacted and cured, with low w/c), $6 < K < 9$ for concrete of medium porosity and $K > 9$ for highly porous concrete (poorly compacted and cured, with high w/c).

From Figure 5.7, which shows the progress in time of the carbonation depth for different values of K, it can be seen that the carbonation depth is less than 20 mm (minimum thickness of concrete cover in many existing structures) after 50 years, only if K is less than 2.82. This means that in areas sheltered from rain, 20 mm of concrete cover lead to an initiation time of over 50 years only in dense concrete.

5.2.2
Other Formulas

Some other empirical models have been proposed to take into direct account the effect of several parameters that influence the carbonation rate. Although these models might be used in the design for durability, they require as input some experimental constants whose evaluation is difficult in practice. For instance Tuutti [4] has proposed a formula based on a nonstationary process of diffusion,

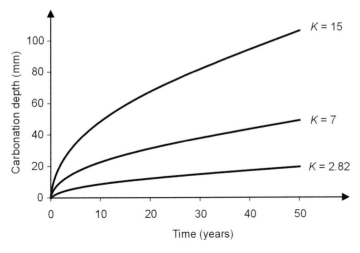

Figure 5.7 Depth of carbonation, calculated by the simplified function $s = K \cdot t^{1/2}$, in relation to time and to K (constant relative humidity, no wetting of concrete).

which evaluates the carbonation depth at a given time on the basis of the concentration of CO_2 in air, the quantity of CO_2 that reacts with the concrete, and the diffusion coefficient of CO_2. Bakker [12] suggests a formula that takes into consideration the duration of wetting–drying cycles on the hypothesis that the progress of carbonation is negligible during wetting. Parrott's formula [19] correlates the depth of carbonation at time t to the concrete's permeability to air, the CaO content in the unhydrated cement and the humidity of the environment.

5.3
Corrosion Rate

Once the carbonation front has reached the reinforcement, steel is depassivated and corrosion can occur if oxygen and water are present. If conditions of complete and permanent saturation of the concrete are excluded, a quantity of oxygen sufficient to permit the corrosion process can always reach the surface of the steel.

5.3.1
Carbonated Concrete without Chlorides

In this case, the corrosion rate is governed by the resistivity of concrete (i.e., corrosion is under ohmic control, as described in Section 7.3). Figure 5.8 depicts the typical relationship between the corrosion rate of steel and the electrical resistivity of carbonated concrete [20].

The moisture content is the main factor in determining the resistivity of carbonated concrete (Section 2.5.3). Of secondary importance, even though significant

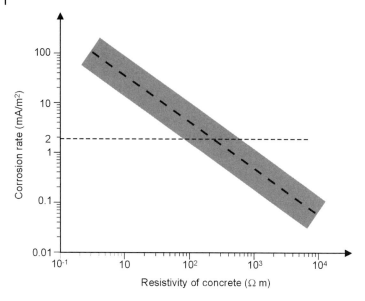

Figure 5.8 Schematic representation of the corrosion rate in carbonated concrete as a function of the resistivity of concrete (after [20]).

for the range of relative humidity from 60 to 90%, is the microstructure of the concrete and the factors determining it (type of cement, w/c ratio, curing, etc.) that are, however, important in affecting the carbonation rate and the initiation time.

Since, at least in high-quality concrete, the rate of corrosion is negligible for relative humidities below 80%, the *time of wetness* (ω, usually defined as the fraction of time in which the relative humidity is greater than 80%) can be introduced to evaluate the penetration of attack (p) in time t. It is then assumed that corrosion propagates only while concrete is wet (i.e., R.H. > 80%). It has also been observed that the corrosion rate tends to decrease with time. In fact, corrosion products, although not able to passivate the reinforcement, can reduce the corrosion rate [2].

In Figure 5.9 the corrosion rate in carbonated concrete is shown as a function of relative humidity of the environment; these data are based on thousands of readings taken mainly in Spain [2]. It can be seen how the maximum corrosion rates, on the order of 100–200 μm/year, will only be reached in very wet environments with relative humidity approaching 100%. Instead, for typical conditions of atmospheric exposure, that is, R.H. = 70–80%, maximum values are between 5 and 50 μm/year (about 5–50 mA/m^2). The average values are one order of magnitude lower.

From the values shown it is evident that the corrosion rate can be considered negligible, unless the humidity is high, or the duration and frequency of periods of condensation of water on the concrete surface can cause variations in the moisture content at the depth of the reinforcement. For example, in structures exposed to indoor environments, or those sheltered from rain, the fact that the concrete is

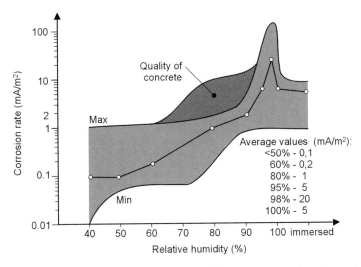

Figure 5.9 Maximum and minimum values of the corrosion rate in carbonated concrete as a function of environmental humidity [2].

carbonated at the depth of the reinforcement rarely presents any problem, since condensation or wetting of the concrete surface for a short period will not result in an increase in moisture at the depth of the reinforcement. It should be pointed out that if, for any reason (such as leakage or infiltration) water penetrates the carbonated concrete, the corrosion rate will no longer be negligible.

A comparison between Figures 5.1 and 5.9 shows that when the carbonation rate is maximum, the rate of corrosion is modest, and vice versa. For this reason the worst situations are those characterized by long alternating periods of low humidity and high humidity. It depends on the climate, whether long dry and wet periods occur. In the "coastal" climate in Northern Europe, dry periods typically are as short as a week. In Southern Europe, however, long dry seasons (summer) and long wet seasons (winter) occur. The latter climate is more probable to cause corrosion damage due to carbonation with regard to atmospheric structures. It should also be noted that the observations mentioned above do not apply for very low cover depths, in particular those well below 15 to 20 mm. It is observed in practice that concrete may be damaged by corrosion of steel at such low cover depths. It appears that even in climates with frequent wetting and drying, dry periods can be so long as to allow carbonation down to those shallow depths in 10 to 20 years, while during wet periods enough water may reach the reinforcement to cause significant corrosion and subsequent cracking and spalling of the cover. Another example may be in specific types of structures. For instance, in tunnels long dry periods and long wet periods (due to condensation) may occur. Consequently, the risk of carbonation-induced corrosion damage can be significant.

5.3.2
Carbonated and Chloride-Contaminated Concrete

The situation is much more serious than the one described above if chlorides are present in the concrete, even in such small quantities that in themselves they would not give rise to corrosion. The presence of a small amount of chlorides in concrete may be due to the use of raw materials (water, aggregates) containing these ions or to the penetration of chlorides from the external environment (seawater, deicing salts). Many buildings of the 1960s and 1970s, particularly in Northern Europe, have been found to contain relatively small quantities of chlorides because, during the cold months, calcium chloride was added as accelerating admixture.

Figure 5.10 shows the corrosion rate of steel in artificially carbonated mortars in the absence and in the presence of chlorides. It can be clearly seen that the corrosion rate will be negligible only in conditions of external relative humidity below 75%, 60%, and even 40% as the chloride content increases from nil to 1% by cement mass [21]. This is at least partially due to the more hygroscopic nature of chloride-containing concrete. In a given humidity, chloride-containing concrete attracts more moisture and thus has a lower resistivity compared to chloride-free concrete.

Figure 5.11 shows the corrosion rate measured on steel reinforcement in carbonated concrete specimens with different composition, exposed to a very wet environment (98% R.H.) and to a dry environment (85% R.H.) [22]. The corrosion rate of steel was mainly determined by the exposure condition and the presence of chlorides, even in small amount, while the w/c ratio did not play any significant

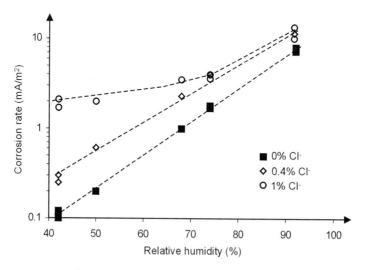

Figure 5.10 Relationship between relative humidity and corrosion rate in carbonated mortar both without chloride and in the presence of chlorides [21].

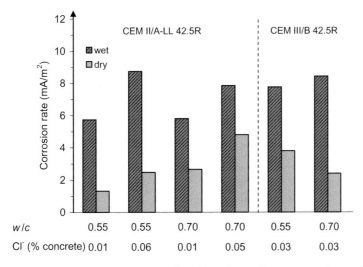

Figure 5.11 Average corrosion rate of steel in carbonated concrete specimens of different composition, in wet (98% R.H.) and dry (85% R.H.) environments [22].

role. In the presence of chlorides the corrosion rate of steel was not negligible (i.e., average values around 5 mA/m^2) even in the dry environment, while in the wet environment average values were always higher than 5 mA/m^2 for all the concrete mixes considered. It should be observed that while for CEM II/A-LL concrete chlorides were added to the mix, for CEM III/B they were present in the cement.

References

1 Neville, A.M. (1995) *Properties of Concrete*, 4th edn, Longman Group Limited, Harlow.
2 Alonso, C. and Andrade, C. (1994) Life time of rebars in carbonated concrete, in *Progress in Understanding and Prevention of Corrosion* (eds J.M. Costa and A.D. Mercer), Institute of Materials, London, p. 624.
3 Sergi, G. (1986) *Corrosion of steel in concrete: cement matrix variables*, PhD Thesis, University of Aston, Birmingham.
4 Tuutti, K. (1982) *Corrosion of Steel in Concrete*, Swedish Foundation for Concrete Research (CBI), Stockholm.
5 Papadakis, V.G., Vayenas, C.G., and Fardis, M.N. (1989) A reaction engineering approach to the problem of concrete carbonation. *AIChE Journal*, **35** (10), 1639–1650.
6 Thiery, M., Villain, G., Dangla, P., and Platret, G. (2007) Investigation of the carbonation front shape on cementitious materials: effects of the chemical kinetics. *Cement and Concrete Research*, **37** (7), 1047–1058.
7 Peter, M.A., Muntean, A., Meier, S.A., and Böhm, M. (2008) Competition of several carbonation reactions in concrete: a parametric study. *Cement and Concrete Research*, **38** (12), 1385–1393.
8 Castellote, M., Andrade, C., Turillas, X., Campo, J., and Cuello, G.J. (2008) Accelerated carbonation of cement paste *in situ* monitored by neutron diffraction.

9 Copuroglu, O. (2006) *Frost salt scaling of cement-based materials with a high slag content*, PhD Thesis, Delft University of Technology.
10 Wierig, H.J. (1984) Long-time studies on the carbonation of concrete under normal outdoor exposure, RILEM Seminar on Durability of concrete structures under normal outdoor exposure, Hannover, 26–27 March 1984.
11 Parrott, L.J. (1992) Carbonation, moisture and empty pores. *Advances in Cement Research*, **4** (15), 111–118.
12 Bakker, R.F.M. (1994) Prediction of service life of reinforcement in concrete under different climatic conditions at given cover. Int. Conf. on Corrosion and corrosion protection of steel in concrete, Sheffield.
13 Hemelop, W. and de Vries, J. (2004) *Monitoring internal environment of RWS tunnels*, DARTS report R2.11.
14 Ho, D.W.S. and Lewis, R.K. (1987) Carbonation of concrete and its prediction. *Cement and Concrete Research*, **17** (3), 489–504.
15 Al-Kadhimi, T.K.H., Banfill, P., Millard, S.G., and Bungey, J.H. (1996) An accelerated carbonation procedure for studies on concrete. *Advances in Cement Research*, **8** (30), 47–59.
16 Skjolsvold, O. (1986) Carbonation depths of concrete with and without condensed silica fume, in *Proc. of the 2nd Int. Conf. Fly Ash, Silica Fume, Slag, and Natural Pozzolans in Concrete*, vol. II (ed. V.M. Malhotra), ACI SP91, Canmet/ACI, pp. 1031–1048.
17 Page, C.L. (1992) Nature and properties of concrete in relation to reinforcement corrosion, in *Corrosion of Steel in Concrete*, Aachen, 17–19 February 1992.
18 Bier, T.A. (1986) Influence of type of cement and curing on carbonation progress and pore structure of hydrated cement paste. *Materials Research Society Proceedings*, **85**, 123–134.
19 Parrott, L.J. (1994) Design for avoiding damage due to carbonation-induced corrosion. Proc. of Canmet/ACI Int. Conf. Durability of concrete, Nice, 283.
20 Alonso, C., Andrade, C., and Gonzales, J.A. (1988) Relation between resistivity and corrosion rate of reinforcements in carbonated mortar made with several cement types. *Cement and Concrete Research*, **18** (5), 687–698.
21 Glass, G.K., Page, C.L., and Short, N.R. (1991) Factors affecting the corrosion of steel in carbonated mortars. *Corrosion Science*, **32** (12), 1283–1294.
22 Redaelli, E. and Bertolini, L. (2011) Electrochemical repair techniques in carbonated concrete. Part II: cathodic protection. *Journal of Applied Electrochemistry*, **41** (7), 829–837.

6
Chloride-Induced Corrosion

Chloride contamination of concrete is a frequent cause of corrosion of reinforcing steel [1]. Modern design codes for reinforced and prestressed concrete structures impose restrictions on the amount of chloride that may be introduced from raw materials containing significant amounts of chlorides. According to the European standard EN 206, the maximum allowed chloride contents are 0.2–0.4% chloride ion by mass of binder for reinforced and 0.1–0.2% for prestressed concrete (Chapter 11). These restrictions are thought to eliminate corrosion due to chloride added in the fresh concrete mix. In some structures built in the past, chlorides have been added in the concrete mix, unknowingly or deliberately, through contaminated mixing water, aggregates (for instance by using sea-dredged sand and gravel without washing them with chloride-free water) or admixtures (calcium chloride, which is now forbidden, in the past was the most common accelerating admixture). Chloride contents from accelerating admixtures in amounts ranging from 0.5% to well over 2% by mass of cement have caused extensive corrosion damage after carbonation and even in alkaline conditions [2–4].[1]

Nowadays, penetration of chloride ions from seawater or deicing salts into concrete and the associated risk for reinforcement corrosion is in many countries regarded as the most important degradation mechanism for reinforced concrete infrastructure. Over several decades chloride-induced corrosion of reinforcement in concrete has been extensively studied under field and laboratory conditions, particularly regarding the so-called *critical chloride content* or *chloride threshold value* [6, 7]. This threshold value corresponds to the chloride content measured at the rebar depth that causes depassivation of the reinforcement. It is affected by numerous interrelated parameters, among them the properties of the steel/concrete

---

1) In some regions in Europe, the presently allowed maximum chloride contents in concrete may be approached by chloride contained in sea-dredged aggregate or in the cement itself. In the former case, no or limited washing of the material is carried out; in the latter, liquid slag from blast furnaces is quenched using seawater [5]. The resulting chloride contents may be between 0.2% and just below 0.4%. Although allowed by standards, this practice does not seem justified from the point of view of long-term corrosion protection. Both carbonation and external chloride may cause corrosion-initiating chloride contents to arrive at the steel. Similarly, steel should be kept free from chloride before casting concrete.

Corrosion of Steel in Concrete: Prevention, Diagnosis, Repair, Second Edition. Luca Bertolini,
Bernhard Elsener, Pietro Pedeferri, Elena Redaelli, and Rob Polder.
© 2013 Wiley-VCH Verlag GmbH & Co. KGaA. Published 2013 by Wiley-VCH Verlag GmbH & Co. KGaA.

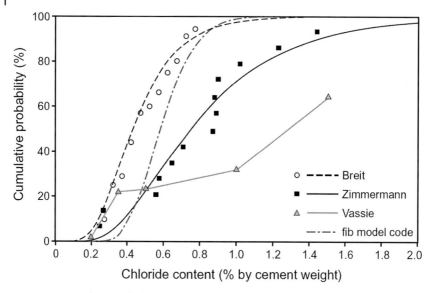

Figure 6.1 Cumulative probability of corrosion initiation versus chloride content, based on laboratory and field data from Refs. [10–16].

interface, pore solution chemistry and the electrochemical potential of the steel that is related to the pH of the pore solution and the amount of oxygen that can reach the surface of the steel. Relatively low levels of chlorides are sufficient to initiate corrosion in structures exposed to the atmosphere, where oxygen can easily reach the reinforcement. Much higher levels of chlorides are necessary in structures immersed in seawater or in zones where the concrete is water saturated, so that oxygen supply is hindered and thus the potential of the reinforcement is rather low [1, 8, 9]. However, even in atmospherically exposed structures such as bridge decks, considerable scatter is present in "threshold" values [6, 7]. Experiments in the laboratory with a large number of samples and field data obtained over larger numbers of structures have established a statistical relationships expressing the percentage of corrosion as a function of chloride content, as shown in Figure 6.1 [10–16]. The chloride threshold will be further discussed in a subsequent section.

6.1
Pitting Corrosion

Chlorides lead to a local breakdown of the protective oxide film on the reinforcement in alkaline concrete, so that a subsequent localized corrosion attack takes place. Areas no longer protected by the passive film act as anodes (active zones) with respect to the surrounding still passive areas where the cathodic reaction of oxygen reduction takes place. The morphology of the attack is that typical of *pitting*

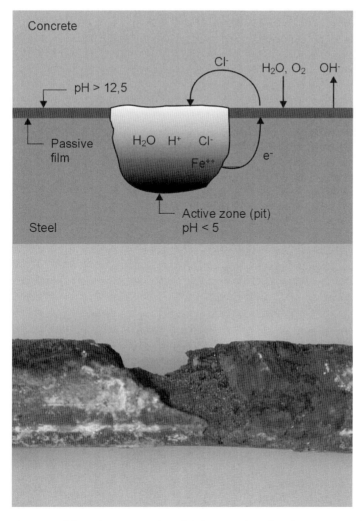

Figure 6.2 Schematic representation of pitting corrosion of steel in concrete and example of steel bar suffering localized corrosion attack [17].

shown in Figure 6.2 [17]. If very high levels of chlorides reach the surface of the reinforcement, the attack may involve larger areas, so that the morphology of pitting will be less evident. The mechanism, however, is the same.

As shown in Figure 6.2, once corrosion has initiated, a very aggressive environment will be produced inside pits. In fact, current flowing from anodic areas to surrounding cathodic areas both increases the chloride content (chlorides, being negatively charged ions, migrate to the anodic region) and lowers the alkalinity (acidity is produced by hydrolysis of corrosion products inside pits). On the contrary, the current strengthens the protective film on the passive surface since it tends to eliminate the chlorides, while the cathodic reaction produces alkalinity. Consequently, both the anodic behavior of active zones and the cathodic behavior of passive zones are stabilized. Corrosion is then accelerated (autocatalytic mechanism of pitting) and can reach very high rates of penetration (up to 1 mm/year) that can quickly lead to a marked reduction in the cross section of the reinforcement without being manifested by cracking or spalling and is thus difficult to detect by visual inspection. This mechanism will be further discussed in Section 7.4.2.

Consequences of pitting corrosion may be very serious in high-strength prestressing steel, where hydrogen embrittlement can be promoted; this is discussed in Chapter 10.

6.2
Corrosion Initiation

Initiation of pitting corrosion takes place when the chloride content at the surface of the reinforcement reaches the *critical chloride content*. From a fundamental point of view the initiation is considered to occur as a sequence of the distinct stages: pit nucleation, metastable pitting and stable pit growth [18], thus a certain time is required from the breakdown of the passive film to the formation of the first stable pit, according to the mechanism of corrosion described above. From a more practical point of view, the initiation time has been considered as the time when the reinforcement in concrete that contains substantial moisture and oxygen has reached the propagation state, characterized by an averaged sustained corrosion rate higher than $2 \, mA/m^2$ [19]. The chloride threshold can be defined as the chloride content required for depassivation (drop in the measured corrosion potential) or the chloride content necessary to reach this corrosion rate. In any case the critical chloride content has to be considered as a statistically distributed value (Figure 6.1).

When chlorides originate from the environment, the initiation time of corrosion will depend on the rate of penetration of chloride ions through the concrete cover. The knowledge of both the chloride threshold and the kinetics of penetration of chlorides into concrete is essential for the assessment of the initiation time of corrosion of reinforced concrete structures exposed to chloride environments (Chapter 11). In this regard, the influence of several parameters related both to the concrete (e.g., type of cement, w/c ratio, moisture content, etc.) and the environment (e.g., chloride concentration, temperature, etc.) has to be considered.

6.2.1
Chloride Threshold

The chloride threshold for the initiation of pitting corrosion for a given structure depends on numerous interrelated factors [1, 7, 11, 19–21]. Major factors have been identified in the pH of concrete, that is, the concentration of hydroxyl ions

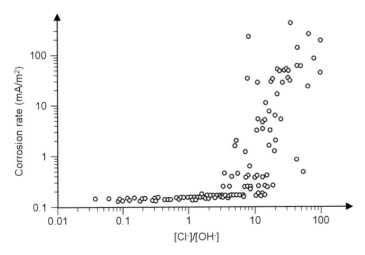

Figure 6.3 Example of relationship between the molar ratio of Cl⁻ and OH⁻ in the pore solution and the corrosion rate of steel [23, 24].

in the pore solution, the electrochemical potential of the steel and the presence of voids at the steel/concrete interface.

The hydroxyl ion concentration in the pore solution mainly depends on the type of cement and additions (Section 2.1.1). It was suggested that pitting corrosion can take place only above a critical ratio of chloride and hydroxyl ions [22]. For instance, Figure 6.3 shows that the risk of significant corrosion increases as the [Cl⁻]/[OH⁻] ratio in the pore solution rises above a certain value [23, 24]. Although several authors confirmed the dependence of corrosion initiation on the [Cl⁻]/[OH⁻] ratio, a great variability was found for the threshold value [7, 11, 21].

The great variability of the [Cl⁻]/[OH⁻] threshold is, first of all, a consequence of the stochastic nature of the initiation of pitting corrosion: the probability that a pit is formed depends on various factors related to the metallurgical properties of the steel, voids at the steel/concrete interface [25] and possibly also on the size of the sample [12]. It has become clear that the chloride threshold can only be defined on a statistical basis [9, 11, 13, 14, 16].

The electrochemical potential of the steel, which is primarily related to the pH of the pore solution and the oxygen availability at the steel surface and thus to the moisture content of concrete, is a second factor that has a marked influence on the chloride threshold. In fact, as the potential of steel decreases, the chloride threshold may increase by more than one order of magnitude [26]; this will be discussed in Section 7.4.

The chloride threshold has been found to be dependent on the presence of macroscopic voids in the concrete near the steel surface [25, 27]. Voids that can be normally found in real structures due to incomplete compaction may weaken the layer of cement hydration products deposited at the steel/concrete interface,

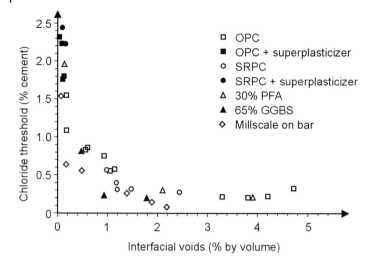

Figure 6.4 Chloride threshold for corrosion initiation as a function of content of interfacial voids (concretes made with different types of cement: OPC = ordinary portland cement, SRPC = sulfate-resistant portland cement) [27].

and thus may favor local acidification that is required for sustained propagation of pits (Section 7.4). For instance, it was shown that by decreasing the volume of entrapped air in the steel/concrete interfacial zone from 1.5% to 0.2% (by volume), the chloride threshold increased from 0.2% to 2% by mass of cement, as illustrated in Figure 6.4 [28]. The presence of air voids, as well as crevices or microcracks, can also be an explanation for the lower values of chloride threshold that are normally found in real structures compared with those found in (usually well-compacted) laboratory specimens with similar materials [29].

Several other factors, such as the temperature, the composition of the cement, the composition or surface roughness of the steel reinforcement, or the polarization with anodic or cathodic current may affect the chloride threshold [30]. Some of these factors will be discussed later. Here, it will be shown how the chloride threshold depends on properties of concrete and on environmental exposure (that, as discussed in Chapter 7, determines the potential of the steel).

Chloride Binding The chloride content in the concrete, and thus the chloride threshold, can be expressed in several ways, either referring to the chloride concentration in the pore solution (i.e., to *free chlorides*) or to the *total chloride content* in the concrete, that is, including chlorides bound to constituents of the cement paste [11]. It is generally believed that only the chloride ions dissolved in the pore solution can promote pitting corrosion, while those chemically bound to constituents of the cement paste, such as chlorides adsorbed on C–S–H or bound to tricalcium aluminate (e.g., as Friedel's salt, $3CaO \cdot Al_2O_3 \cdot CaCl_2 \cdot 10H_2O$) do not.

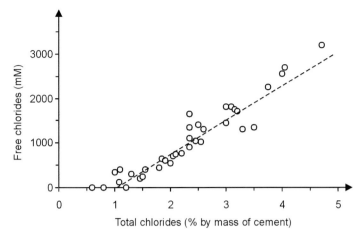

Figure 6.5 Relationship between free chlorides and total chloride content found in an ordinary portland concrete subjected to chloride penetration [23].

However, from a study of chemical aspects of binding it was suggested that bound chloride may also play a role in corrosion initiation. Glass and Buenfeld [21] state that a large part of the bound chloride is released as soon as the pH drops to values below 12, which may happen locally in voids at the steel/concrete interface. The bound chloride dissolves and may subsequently be involved in initiation of corrosion (see also the previous discussion of the role of air voids with respect to the threshold).

Equilibrium conditions tend to establish between free chloride ions and bound chlorides, depending on the composition of the cement and its binding capacity (Figure 6.5). Therefore, it is possible that the concentration of free chloride ions in the pore solution of different concrete varies, even if the total chloride content is the same. Binding in ordinary portland cement concrete depends on the content of the tricalcium aluminate phase. For example, the risk of corrosion in sulfate-resistant portland cements, which are characterized by a low content of tricalcium aluminate, is higher than in normal portland cement, considering an equal content of total chloride [31]. Similarly, slag, silica fume and fly ash blended cements have lower free chloride concentrations than the original portland cement, although their pH is slightly lower [31]. Adsorption of chloride ions in the C–S–H may be even more important than chemical binding, and this could explain the higher binding capacity observed in blended cements, which tend to form a finer microstructure of the hydration products [20] (Section 1.3).

Analyzing data on bound and free chloride from the literature using neural network modeling, Glass and coworkers [32] found that among the factors influencing bound versus free chloride were: C_3A content, silica fume addition, addition of blast furnace slag and addition of PFA. Addition of 65% slag decreased the free chloride content by about 50% and addition of 35% PFA decreased the free chloride content by about 9%. Addition of 20% silica fume increased the free chloride content by about 50%.

The importance of adsorption for binding is illustrated from results reported in ref [33]. No chloride-containing crystalline phases were detected by X-ray diffraction in a blast furnace slag cement specimen after submersion in the North Sea for 16 years, while containing 2 to 4% of chloride ion by mass of cement in the outer layers. It follows that binding in (crystalline) chloro-aluminates does not always take place in considerable amounts, even if rather large amounts of chloride are removed from the solution in solid form. Adsorbed chloride would not show in X-ray diffraction due to its noncrystalline nature. In the same samples, ettringite was found to be present. Speculating, the sulfate from the seawater may have reacted with the aluminates and kept chloride from chemical binding.

In practice, since the total chloride content can be measured much easier than the free chloride concentration (Section 16.3.2), the chloride threshold is expressed as a critical total chloride content. The critical value is usually given as a percentage of chlorides with respect to the mass of cement, since the amount of chlorides that can be tolerated increases as the cement content in the concrete increases.

Atmospherically Exposed Structures In structures exposed to the atmosphere, oxygen can easily reach the steel surface through the air-filled pores and the electrochemical potential of the reinforcement is quite positive, around 0 V SCE. In practice, the risk of corrosion in noncarbonated concrete obtained with normal portland cement (for which pH > 13) is considered low for chloride contents below 0.4% by mass of cement and high for levels above 1%. The threshold value tends to be higher in concretes of low permeability. The influence on the corrosion threshold of blended cements with additions of pozzolana or blast furnace slag is subject to controversy. There is laboratory evidence that blast furnace slag cement has a higher threshold for corrosion initiation [34]. On the other hand, marine exposure data suggest that thresholds with slag or fly ash cements are similar to or even lower than with portland cement [35]. However, in the case of higher additions of silica fume, the threshold is significantly lower due to the lower pH of the pore solution [36].

For sulfate-resisting cements with low C_3A content, the critical level of chlorides to initiate corrosion can be found among the lower values mentioned above, or at values even below those (e.g., 0.2% by mass of cement).

Submerged Structures When reinforced concrete structures are submerged in water, or in any case the moisture content of concrete is near the saturation level, the transport of oxygen to the steel is low and the reinforcement reaches very negative potentials, for example, between −400 and −600 mV SCE [1]. In that case, the chloride threshold is much higher, sometimes even reaching values one order of magnitude greater than that of structures exposed to the atmosphere (Section 7.4). However, for hollow (air-filled) structures submerged in seawater (e.g., offshore platform legs, tunnels), macrocells between outer and inner reinforcement may cause significant corrosion (Section 8.2).

6.2.2
Chloride Penetration

Chloride penetration from the environment produces a profile in the concrete characterized by a high chloride content near the external surface and decreasing contents at greater depths. The experience on both marine structures and road structures exposed to deicing salts has shown that, in general, these profiles can be approximated by means of relation (2.4) obtained from Fick's second law of diffusion [20]. This relation theoretically describes the kinetics of a nonstationary diffusion process, under the assumptions described in Section 2.2.2 [37]. Indeed, these assumptions are rarely met in real structures. Only in concrete completely and permanently saturated with water, can chloride ions penetrate by pure diffusion. In most situations, other transport mechanisms contribute to chloride penetration. For instance, when dry concrete comes into contact with salt water, initially capillary suction of the chloride solution occurs. In many cases wetting and drying occur, at least in the outer zones of the concrete, so that evaporation of water leads to enrichment of chloride ions (Section 2.6). Furthermore, binding due to adsorption or chemical reaction with constituents of the cement paste alters the concentration of chloride ions in the pore solution.

Nevertheless, experience shows that, even in these cases, chloride profiles can be mathematically modeled with good approximation using an equation formally identical to (2.4). In this equation, the total content of chlorides is usually considered, hence including bound chlorides:

$$C_x = C_s \left(1 - \mathrm{erf} \frac{x}{2\sqrt{D_{app} \cdot t}} \right) \quad (6.1)$$

where: C_x = total chloride content (% by mass of cement or concrete) at time t (s) and at the depth x from the surface of the concrete (m); D_{app} = apparent diffusion coefficient for chloride (m²/s); C_s = surface chloride content (% by mass of cement or concrete). The apparent diffusion coefficient and the surface chloride content are calculated by fitting the experimental data to Eq. (6.1) and are often used to describe chloride profiles measured on real structures [33, 38–43].

The apparent diffusion coefficient, obtained from real structures or laboratory tests, is often also used as the parameter that describes the resistance of concrete to chloride penetration, for example, when performances of different materials are compared. The lower D_{app} is, the higher the resistance to chloride penetration is. It should, however, be observed that, while the diffusion coefficient obtained from pure (steady-state) diffusion tests can be considered as a property of the concrete, the apparent diffusion coefficient obtained from real structures also depends on other factors (such as the exposure conditions or the time of exposure). Therefore, results obtained under particular conditions, especially during short-term laboratory tests, may not be applicable to other exposure conditions.

Furthermore, Eq. (6.1) is also used for the prediction of long-term performance of structures exposed to chloride environments, for example, during the design stage or the evaluation of the residual life of existing structures. In principle, if

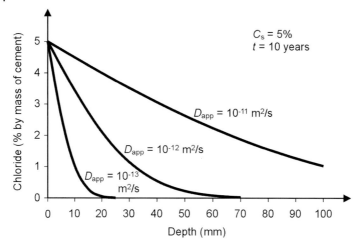

Figure 6.6 Examples of profiles of diffusion for chlorides in concrete for different values of the apparent diffusion coefficient (D_{app}) after 10 years of exposure.

D_{app} and C_s are known and can be assumed to be constant in time, it is possible to evaluate the evolution with time of the chloride profile in the concrete and then to estimate the time t at which a particular chloride content will be reached. Comparing with the chloride threshold value the time of corrosion initiation can be calculated. For example, Figure 6.6, plots chloride profiles calculated for a surface chloride content of 5% by mass of cement and a time of 10 years as a function of the apparent diffusion coefficient. Figure 6.7 shows the initiation times that can be estimated from those profiles as a function of the thickness of the concrete cover when a fixed chloride threshold of 1% by mass of cement is assumed.

Although this approach to the evaluation of the corrosion initiation time is simple, it should be observed that its reliability strongly depends on the reliability of the parameters utilized, that is, not only D_{app} and C_s but also the chloride threshold. Factors affecting the chloride threshold were illustrated in Section 6.2.1. D_{app} and C_s, in general, cannot be assumed as constants in the case of real structures where binding as well as processes other than diffusion take place. In particular, the apparent diffusion coefficient changes in time and as a function of the exposure conditions, among others due to hydration of slowly reacting cement constituents, in particular blast furnace slag or fly ash. The following paragraphs deal with the main factors that affect the surface content and the apparent diffusion coefficient. Use of probabilistic approaches that consider the statistical variability of the parameters involved (Cl_{th}, D_{app}, C_s) may allow the calculation of the probability of initiation of corrosion. The application of the "error-function-solution" Eq. (6.1) to the prediction of the service life of reinforced concrete structures exposed to chloride environments will be discussed in Chapter 11.

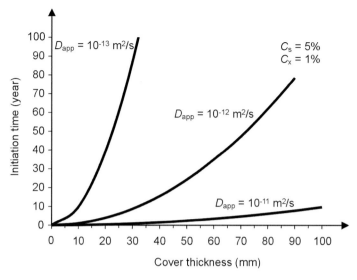

Figure 6.7 Initiation time of corrosion as a function of cover thickness and apparent diffusion coefficient assuming that the surface chloride concentration is 5% and the critical threshold is 1% by mass of cement.

6.2.3
Surface Content (C_s)

Analysis of chloride profiles from real structures shows that the surface chloride content is different in different structures, but may also vary in time [20]. For structures exposed to a marine environment it was observed that the value of C_s reached in a few months' time tends to remain constant [44], although there is no general agreement on this point (see ref. [45]). In marine environments, several transport processes may interact (Chapter 2) like capillary absorption and diffusion, depending on the relative position with respect to the mean water level, wave height, tidal cycle and so on (Figure 6.8). Moreover, cyclic wetting and drying (with different cycle lengths for tidal and splash zones) may cause accumulation of chloride; exposure to prevailing wind and precipitation may wash out previously absorbed chloride, and carbonation will release bound chloride. Most of these factors also depend on the concrete composition (cement type, chloride binding, absorption, permeability for water vapor). The effect is that chloride penetration is a complex function of position, environment, and concrete. As an example, Figure 6.9 shows chloride penetration contours. The effect of height and in particular of the position with respect to the mean seawater level is clear [46]. It can be seen that the highest surface contents occur at relatively high position, where the accumulating effect of drying out will be strong. At somewhat lower positions, the concrete is wetter and transport towards the inner parts will be faster and consequently the penetration will be deepest. In the submerged part, accumulation

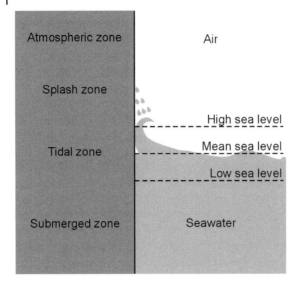

Figure 6.8 Different zones of a marine structure, in relation to chloride penetration and corrosion of reinforcement.

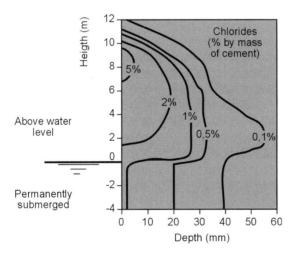

Figure 6.9 Example of chloride penetration contours in a marine structure as a function of the height above the seawater (portland cement, $w/c = 0.5$, $C_3A = 10\%$; after Nürnberger [46]).

due to wetting and drying is absent; here, equilibrium between bound and free chloride governs the total chloride content, together with relatively fast transport to the inner parts.

The value of C_s will, however, depend on the composition of concrete, the position of the structure, the orientation of its surface and the microenvironment, the

Table 6.1 Composition, surface chloride content, apparent diffusion coefficient, corrosion rate and resistivity data for three mixes in marine splash zone [38, 47] and motorway exposure [48] and strength from [38].

South East UK coast, splash zone (Folkestone), 6 years [38, 47]

	OPC	30% fly ash	70% blast furnace slag
Binder content (kg/m^3)	288	325	365
w/c	0.66	0.54	0.48
28-day compressive cube strength (MPa)	39	50	38
C_s (% by mass of cement)	1.7–3.5	3.7–5.2	2.8–3.8
D_{app} (10^{-12} m^2/s)	6.4–14.5	0.34–0.47	0.40–1.4
Corrosion rate of bars at 20 mm cover (μm/year)	11	0.6	1.2
Chloride content at 20 mm depth (% by mass of cement)	2.4	0.5	1.0
Resistivity (Ω m)	60	575	400–600

UK motorway (Middlesbrough), 9 years [48]

	OPC	30% fly ash	50% blast furnace slag
C_s (% by mass of cement)	2.8	4.1	3.7
D_{app} (10^{-12} m^2/s)	5.21	0.85	1.7
Resistivity (Ω m) after two years	100	650	550

chloride concentration in the environment and the general conditions of exposure with regard to prevailing winds and rain. The highest values of C_s were found in the splash zone, where evaporation of water leads to an increase in the chloride content at the concrete surface. A dependence of C_s on the cement content was also observed. Maximum values of C_s measured on concrete blocks in the splash zone of marine environments ranged from 3.5 to 5.2% (by mass of cement), depending on the type of binder, with higher values for cement containing fly ash or slag, as shown in Table 6.1 [38]. The curing was found to influence C_s as well. Rather stable C_s values were reached after one to two years.

An indepth survey of coastal structures has shown some of the effects mentioned [49]. Seven structures on the North Sea coast in The Netherlands, aged between 8 and 41 years were sampled at different heights above mean sea level (one structure at two points in time). In all cases reported here, the concrete was made using blast furnace slag cement with about 70% slag (comparable to modern CEM III/B 42.5 N) and roughly similar compositions; the execution quality may have varied, however. Most sampled faces were roughly oriented towards the North Sea. Chloride surface contents and apparent diffusion coefficients were obtained by fitting Eq. (6.1). Chloride surface contents are reported in Table 6.2. At heights between mean sea level and +7 m, the surface content varied randomly between 2% and 4% by mass of cement. This range appeared to be established after about ten years. Apparently, heights up to 7 m constitute the splash zone. At

Table 6.2 Chloride surface contents of slag cement concrete in seven coastal structures in The Netherlands in relation to height above mean sea level [49].

Structure	Test area, age (year)	Chloride surface content (% by mass of cement)		Height (m above mean sea level)	Note
		μ	σ		
Pier Scheveningen	Deck, underside, 41	3.2	1.3	+5	Sheltered from rain
Haringvliet	Pier, vertical, 40	2.8	2.0	+1	
Haringvliet	Pier, vertical, 40	0.4	0.0	+9	Fully exposed to rain
Haringvliet	Pier, vertical, 40	0.7	0.2	+14	Fully exposed to rain
Caland canal	Quay wall, vertical, 35	3.9	1.9	+1	
Hartel harbor	Quay wall, vertical, 30	2.9	0.3	+1	
Europa harbor	Quay wall, vertical, 19	3.9	1.3	+1	
Eastern scheldt barrier	Pier, vertical, 22	2.2	0.6	+1	
Eastern scheldt barrier	Upper beam, vertical, 16	4.1	0.3	+4	
Eastern scheldt barrier	Bridge element, overhang slope, 18	5.3	1.5	+9	Sheltered from rain
Noordland	Quay wall, vertical, 8	1.8	0.3	+1	
Noordland	Quay wall, vertical, 18	2.2	0.1	+1	

larger heights, surfaces exposed to rain (Haringvliet) showed lower C_s (0.4–0.7% by mass of cement), while surfaces sheltered from rain (but not from seawater splash) in the Eastern Scheldt bridge element had the highest C_s of all, about 5%.

Under other environmental conditions, C_s can have different values. Bamforth [48] has reported on exposure of concrete blocks made with similar compositions and binders as described above along a motorway in the UK, subjected to deicing salt application. The results (Table 6.1) show that C_s increases over several years, taking values between 0.3% and 0.6% chloride by mass of concrete (corresponding to 3–4% by mass of cement) after 9 years.

6.2.4
Apparent Diffusion Coefficient

Values of D_{app} usually vary from $10^{-13}\,\text{m}^2/\text{s}$ to $10^{-10}\,\text{m}^2/\text{s}$ in relation to the properties of the concrete and the conditions of exposure. D_{app} in particular depends on the

6.2 Corrosion Initiation

Table 6.3 Examples of stationary diffusion coefficient of chlorides, D, in cement pastes with $w/c = 0.5$, cured 60 days, made of portland cement and cement with additions of fly ash (PFA) or ground granulated blast furnace slag (GGBS) [31].

Type of cement	D (10^{-12} m²/s)
Portland	4.47
Portland + 30% PFA	1.47
Portland + 65% GGBS	0.41

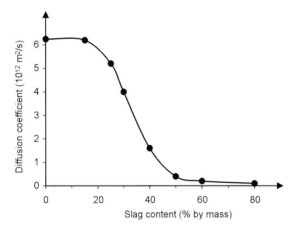

Figure 6.10 Influence of the replacement by blast furnace slag of portland cement on the diffusion coefficient of cement pastes with $w/c = 0.6$ at 21 °C (Brodersen, from [48]).

pore structure of the concrete and on all the factors that determine it, such as: w/c ratio, compaction, curing,, and the presence of microcracks. The type of cement has also a considerable effect: in passing from concrete made with portland cement to concrete made with increasing addition of pozzolana or blast furnace slag, D_{app} can be drastically reduced (Table 6.3). Of particular interest is the addition to portland cement of elevated percentages of blast furnace slag which, as shown in Figure 6.10, may reduce the diffusion coefficient by more than one order of magnitude.

The exposure of concrete blocks made with different binders in the marine splash zone or in a motorway environment described above show a wide range of apparent diffusion coefficients after 6 and 9 years, respectively, as shown in Table 6.1, ranging from 0.3 to 14 × 10^{-12} m²/s, with lower values for fly-ash- and slag-containing binders and higher values for pure portland cement [38, 48, 50].

From Eq. (6.1) it is clear that the penetration depth x for a certain content C_x depends on the product $D_{app} \cdot t$. Consequently, other factors being constant, reducing the apparent diffusion coefficient by a factor of 10 increases the initiation time

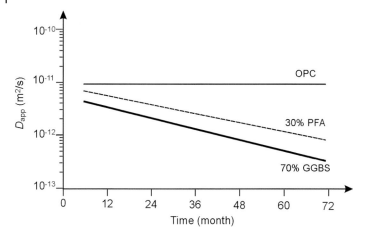

Figure 6.11 Schematic of the apparent diffusion coefficient for chloride (D_{app}) as a function of time and type of cement for portland cement, 30% fly ash and 70% granulated blast furnace slag during marine splash zone exposure, after [33].

by a factor of 10. Advantages of using blended cements or mineral additions in the design of durable structure will be illustrated in Section 12.6.1.

When the apparent diffusion coefficient is used for prediction purposes, it should be considered that it tends to reduce in time [20, 38], especially for blended cement. For example, Figure 6.11 sketches the variation in time of the apparent diffusion coefficient obtained from long-term field tests on concretes of various composition exposed to a marine environment [38]. Modeling of this process will be dealt with in Chapter 11.

6.3
Corrosion Rate

On structures exposed to the atmosphere, the corrosion rate can vary from several tens of μm/year to localized values of 1 mm/year as the relative humidity rises from 70% to 95% and the chloride content increases from 1% by mass of cement to higher values. These high corrosion rates have been observed in particular on heavily chloride containing structures such as bridge decks, retaining walls and pillars in the Swiss Alps under macrocell conditions.

Moving from temperate climates to tropical ones, the corrosion rate further increases. Therefore, once the attack begins in chloride-contaminated structures, in a relatively short time it can lead to an unacceptable reduction in the cross section of the reinforcement, even under conditions of normal atmospheric exposure.

The lower limits of relative humidity near which the corrosion rate becomes negligible depend on the characteristics of the concrete, on the amount of chlorides in the concrete and the type of salt they originate from. In any case, this limit is at much lower relative humidity than that which makes carbonation-induced corrosion negligible. In the presence of high chloride contents, above all with hygroscopic salts admixed like calcium or magnesium chloride, even for relative humidities of 40–50%, the corrosion rate can be up to $2\,\mu m/year$.

Temperature and humidity affect the corrosion rate through their influence on the electrochemical reactions at the steel/concrete interface and through their influence on ion transport between anodes and cathodes. Although the mechanisms are not fully understood, it appears that the concrete resistivity is strongly related to the corrosion rate at moderate or low temperature [51–54]. Variation in resistivity due to variation of humidity (at constant temperature) caused an inversely proportional variation of corrosion rate in carbonated mortar and concrete with low amounts of chloride or without chloride. Variation of temperature (at constant humidity) caused a similarly varying corrosion rate.

In a given set of conditions in terms of humidity and temperature and provided corrosion has initiated, the higher resistivity of blended cements results in a lower corrosion rate than for portland cement. As an illustration, the resistivities and corrosion rates measured on the previously mentioned concrete blocks exposed in a marine splash zone have been shown in Table 6.3. Bars at 20 mm depth in fly ash and blast furnace slag concrete show low corrosion rates although the chloride contents suggest corrosion initiation has taken place; both have a high resistivity ($>500\,\Omega m$). Bars at 20 mm in portland cement blocks show much higher corrosion rates, associated with low resistivity ($<100\,\Omega m$) [47].

Exceptions Only for structures completely and permanently submerged in water, even if conditions for the initiation of corrosion are satisfied (though this does not usually happen thanks to the low potential of the steel), the very low supply of oxygen reaching the reinforcement keeps the corrosion rate so low that corrosive attack is negligible even after long periods [1, 33]. This implies, however, that in hollow structures with air inside, corrosion of bars in the outer parts may occur [55]; this is due to macrocell corrosion, which will be treated in Chapter 8. Finally, gross defects like honeycombs or wide cracks in the cover of submerged structures may allow sufficient oxygen dissolved in the seawater to reach the bars and consequently sustain significant corrosion.

References

1 Arup, H. (1983) The mechanisms of the protection of steel by concrete, in *Corrosion of Reinforcement in Concrete Construction* (ed. A. P. Crane), Ellis Horwood Ltd., Chichester, pp. 151–157.

2 Polder, R.B. (1998) Cathodic protection of reinforced concrete structures in The Netherlands – experience and developments, in *Corrosion of Reinforcement in Concrete – Monitoring, Prevention and Rehabilitation* (eds J. Mietz, B. Elsener, and R.B. Polder), The European Federation of Corrosion Publication number 25, The Institute of Materials, London, pp. 172–184.

3 Polder, R.B. (1998) Cathodic protection of reinforced concrete structures in The Netherlands–experience and developments. *Heron*, **43** (1), 3–14.
4 Schuten, G., Leggedoor, J., and Polder, R.B. (2000) Cathodic protection of concrete ground floor elements with mixed in chloride, in *Corrosion of Reinforcement in Concrete, Corrosion Mechanisms and Corrosion Protection* (eds J. Mietz, R.B. Polder, and B. Elsener), The European Federation of Corrosion Publication number 31, The Institute of Materials, London, pp. 85–92.
5 Bertolini, L., Gastaldi, M., Carsana, M., and Berra, M. (2004) Comparison of resistance to chloride penetration of concretes and mortars for repair. Proc. 3rd RILEM Workshop Testing and modelling chloride ingress into concrete (eds C. Andrade and J. Kropp), RILEM PRO 38, 165–178.
6 Glass, G.K. and Buenfeld, N.R. (1997) Chloride threshold levels for corrosion induced deterioration of steel in concrete. Proc. RILEM Int. Workshop Chloride penetration into concrete (eds L.O. Nilsson and J.P. Olivier), Paris, 429.
7 Angst, U., Elsener, B., Larsen, C.K., and Vennesland, Ø. (2009) Critical chloride content in reinforced concrete–a review. *Cement and Concrete Research*, **39**, 1122–1138.
8 Pedeferri, P. (1996) Cathodic protection and cathodic prevention. *Construction and Building Materials*, **10**, 391–402.
9 Alonso, C., Castellote, M., and Andrade, C. (2000) Dependence of chloride threshold with the electrical potential of reinforcements. Proc. 2nd International RILEM Workshop Testing and modelling the chloride ingress into concrete (eds C. Andrade and J. Kropp), RILEM PRO 19, 415–425.
10 Vassie, P.R. (1984) Reinforcement corrosion and the durability of concrete bridges. *Proceedings of the Institution of Civil Engineers*, **76** (3), 713–723.
11 Alonso, M.C. and Sanchez, M. (2009) Analysis of the variability of chloride threshold values in the literature. *Materials and Corrosion*, **60** (8), 631–637.
12 Angst, U., Ronnquist, A., Elsener, B., Larsen, C.K., and Vennesland, Ø. (2011) Probabilistic considerations on the effect of specimen size on the critical chloride content in reinforced concrete. *Corrosion Science*, **53**, 177–187.
13 Polder, R.B. (2009) Critical chloride content for reinforced concrete and its relationship to concrete resistivity. *Materials and Corrosion*, **60** (8), 623–630.
14 Breit, W. (2001) *Critical Corrosion Inducing Chloride Content–State of the Art and New Investigation Results*, Verein Deutscher Zementwerke e.V., Verlag Bau + Technik, Düsseldorf, pp. 145–167.
15 Zimmermann, L. (2000) Korrosionsinitiierender Chloridgehalt von Stahl in Beton, Dissertation ETH Nr. 13870, ETH Zürich.
16 *fib*, International Federation for Structural Concrete (2006) Model code for service life design, Bulletin 34.
17 Bertolini, L. (2008) Steel corrosion and service life of reinforced concrete structures. *Structure and Infrastructure Engineering*, **4** (2), 123–137.
18 Angst, U., Elsener, B., Larsen, C.K., and Vennesland, Ø. (2011) Chloride induced reinforcement corrosion: rate limiting step of early pitting corrosion. *Electrochimica Acta*, **56**, 5877–5889.
19 Andrade, C. (2003) Determination of the chloride threshold in concrete, in *COST Action 521, Corrosion of Steel in Reinforced Concrete Structures*, Final report (eds R. Cigna, C. Andrade, U. Nürnberger, R. Polder, R. Weydert, and E. Seitz), European Communities, Luxembourg, Publication EUR 20599, pp. 101–111.
20 Frederiksen, J.M. (ed.) (1996) HETEK–chloride penetration into concrete. State of the art. Transport processes, corrosion initiation, test methods and prediction models, The Road Directorate, Report No. 53, Copenhagen.
21 Glass, G.K. and Buenfeld, N.R. (1997) Chloride threshold level for corrosion of steel in concrete. *Corrosion Science*, **39**, 1001–1013.
22 Hausmann, D.A. (1967) Steel corrosion in concrete, how does it occur? *Materials Protection*, **11**, 19–23.
23 Page, C.L., Lambert, P., and Vassie, P.R.W. (1991) Investigation of reinforcement corrosion: 1. The pore electrolyte phase in chloride-

contaminated concrete. *Materials and Structures*, **24**, 243–252.
24. Lambert, P., Page, C.L., and Vassie, P.R.W. (1991) Investigation of reinforcement corrosion: 2. Electrochemical monitoring of steel in chloride-contaminated concrete. *Materials and Structures*, **24**, 351–358.
25. Page, C.L. (2009) Initiation of chloride-induced corrosion of steel in concrete: role of the interfacial zone. *Materials and Corrosion*, **60** (8), 586–592.
26. Bertolini, L., Bolzoni, F., Gastaldi, M., Pastore, T., Pedeferri, P., and Redaelli, E. (2009) Effects of cathodic prevention on the chloride threshold for steel corrosion in concrete. *Electrochimica Acta*, **54** (5), 1452–1463.
27. Glass, G.K. and Buenfeld, N.R. (2000) The inhibitive effects of electrochemical treatment applied to steel in concrete. *Corrosion Science*, **42**, 923–927.
28. Glass, G.K. and Buenfeld, N.R. (1997) The Presentation of the chloride threshold level for corrosion of steel in concrete. *Corrosion Science*, **39**, 1001–1013.
29. Page, C.L. (2002) Advances in understanding and techniques for controlling reinforcement corrosion. 15th International Corrosion Congress, Granada, 22–27 September.
30. Bertolini, L. and Redaelli, E. (2009) Depassivation of steel reinforcement in case of pitting corrosion: detection techniques for laboratory studies. *Materials and Corrosion*, **60** (8), 608–616.
31. Page, C.L., Short, N.R., and Holden, W.H.R. (1986) The influence of different cements on chloride-induced corrosion of reinforcing steel. *Cement and Concrete Research*, **16**, 79–86.
32. Glass, G.K., Hassanein, N.M., and Buenfeld, N.R. (1997) Neural network modelling of chloride binding. *Magazine of Concrete Research*, **49**, 323–335.
33. Polder, R.B. and Larbi, J.A. (1995) Investigation of concrete exposed to North Sea water submersion for 16 years. *Heron*, **40** (1), 31–56.
34. Bakker, R.F.M., Wegen, G., and Van der Bijen, J. (1994) Reinforced concrete: an assessment of the allowable chloride content. Proc. of Canmet/ACI Int. Conf. on Durability of concrete, Nice.
35. Thomas, M.D.A. (1996) Chloride thresholds in marine concrete. *Cement and Concrete Research*, **26**, 513–519.
36. Manera, M., Vennesland, Ø., and Bertolini, L. (2008) Chloride threshold for rebar corrosion in concrete with addition of silica fume. *Corrosion Science*, **50** (2), 554–560.
37. Collepardi, M., Marcialis, A., and Turriziani, R. (1972) Penetration of chloride ions into cement pastes and concretes. *Journal of American Ceramic Society*, **55**, 534.
38. Bamforth, P.B. and Chapman-Andrews, J. (1994) Long term performance of RC elements under UK coastal exposure condition, in *Proc. Int. Conf. on Corrosion and Corrosion Protection of Steel in Concrete* (ed. R.N. Swamy), Sheffield Academic Press, 24–29 July, pp. 139–156.
39. Nilsson, L.O., Andersen, A., Tang, L., and Utgenannt, P. (2000) Chloride ingress data from field exposure in a Swedish environment. Proc. 2nd Int. RILEM Workshop Testing and modelling the chloride ingress into concrete (eds C. Andrade and J. Kropp), RILEM PRO 19, 69–83.
40. Lindvall, A., Andersen, A., and Nilsson, L.O. (2000) Chloride ingress data from Danish and Swedish road bridges exposed to splash from de-icing salt. Proc. 2nd Int. RILEM Workshop Testing and modelling the chloride ingress into concrete (eds C. Andrade and J. Kropp), RILEM PRO 19, 85–103.
41. Tang, L. and Andersen, A. (2000) Chloride ingress data from five years field exposure in a Swedish marine environment. Proc. 2nd Int. RILEM Workshop Testing and modelling the chloride ingress into concrete (eds C. Andrade and J. Kropp), RILEM PRO 19, 105–119.
42. Andrade, C., Sagrera, J.L., and Sanjuán, M.A. (2000) Several years study on chloride ion penetration into concrete exposed to Atlantic ocean water. Proc. 2nd Int. RILEM Workshop Testing and modelling the chloride ingress into concrete (eds C. Andrade and J. Kropp), RILEM PRO 19, 121–134.

43 Izquierdo, D., Andrade, C., and de Rincon, O. (2000) Statistical analysis of the diffusion coefficients measured in the piles of Maracaibo's bridge. Proc. 2nd Int. RILEM Workshop Testing and modelling the chloride ingress into concrete (eds C. Andrade and J. Kropp), RILEM PRO 19, 135–148.

44 Bamforth, P.B. (1993) Concrete classification for R.C. structures exposed to marine and other salt-laden environments. Proc. of Structural faults and repair–93, Edinburgh, 29 June–1 July.

45 Swamy, R.N., Hamada, H., and Laiw, J.C. (1994) A critical evaluation of chloride penetration into concrete in marine environment, in *Proc. Int. Conf. on Corrosion and Corrosion Protection of Steel in Concrete* (ed. R.N. Swamy), Sheffield Academic Press, 24–29 July, pp. 409–419.

46 Nürnberger, U. (1995) *Korrosion Und Korrosionsschutz Im Bauwesen*, Bauverlag GmbH, Wiesbaden/Berlin.

47 Polder, R.B., Bamforth, P.B., Basheer, M., Chapman-Andrews, J., Cigna, R., Jafar, M.I., Mazzoni, A., Nolan, E., and Wojtas, H. (1994) Reinforcement corrosion and concrete resistivity–state of the art, laboratory and field results, in *Proc. Int. Conf. on Corrosion and Corrosion Protection of Steel in Concrete* (ed. R.N. Swamy), Sheffield Academic Press, 24–29 July, pp. 571–580.

48 Bamforth, P.B. (1997) Corrosion of reinforcement in concrete caused by wetting and drying cycles in chloride-containing environments – results obtained from RC blocks exposed for 9 years adjacent to bridge piers on the A19 near Middlesbrough, Taywood Engineering Ltd report PBB/BM/1746.

49 Polder, R.B. and de Rooij, M.R. (2005) Durability of marine concrete structures–field investigations and modelling. *Heron*, **50** (3), 133–143.

50 Bijen, J. (1996) Benefits of slag and fly ash. *Construction and Building Materials*, **10**, 309–314.

51 Glass, G.K., Page, C.L., and Short, N.R. (1991) Factors affecting the corrosion rate of steel in carbonated mortars. *Corrosion Science*, **32**, 1283–1294.

52 Fiore, S., Polder, R.B., and Cigna, R. (1996) Evaluation of the concrete corrosivity by means of resistivity measurements, in *Proc. Fourth Int. Symp. on Corrosion of Reinforcement in Concrete Construction* (eds C.L. Page, P.B. Bamforth, and J.W. Figg), Society of Chemical Industry, Cambridge, UK, 1–4 July, pp. 273–282.

53 Bertolini, L. and Polder, R.B. (1997) Concrete resistivity and reinforcement corrosion rate as a function of temperature and humidity of the environment, TNO report 97-BT-R0574.

54 Alonso, C., Andrade, C., and Gonzalez, J. (1988) Relation between resistivity and corrosion rate of reinforcements in carbonated mortar made with several cement types. *Cement and Concrete Research*, **18**, 687–698.

55 Della Pergola, A., Lollini, F., Redaelli, E., and Bertolini, L. (2013) Numerical modeling of initiation and propagation of corrosion in hollow submerged marine concrete structures, submitted.

7
Electrochemical Aspects

7.1
Electrochemical Mechanism of Corrosion

The corrosion process of steel can be summarized with the following reaction:

$$\text{iron} + \text{oxygen} + \text{water} \rightarrow \text{corrosion products} \qquad (7.1)$$

Actually, this is an electrochemical reaction [1–4] and is composed of four partial processes (Figure 7.1):

- the oxidation of iron (*anodic process*) that liberates electrons in the metallic phase and gives rise to the formation of iron ions (Fe $\rightarrow$ Fe^{2+} + 2e) whose hydrolysis produces acidity (Fe^{2+} + 2H$_2$O $\rightarrow$ Fe(OH)$_2$ + 2H$^+$);

- the reduction of oxygen (*cathodic process*) that consumes these electrons and produces alkalinity: O$_2$ + 2H$_2$O + 4e $\rightarrow$ 4OH$^-$;

- the transport of electrons within the metal from the anodic regions where they become available to the cathodic regions where they are consumed (since the electrons carry a negative charge, this gives rise to a nominal electrical current flowing in the opposite direction);

- finally, in order for the circuit to be complete, the flow of current inside the concrete from the anodic regions to the cathodic ones, transported by ions in the pore solution.

These four processes are complementary, which is to say that they occur at the same rate. In fact, the anodic current I_a (i.e., the number of electrons liberated by the anodic reaction in a unit of time), the cathodic current I_c (i.e., the number of electrons that are consumed in the cathodic reaction in a unit of time), the current that flows inside the reinforcement from the cathodic region to the anodic (I_m), and finally the current that circulates inside the concrete from the anode to the cathode (I_{con}), should all be equal:

Corrosion of Steel in Concrete: Prevention, Diagnosis, Repair, Second Edition. Luca Bertolini,
Bernhard Elsener, Pietro Pedeferri, Elena Redaelli, and Rob Polder.
© 2013 Wiley-VCH Verlag GmbH & Co. KGaA. Published 2013 by Wiley-VCH Verlag GmbH & Co. KGaA.

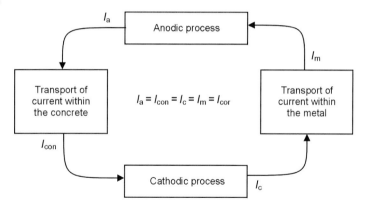

Anodic process
iron + water → corrosion products of iron + acidity + electrons (within the metal phase)

Cathodic process
oxygen + water + electrons (from the metal phase) → alkalinity

Transport of current within concrete
ion movement in the presence of water; enhanced by increase in pH and presence of chloride

Transport of current in the metal
electrons move from the anode, where they are produced, towards the cathode, where they are consumed

Figure 7.1 Electrochemical mechanism of corrosion of steel in concrete [5].

$$I_a = I_c = I_m = I_{con} = I_{cor} \qquad (7.2)$$

The common value of all these currents (I_{cor}) is, in electrochemical units, the rate of the overall process of corrosion.[1] The corrosion rate will thus be determined by the slowest of the four partial processes.

In reality, the electrical resistance of the reinforcement is always negligible with respect to that of the concrete. Therefore, the transport of current within the reinforcement is never a slow process and thus never contributes to reducing the rate of corrosion. Under particular conditions inside the concrete, each of the other three processes can take place at negligible rate and thus become the kinetically controlling one. More precisely, the corrosion rate is negligible when one of the following conditions exists:

1) The relationship between the anodic current I_a and the corrosion rate expressed as loss of mass in time t can be obtained from Faraday's first law of electrochemical stoichiometry: $\Delta m = e_{ech}\, q = e_{ech} \cdot I_a \cdot t$ where: Δm is the loss of mass at the anode following the passage of charge q; e_{ech} is the electrochemical equivalent of a metal which corrodes, $e_{ech} = M/(z \cdot F)$ with M = molar mass of the metal (55.8 g/mol for iron), $F = 96\,490$ C/mol (Faraday's constant) and z = valence of the ion formed following the anodic reaction ($z = 2$ for the reaction Fe → $Fe^{2+} + 2e$).

- the anodic process is slow because the reinforcement is passive, as when the concrete is not carbonated and does not contain chlorides;
- the cathodic process is slow because the rate at which oxygen reaches the surface of the reinforcement is low, as in the case of water-saturated concrete;
- the electrical resistivity of the concrete is high, as in the case of structures exposed to environments that are dry or low in relative humidity.

The first case is referred to as *passive control*, the second as *oxygen diffusion control*, the third as *ohmic control*.

On the other hand, the corrosion rate is high in those cases where the three following conditions are present simultaneously: (i) the reinforcement is no longer in the condition of passivity, (ii) oxygen can reach the surface of steel, (iii) the resistivity of the concrete is low (i.e., below $1000\,\Omega\,m$).

The moisture content in concrete is the main factor controlling the corrosion rate. When concrete is in equilibrium with the atmosphere, the moisture content can be correlated to the relative humidity of the environment (Section 2.1.2). Actually, in real structures this condition normally occurs only at the concrete surface. In fact, often concrete is periodically wetted and, since it tends to absorb water more quickly than it releases it, the moisture content at the depth of the reinforcement tends to be higher than that of equilibrium with relative humidity of the environment.

In concrete of low porosity, the maximum values of corrosion rate can be found for moisture content equivalent to the equilibrium with a relative humidity in the atmosphere (R.H.) of about 95%. For less-dense concrete, it corresponds to equilibrium with atmospheres with a slightly higher R.H. Moving away from these values of humidity in either direction, the corrosion rate decreases (Figure 7.2) [6]. Concrete with a high water content (that is, near saturation) is characterized by low resistivity, but the oxygen diffusion is slow (Chapter 2). The controlling process is then the cathodic process, and the corrosion rate will decrease as the water content increases, until it nearly becomes zero in conditions of saturation. Conversely, in concrete of lower water content, although oxygen diffusion can take place unhindered, the resistivity is high and it increases as the moisture content decreases. The corrosion rate then depends on the resistivity of the concrete; the lower the moisture content, the lower the corrosion rate.

Polarization Curves The rate at which the anodic or the cathodic process takes place depends on the electrochemical potential (E). The corrosion behavior of the reinforcement can be described by means of polarization curves that relate the potential and the anodic or cathodic current density. Unfortunately, determination of polarization curves is much more complicated for metals (steel) in concrete than in aqueous solutions, and often curves can only be determined indirectly, using solutions that simulate the solution in the pores of cement paste. This is only partly due to the difficulty encountered in inserting reference electrodes into the concrete and positioning them in such a way as to minimize errors of

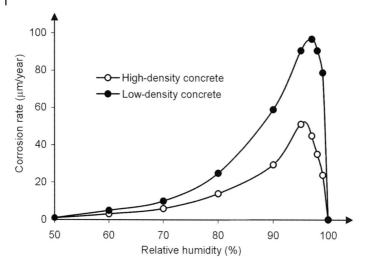

Figure 7.2 Corrosion rate as a function of external relative humidity in the case of initiation due to chloride for concrete with low and high density (after [6]).

measurement. The main problem is that diffusion phenomena in the cement paste are slow (Chapter 2). So when determining polarization curves, pH, and ionic composition of the electrolyte near the surface of the reinforcement may actually be altered.

7.2
Noncarbonated Concrete without Chlorides

7.2.1
Anodic Polarization Curve

In noncarbonated concrete without chlorides, steel is passive and a typical anodic polarization curve is shown in Figure 7.3. The potential is measured versus the saturated calomel reference electrode (SCE), whose potential is +244 mV versus the standard hydrogen electrode (SHE). Other reference electrodes used to measure the potential of steel in concrete are: Ag/AgCl, Cu/CuSO$_4$, MnO$_2$, and activated titanium types. From this point on in the text, unless explicitly stated otherwise, potentials are given versus the SCE electrode.

Iron has a tendency to oxidation at potentials more positive than the equilibrium potential of the reaction Fe → Fe^{2+} + 2e, which is about −1 V SCE. Therefore, below −1 V steel is in a condition of *immunity*.

In the range of potentials between −800 and +600 mV SCE, the anodic current is very low (0.1 mA/m^2) because the steel is covered by a very thin film of iron oxide that protects it completely (passive film). Thus, in this interval of potentials

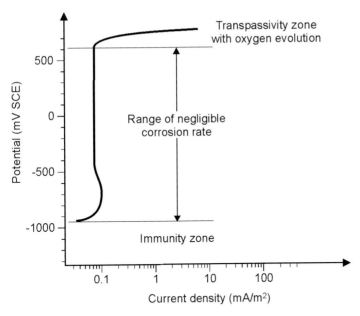

Figure 7.3 Schematic anodic polarization curve of steel in noncarbonated concrete without chlorides.

the dissolution rate of iron is negligible ($\approx 0.1\,\mu m$/year).[2] This is known as the condition of *passivity* and it exists in the interval of potentials known as the *passivity range*.

In the interval of potentials between equilibrium and about $-800\,mV$ SCE, the protective film does not form spontaneously. In this condition, called *activity*, steel can theoretically corrode. Nevertheless, given the proximity to equilibrium conditions, the rate of the anodic process is still negligible. To emphasize the fact that these conditions of activity are characterized by low corrosion rates because of this proximity to equilibrium, they are also called *quasi-immunity* conditions.

Above the passivity range, that is for potentials above about $+600\,mV$ SCE, the steel is brought to conditions known as *transpassivity*: oxygen may be produced on its surface according to the anodic reaction of oxygen evolution: $2H_2O \rightarrow O_2 + 4H^+ + 4e$, which produces acidity. Steel reaches these conditions only in the presence of an external polarization (e.g., in the presence of stray currents). Since the anodic reaction is oxygen evolution, dissolution of iron and consequent corrosion of the steel does not take place (i.e., the passive film is not destroyed). Nevertheless, if these conditions persist until the quantity of acidity produced is sufficient to neutralize the alkalinity in the concrete in contact with the steel, the passive

---

2) A dissolution rate of $0.1\,mA/m^2$ corresponds to a penetration rate of about $0.1\,\mu m$/year. Such low corrosion rate can only be measured electrochemically.

7.2.2
Cathodic Polarization Curve

The kinetics of oxygen reduction is illustrated by the cathodic polarization curves *a* and *b* shown in Figure 7.4. Even if the equilibrium potential for oxygen reduction within the concrete (pH about 13) has a value of approximately +200 mV SCE, the reaction takes place at significant rates only at potentials below 0 mV SCE. The rate of the cathodic process for potentials lower than 0 mV SCE depends on the oxygen availability at the surface of the steel. For a given potential it decreases as the moisture content in concrete increases and is reduced by 2–3 orders of magnitude in passing from concrete in equilibrium with atmospheres of relative humidity (R.H.) of about 70% ("semidry") to saturated concrete (in which the current density reaches values as low as about 0.2–2 mA/m², in relation to the thickness of the concrete cover and the quality of the concrete).

For potential values more negative than −900 mV SCE, along with the process of oxygen reduction there is also that of hydrogen evolution ($H_2O + e^- \rightarrow H_{ad} + OH^-$), so that the cathodic current density increases again.

If the concrete is completely saturated with water, and thus there is no oxygen, the only cathodic process possible is hydrogen evolution and the cathodic polarization curve is curve *c*.

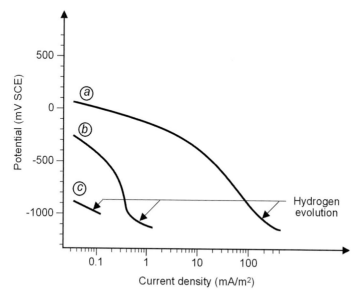

Figure 7.4 Schematic cathodic polarization curves in alkaline concrete: (a) aerated and semidry; (b) wet; (c) completely saturated with water.

7.2.3
Corrosion Conditions

From the cathodic and anodic polarization curves, the conditions of corrosion of reinforcement in various situations can be determined.

If the ohmic drop due to the current passing from the anodic to the cathodic areas is negligible, the corrosion rate (i_{corr}) and corrosion potential (E_{corr}) are determined by the intersection of anodic and cathodic curves. Figure 7.5 shows the intersections of these curves in concrete that is exposed to the atmosphere. In this condition the reinforcement generally has a corrosion potential between +100 and −200 mV SCE.

For concrete immersed in water, or in any way saturated with water, the diminished supply of oxygen to the surface of the steel can bring the potential down to values below −400 mV SCE. Finally, when oxygen is totally lacking (a very difficult condition to achieve, even in the laboratory), the potential may even drop to values below −900 mV SCE and the cathodic process will lead to hydrogen evolution. Under all of these conditions, embedded steel is subjected to a corrosion rate that is practically nil. Consequently, the cathodic current density is also very small.

The corrosion potential of passive reinforcement (E_{corr}) is determined by the availability of oxygen at the surface of the rebars. The maximum and minimum values of potential that characterize passive reinforcement under different environmental conditions are, respectively, +100 mV SCE in aerated concrete, and −1 V SCE in the total absence of oxygen. This means that in concrete the reinforcement

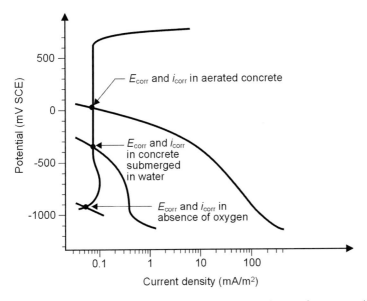

Figure 7.5 Schematic representation of the corrosion conditions of passive steel in concrete, under different conditions of moisture content.

does not reach conditions of immunity nor of transpassivity, unless it is polarized by imposing an external current.

7.3
Carbonated Concrete

The almost neutral environment of carbonated concrete hinders the formation of a protective film on the steel and thus the basis for conditions of passivity. The anodic curve therefore shows a progressive increase as shown in Figure 7.6, curve a. On a semilogarithmic scale (E vs log i) the anodic curve is a straight line with a slope of between 60 and 120 mV/decade over a wide range of current densities.

The cathodic polarization curve starts from the equilibrium potential of oxygen reduction, which is approximately 200 mV higher compared to alkaline concrete, due to the lower pH of carbonated concrete. The moisture content in carbonated concrete is expected to influence the cathodic curve in a similar way as in alkaline concrete. However, other parameters, such as the cement paste porosity (which is reduced by carbonation), may play a role in determining the amount of oxygen that is able to reach the steel surface. In this regard, the denser pore structure of carbonated concrete compared to alkaline concrete may hinder oxygen access to the steel surface and, therefore, lower the polarization curve towards more negative potential values, compensating for the effect of pH that would shift the curve

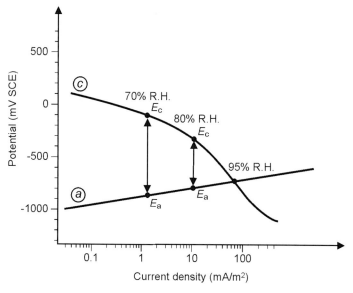

Figure 7.6 Schematic representation of the corrosion conditions of steel in carbonated concrete in equilibrium with environments of different relative humidity (R.H.).

towards more positive potential values. The determination of cathodic curves in concrete is rather difficult to achieve and previous considerations stem from theory rather than experimental tests. Correlations between free corrosion potential and corrosion rate of steel in alkaline and carbonated concrete do not seem to highlight substantial differences in the cathodic curves.

In any given set of environmental circumstances, the combination of anodic and cathodic curves determines the conditions of corrosion, as shown in Figure 7.6 (for simplicity changes in the cathodic curve due to changes in humidity have been neglected). When the water content is above that of equilibrium with environments with relative humidity greater than 95%, the corrosion rate of steel is determined almost exclusively by the rate of oxygen supply and can reach several tens of mA/m^2, in correspondence with the water content mentioned above. It will diminish if the water content increases beyond this value, because the cathodic curve will change in a similar way as shown in Figure 7.4, and it becomes negligible once conditions of saturation are reached. The potential of embedded steel is normally maintained between −300 and −500 mV SCE, and leads to lower values only if conditions of water saturation are reached.

When the moisture content in the concrete falls below these very high values, then the effects of ohmic drop, due to current circulating from anodic to cathodic areas through the concrete, must be taken into account.[3] As the moisture content in concrete decreases, the ohmic drop contribution increases. Conditions of corrosion can still be obtained from the polarization curves. The corrosion rate is given by the current density corresponding to the difference between the cathodic and anodic potentials, which is equal to the ohmic drop. In practice, in passing from a moisture content that gives rise to the maximum corrosion rate (defined by equilibrium with relative humidities between 95% and 98%) to lower levels of moisture corresponding to relative humidities between 80% and 90%, and to even lower levels corresponding to relative humidities below 70%, the corrosion rate indicatively falls, respectively, by 1 and 2 orders of magnitude.

In this case, the potentials of the anodic and cathodic regions are different. The potential of the steel (i.e., the value measured against a reference electrode placed in the concrete) has an intermediate value between the anodic and cathodic potentials, within the range of 0 to −600 mV SCE. Experience shows that the corrosion potential depends on the moisture content in the concrete, with more positive values being measured in dry concrete. Furthermore, a correlation between corrosion rate and corrosion potential can be found in carbonated concrete. To explain the change in corrosion potential as a function of corrosion rate it has been suggested that the ohmic contributions are localized at the anode [7]. This mechanism, termed *anodic resistance control*, is based on the hypothesis that the corrosion rate is under anodic control, that is, the anodic reaction determines the corrosion rate, but the anodic reaction is controlled by the resistivity of concrete, as shown in

---

3) Ohmic drop contributions cannot be neglected as in the case of passive steel discussed earlier because the current circulating in actively corroding steel is higher and, furthermore, carbonated concrete has a greater resistivity than alkaline concrete.

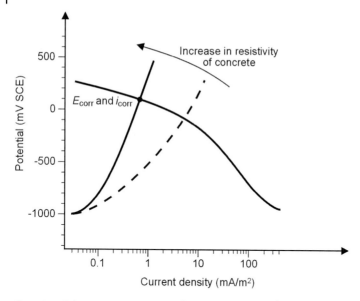

Figure 7.7 Schematic representation of the anodic resistance control mechanism proposed for steel in carbonated concrete [7].

Figure 7.7. This hypothesis is supported by the observation from practice and laboratory studies that a more negative potential corresponds to a higher corrosion rate.

7.4
Concrete Containing Chlorides

7.4.1
Corrosion Initiation and Pitting Potential

The presence of chloride ions in concrete leads to variations in the anodic behavior of steel, modifying the anodic polarization curve as shown in Figure 7.8. The passivity range is reduced because its upper limit, E_{pit}, known as the *breakdown potential* or *pitting potential* decreases as the chloride content increases [8]: it passes from values of about +600 mV SCE in uncontaminated concrete to values below −500 mV in concrete with a high content of chloride.

The presence of chloride ions produces, at more positive potentials than E_{pit}, localized breakdown of the passive film and thus allows attack on the underlying metal. For potentials below E_{pit}, the action of chloride is, at first approximation, negligible. The chloride content being equal, E_{pit} decreases as the pH of the pore solution in concrete decreases and as temperature and porosity increase. E_{pit} is

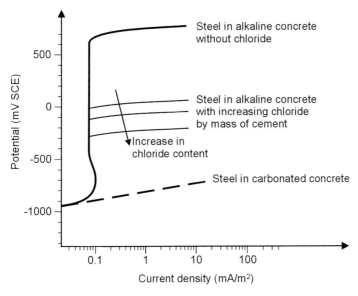

Figure 7.8 Schematic representation of the anodic polarization curve of steel in concrete with different chloride contents.

difficult to measure with laboratory measurements since during the measurement significant variations of pH and chloride level in the concrete near the surface of the steel can be introduced, altering the result [9, 10].

For a given potential of the steel, the highest content of chlorides compatible with conditions of passivity is the *critical chloride content* (or chloride threshold) at that potential. As already discussed in Chapter 6, for structures exposed to the atmosphere (whose reinforcement operates at a potential around 0 V SCE) the critical level is usually considered to be in the range of 0.4% to 1% of the cement content. For structures immersed in water (whose reinforcement operates instead at a much lower potential, around −400 to −500 mV SCE) or for reinforcement that is cathodically polarized for any reason, the chloride threshold is much higher.

The correlation between the critical chloride content and the potential of steel is expressed by Pedeferri's diagram (Figure 7.9) [11–13], which identifies, in a potential versus chloride content graph, areas representative of pitting corrosion and of passivity (perfect passivity and imperfect passivity, see Section 7.4.3). The boundary line between pitting corrosion and passivity can either be considered to represent the pitting potential for a given chloride content or the chloride threshold for a given steel potential. The determination of such a correlation is rather difficult due to the stochastic nature of pitting corrosion and the experimental difficulties in controlling all the parameters that affect it. As a consequence, both the pitting potential and the chloride threshold are characterized by a great variability.

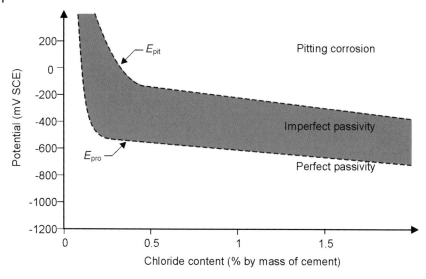

Figure 7.9 Simplified Pedeferri's diagram indicating regions of pitting corrosion and of perfect and imperfect passivity as a function of steel potential and chloride content (modified from [11–13]).

7.4.2
Propagation

Once the attack has initiated, acidity is produced progressively in the anodic zone and the chloride level increases until stable conditions are reached. As shown in Figure 6.2, current that circulates from the anodic zones (which corrode) to the cathodic zones (passive) induces the transport of chlorides in the opposite direction (since they are negatively charged ions). Chlorides are thus concentrated in the area where attack occurs. In addition, because of hydrolysis of anodic products, acidity is created in the same zone (pH to levels even below 3 can be reached in some cases, as confirmed by laboratory experiments [14]). Consequently, the local environment becomes more and more aggressive. In time, a condition of "stable propagation" is reached, in correspondence with which there is equilibrium between chlorides carried by the current and those that move away by diffusion, and between hydrogen ions produced in the anodic zone and those that move away and/or inflow of hydroxyl ions (which move in the same direction of chloride ions).

In practice, the potential in structures exposed to the atmosphere is about −500 to −600 mV SCE in the anodic zones and about −100 to −200 mV SCE in the cathodic zones.

Corrosion penetration may even exceed 1 mm/year in the most critical situations, which occur with high levels of chlorides and a moisture content near saturation. For lower moisture contents, the ohmic resistance increases and the corrosion rate slows down until it becomes negligible, when humidity is less than that of equilibrium with an atmosphere of 40–50% R.H.

7.4.3
Repassivation

Once pitting has initiated, circulation of current in the anodic region provokes and maintains an increase in acidity and chloride content, so that propagation may take place even if the potential of the steel is reduced, for example, owing to an external cathodic polarization.

This behavior is clarified by Figure 7.10 from Pourbaix [1]. It shows the progress of current exchanged anodically during cyclical polarization in which the potential of the steel (by external polarization) is first raised above E_{pit} to initiate localized attack and then lowered until conditions of passivity are established again.

It can be seen how, to stop the attack, it is necessary to reach a potential value, called the *protection potential* (E_{pro}) more negative than E_{pit}. Thus, the interval of potential included between E_{pit} and E_{pro} is characterized by the fact that it does not initiate the attack, but if the attack has already begun, it permits propagation of the attack. E_{pro} varies, as does E_{pit}, with chloride level, pH, and temperature, and the difference between them is of the order of 300 mV (Figure 7.11).

This effect is also considered in Pedeferri's diagram (Figure 7.9), in which the imperfect passivity and perfect passivity regions are plotted. These matters are further explored in Chapter 20.

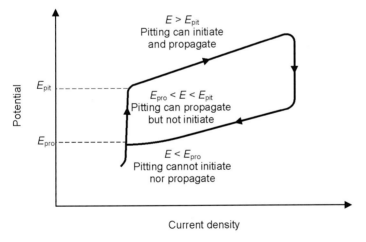

Figure 7.10 Schematic representation of a cyclic anodic polarization curve of an active–passive material in a chloride-containing environment: pitting potential (E_{pit}) and protection potential (E_{pro}) are identified [1].

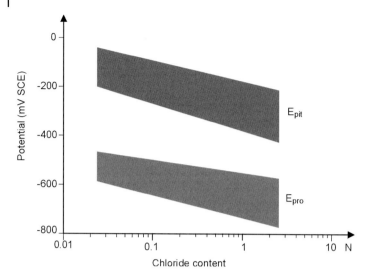

Figure 7.11 Values of the pitting (E_{pit}) and protection (E_{pro}) potentials determined with tests on steel immersed in saturated solutions of calcium hydroxide (pH = 12.6) at various levels of chlorides [11, 15].

7.5
Structures under Cathodic or Anodic Polarization

In the case of steel that exchanges a current I with the concrete from an external source, either anodic or cathodic, the relation $I_a = I_c$ shown in Section 7.1, which expresses equality between the electrons produced and consumed at the surface of the steel in the absence of exchanged current, should be modified. In fact, electrons extracted or provided by the external current also have to be taken into consideration. If the current is exchanged anodically the following condition occurs:

$$I_a = I_c + |I| \qquad (7.3)$$

while for a current exchanged cathodically:

$$I_c = I_a + |I| \qquad (7.4)$$

Consequently, the rates I_a of oxidation (corrosion) and I_c of reduction, which take place on the surface of the steel, are modified according to Figure 7.12.

If the current exchanged passes from the steel to the concrete or vice versa, the above equations are verified for a potential, respectively, more positive or more negative than the corrosion potential, so that the steel is polarized either anodically or cathodically.

Variations of the rate of anodic and cathodic processes, as well as variations of potential due to the external current, depend on the direction and magnitude of

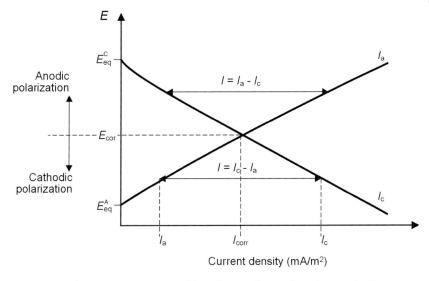

Figure 7.12 Schematic representation of the influence of external anodic or cathodic polarization.

the current and on the anodic and cathodic polarization curves. This will be dealt with in the following chapters.

References

1 Pourbaix, M. (1973) *Lectures on Electrochemical Corrosion*, Plenum Press, New York.
2 Pedeferri, P. (1980) *Corrosion and Protection of Metallic Materials*, Clup, Milan (in Italian).
3 Shreir, L.L., Jarman, R.A., and Burnstein, G.T. (eds) (1994) *Corrosion*, 3rd edn, Buttherworth Heinemann.
4 Revie, R.W. (2000) *Uhlig's Corrosion Handbook*, 2nd edn, John Wiley & Sons, Inc.
5 Pedeferri, P. and Bertolini, L. (2000) *Durability of Reinforced Concrete*, McGrawHill Italia, Milan (in Italian).
6 Tuutti, K. (1982) *Corrosion of Steel in Concrete*, Swedish Foundation for Concrete Research, Stockholm.
7 Glass, G.K., Page, C.L., and Short, N.R. (1991) Factors affecting the corrosion rate of steel in carbonated mortars. *Corrosion Science*, **32** (12), 1283–1294.
8 Page, C.L. and Treadaway, K.W.J. (1982) Aspects of the electrochemistry of steel in concrete. *Nature*, **297**, 109–115.
9 Bertolini, L. and Redaelli, E. (2009) Depassivation of steel reinforcement in case of pitting corrosion: detection techniques for laboratory studies. *Materials and Corrosion*, **60** (8), 608–616.
10 Angst, U.M., Elsener, B., Larsen, C.K., and Vennesland, Ø. (2011) Chloride induced reinforcement corrosion: electrochemical monitoring of initiation stage and chloride threshold values. *Corrosion Science*, **53** (4), 1451–1464.
11 Pedeferri, P. (1993) Cathodic prevention and protection of reinforced and prestressed concrete structures. *L'Edilizia*, **XII** (10), 69–81 (in Italian).

12 Pedeferri, P. (1996) Cathodic protection and cathodic prevention. *Construction and Building Materials*, **10** (5), 391–402.
13 Bertolini, L. (2011) A tribute to Pietro Pedeferri's contribution to the knowledge on corrosion of steel in concrete and its prevention. *Materials and Corrosion*, **62** (2), 96–97.
14 Pacheco, J., Polder, R.B., Fraaij, A.L.A., and Mol, J.M.C. (2011) Short-term benefits of cathodic protection of steel in concrete, in *Proc. Concrete Solutions, Dresden* (eds M. Grantham, V. Mechtcherine, and U. Schneck), Taylor and Francis, pp. 147–156.
15 Cigna, R. and Fumei, O. (1981) On the cathodic protection of steel in reinforced concrete. *L'Industria Italiana del Cemento*, **51** (9), 595–601 (in Italian).

8
Macrocells

Anodic and cathodic processes may take place preferentially on separate areas of the surface of the reinforcement, leading to a macrocell. This can establish, for instance, between active and passive areas of the reinforcement. Current circulating between the former, which are less noble and thus function as anodes, and the latter, which are more noble and thus function as cathodes, accelerates the corrosion attack on active surfaces while further stabilizing the protective state of passive ones. The magnitude of this current, known as the *macrocell current*, increases as the difference in the free corrosion potential between passive and active areas increases, and decreases as the dissipation produced by the current itself at the anodic and cathodic sites and within the concrete increases.

The most frequent type of macrocell in reinforced concrete structures exposed to the atmosphere is the one established between more superficial rebars that have been depassivated by carbonation or chloride penetration, and deeper-lying passive rebars. Another example may be walls where chloride penetrates from one side and oxygen penetrates from the other side, which may occur in hollow structures like tunnels and off-shore platform legs or with ground retaining walls.

For reinforced concrete structures buried in soil or immersed in water, cathodic areas may be due to noble metals present in the environment and electrically connected with the steel embedded in the concrete. For example, this is the case with copper grounding systems.

It should be observed that, because of the current flowing from the anodic area towards the cathodic areas, theoretically some Fe^{2+} ions migrate away from the corroding site and thus they do not precipitate locally as expansive oxides. This could mean that corrosion products due to macrocell action may have less-expansive effect than corrosion products due to microcells.

8.1
Structures Exposed to the Atmosphere

Depassivation of rebars due to carbonation or chloride penetration often does not extend to the whole surface of the reinforcement but, for instance, it is limited to the outer layer of rebars, or to parts where the concrete cover has a lower thickness

---

Corrosion of Steel in Concrete: Prevention, Diagnosis, Repair, Second Edition. Luca Bertolini, Bernhard Elsener, Pietro Pedeferri, Elena Redaelli, and Rob Polder.
© 2013 Wiley-VCH Verlag GmbH & Co. KGaA. Published 2013 by Wiley-VCH Verlag GmbH & Co. KGaA.

or is more porous, or in those areas where the chloride content is higher (e.g., where drainage of chloride-contaminated water occurs or where wetting–drying cycles lead to concentration of chlorides).

On the surface of depassivated steel, in the presence of water and oxygen, corrosion takes place mainly due to the mechanisms described in Chapter 7, that is, anodic and cathodic areas are intrinsically mixed (this is called *microcell corrosion*). On the other hand, if corroding steel is electrically connected to surrounding passive steel, a macrocell can form that concentrates the anodic process on the corroding steel and the cathodic process on the passive steel. The overall increase in the corrosion rate on the active steel induced by macrocell action depends strongly on the ratio between anodic and cathodic sites. An increase in the rate of attack of only 10% has been reported for concrete of high electrical resistivity [1]. On the contrary, in the case of chloride-induced localized corrosion attack in low-resistive concrete, macrocell corrosion is the dominating mechanism as has been shown in field tests in bridge decks [2]. More than 90% of the metal loss was due to macrocell corrosion. Low-resistivity concrete allows cathodic current to flow from relatively large distances (in the range of a meter) to the local anode

Coated Reinforcement Macrocell formation may be important in the case of epoxy-coated rebars (Section 15.4) in chloride-contaminated concrete if there are defects in the coating and the coated bars are electrically connected with uncoated passive steel bars in deeper parts of the structure. Small anodic areas are created at the defect points of coated rebars in contact with chloride-contaminated concrete, while the uncoated passive rebars provide a cathodic surface of much greater size. In these situations the macrocell can result in considerable anodic current densities and can significantly accelerate the attack on corroding sites. This is why coated rebars should be electrically isolated from uncoated bars. Recent laboratory and modeling work by Sagues and coworkers [3] has shown that macrocell effects aggravate corrosion in bridges with epoxy-coated rebars.

Protection Effect Macrocell currents can have beneficial effects on rebars that are polarized cathodically. This is indirectly evident for patch repair of chloride-contaminated structures when only the concrete in the corroding areas is replaced with alkaline and chloride-free mortar, but surrounding concrete containing chlorides is not removed. Before the repair, the corroding rebars behave as an anode with respect to those in the surrounding areas, which are polarized cathodically and thus are protected by the macrocell. After the repair, formerly anodic zones no longer provide protection, and corrosion can initiate in the areas surrounding repaired zones (these have been called "incipient" anodes) [4]. Consequences for repair are discussed in Chapter 18.

Presence of Different Metals Rebars of carbon steel in certain cases can be connected to rebars or facilities made of stainless steel or copper. This type of coupling, which in other electrolytes would provoke a considerable degree of corrosion on carbon steel by galvanic attack, does not cause problems in the case of concrete that are any different from those provoked by coupling with normal passive steel. In fact, the corrosion potential of passive carbon steel in concrete is not much different from that of stainless steel or copper. Therefore, these materials behave more or less like rebars of passive steel. The consequences of coupling of bars of carbon steel and stainless steel are illustrated in Section 15.2.3. Zinc or galvanized steel can exert some protective action on carbon steel.

Other Macrocell Effects A special case of macrocell effects has been observed on structures contaminated by chlorides where an activated titanium mesh anode was installed in order to apply cathodic protection; when the cathodic protection system is installed but is not in operation, localized corrosion on steel can be slightly enhanced by the presence of the distributed anode, which effectively shortcuts part of the concrete between anodes and cathodes and thus reduces the overall circuit resistance [5].

Macrocell currents can also affect electrochemical measurements carried out on reinforcement. In particular, macrocell currents can generate in structures under cathodic protection when the current is switched off during depolarization tests (Section 20.3.8). Different polarization conditions of different parts of the reinforcement (due for instance to uneven distribution of current) can lead to the onset of a macrocell that can alter the result of the potential decay measurement of reinforcement [6].

8.2
Buried Structures and Immersed Structures

The action of macrocells in structures buried in the soil or immersed in water is different from that of structures exposed to the atmosphere: two circumstances promote macrocell effects, while another reduces them. First, concrete is wetter than in aerated structures and its resistivity is lower, particularly in structures immersed in seawater. This reduces the ohmic drop in the concrete and increases the size of the effective cathodic area in relation to the anodic one. Secondly, the soil or the seawater around the concrete is an electrolyte of low resistivity, and the macrocell current can also flow outside the concrete. This further reduces the ohmic resistance between the anodic area and passive reinforcement. Thirdly, there is, however, a mitigating aspect. Oxygen can only diffuse with great difficulty through wet concrete and thus it hardly reaches the surface of the embedded steel. Depletion of oxygen at the surface of the rebar that is observed in this case makes initiation of corrosion very difficult, and, even when corrosion initiates, the driving voltage for the macrocell is very low.

Nevertheless, there are specific situations in which macrocells may form and promote localized attack.

Differential Aeration in Buried Structures A clear example of macrocell action was documented in diaphragm walls in Berlin, illustrated in Figure 8.1 [7]. In this case,

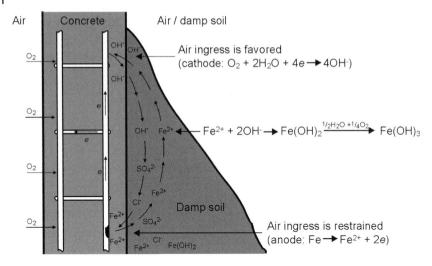

Figure 8.1 Diagrammatic representation of the macrocell formed on a diaphragm wall [7].

anodic areas had formed at the lower, nonaerated parts of the reinforcement at the ground side, while steel on the free side and higher up acted as a cathode. Large amounts of corrosion products were found inside the concrete at various distances from the anodes and in the soil, suggesting that relatively soluble iron oxides had formed that were able to move away from the anodes. Chlorides originated from deicing salts applied on the motorway, but chloride was also enriched at the ground side by the macrocell current. The initiation of anodes probably originated from defects in the concrete and the locally anaerobic conditions.

Structures Immersed in Seawater Macrocells may form between rebars reached by chlorides and passive rebars on which, for any reason, oxygen is available. Macrocell current is then controlled by the amount of oxygen that can be reduced on the passive rebars and the ohmic resistance in the circuit. The galvanic coupling lowers the potential on passive rebars and produces alkalinity on their surface. Therefore, the macrocell contributes to maintaining the steel passive.

Where oxygen access is low, it can be seen that the macrocell current tends to diminish in time because of oxygen depletion at the surface of the passive steel. The potential of passive steel consequently decreases in time until it reaches a value similar to that of corroding bars.

This decrease may not occur in the case of structures subjected to wetting–drying cycles or in conditions where oxygen consumed at the surface of the passive steel is replaced. This may happen in hollow piles of off-shore structures, as depicted in Figure 8.2. Similar conditions may arise in tunnels buried or submersed in a chloride-containing environment. Rebars on the inside of hollow (air-filled) structures may be effective cathodes with noble potentials. They increase the potential of rebars closer to the seawater side of the cross section, stimulating

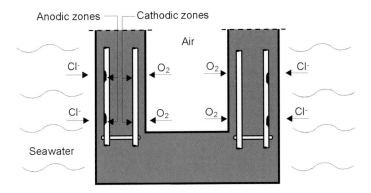

Figure 8.2 Diagrammatic representation of the macrocell formed in a hollow reinforced concrete structure immersed in the sea [8].

corrosion initiation at lower chloride contents than without anodic polarization. Subsequently, they may increase the corrosion rate at the anodes by consuming the electrons produced. The final corrosion rate will be a function of the ratio between anodic and cathodic areas, which is influenced by the concrete resistivity. This problem has been analyzed by numerical modeling (see below).

Rebars Not Entirely Embedded in Concrete Macrocell corrosion can occur when there are macroscopic defects in the concrete (cracks with large width, honeycombs, delaminations, etc.) or when there are metallic parts connected to the rebars that are only partially embedded in the concrete. This case is important for structures immersed in seawater or in aggressive soil. Besides being subjected to direct attack, those parts in direct contact with water or soil may also undergo more severe attack caused by the galvanic coupling with steel embedded in concrete.

Buried Structures Connected with Ground Systems The steel in buried reinforced concrete structures may be in contact with ground systems made of copper, steel or galvanized steel. As long as the steel embedded in concrete remains passive, the galvanic coupling with the ground systems does not influence the corrosion rate. In fact, buried steel or galvanized steel would function anodically, while copper would not produce macrocell effects since copper buried in aerated soil has about the same potential as steel in aerated concrete. If the rebars are in wet concrete, and thus have a low potential because of the lack of oxygen, coupling with buried copper will only take their potential to the same value that they would have if they were in aerated concrete.

Conversely, if, for any reason, steel within concrete is no longer passive, the presence of ground systems may be dangerous, since these connections offer a large and practically unpolarized cathodic area. In addition, since the anodic and cathodic areas are far apart, they may be operating under very different environmental conditions. For example, the anodic area may be in contact with water-saturated concrete (therefore in an anaerobic but highly conductive environment) because it is surrounded by water-saturated soil; at the same time the cathodic area may be found in a soil that is completely aerated.

8.3
Electrochemical Aspects

Electrochemical conditions of the steel bars and electrical resistivity of concrete play an important role in determining the effects of macrocells. To describe the consequences of macrocell corrosion, it will be assumed that inside the concrete there are only a passive rebar and an active one. If these are not electrically connected, the corrosion potential and the corrosion rate of the passive rebar are $E_{p,cor}$ and $I_{p,cor}$ and those of the active rebar are $E_{a,cor}$ and $I_{a,cor}$ (Figure 8.3). A driving voltage for the macrocell is then available and it is equal to:

$$\Delta E = E_{p,cor} - E_{a,cor} \tag{8.1}$$

It may vary from values near zero, in the absence of oxygen, to more than 300 mV for aerated structures. If the two rebars are electrically connected, the driving voltage ΔE produces a macrocell current I. Figure 8.4 helps to understand where the driving voltage is dissipated. The potential of the active rebar, which is anodically polarized by the current I, changes from $E_{a,cor}$ up to the value of E_a, so that:

$$I_{a,a} = I + I_{a,c} \tag{8.2}$$

The potential of the passive rebar is depressed by the cathodic current I, moving from $E_{p,cor}$ to the value E_c, for which the relationship:

$$I_{p,c} = I + I_{p,cor} \tag{8.3}$$

holds. That is in practice, since $I_{p,cor}$ is negligible, $I_{p,c} = I$.

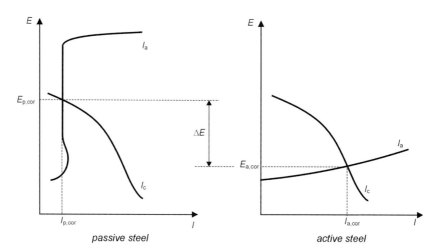

Figure 8.3 Schematic representation of the corrosion conditions of a passive and an active rebar, and determination of the driving voltage (ΔE) available for the macrocell.

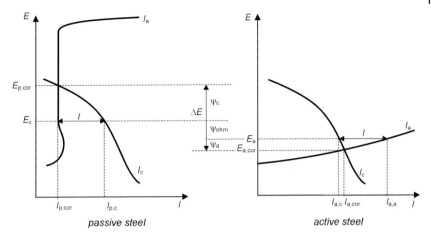

Figure 8.4 Schematic representation of the macrocell between a passive and an active rebar and determination of dissipation occurring at the anode (ψ_a), at the cathode (ψ_c) and in the concrete (ψ_{ohm}).

The increase in potential on the active rebar (ψ_a) and the decrease on the passive rebar (ψ_c), are the polarizations induced, respectively, at the anode and cathode by the current I:

$$\psi_a = E_a - E_{a,cor} \quad \psi_c = E_{p,cor} - E_c \tag{8.4}$$

In addition, current circulating through the concrete between the anodic and cathodic areas generates an ohmic drop (ψ_{ohm}). Therefore, the potential of the passive rebar, E_c, is more positive than that of the active one by an amount that is equal to the ohmic drop:

$$E_c - E_a = \psi_{ohm} \tag{8.5}$$

The driving voltage ΔE is thus dissipated by anodic and cathodic polarization and by the ohmic drop within the concrete:

$$\Delta E = \psi_a + \psi_c + \psi_{ohm} \tag{8.6}$$

It can be seen in Figure 8.4 how the corrosion rate induced on the active rebar (measured by $I_{a,a}$) depends on the value of the macrocell current (I), even though the increase in corrosion rate ($I_{a,a} - I_{a,corr}$) is lower than this current. In fact, the anodic polarization ψ_a causes a decrease in the cathodic current from $I_{a,cor}$ to $I_{a,c}$.

To evaluate polarizations and thus determine conditions of corrosion due to the macrocell, it is necessary to consider current densities exchanged at the anodic and cathodic surfaces as well as the macrocell current, I.

If we call: i the current density at the anodic surface (of area A_a) due to the current I (i.e., $i = I/A_a$), r the ratio between the anodic and cathodic areas ($r = A_a/A_c$), and R the resistance of the concrete between the cathodic and the anodic areas, relationship (8.6) can be written as:

$$\Delta E = \psi_a(i) + \psi_c(i \cdot r) + R \cdot i \cdot A_a \tag{8.7}$$

The size of the surfaces acting as anode and cathode (and thus r, which is their ratio) also depends on the resistivity of the concrete and the geometry of the system. For reinforced concrete structures exposed to the atmosphere, usually concrete has a high resistivity and thus only those areas near the site of active corrosion act as cathode (e.g., r can approach unity). In the case of structures immersed in seawater or buried, the concrete is wet and has a low resistivity and, moreover, the sea or the soil also act as electrolytes. Therefore, even areas very far from the anodic areas can function as cathodes, so that the ratio r can be very small.

The three dissipative contributions of Eq. (8.7) increase as i increases, so that once the driving voltage ΔE is given, there is only one value of i that satisfies that relationship.

Let us consider the opposite cases in which the ratio r tends to unity or to zero. The most frequent and also most favorable case is that in which the ratio r between the anodic and cathodic areas is near unity. The cathodic polarization, $\psi_c(i \cdot r)$, is prevalent with respect to other contributions. In fact, the cathodic polarization of passive steel reaches values of about 200–300 mV even for a current density of about 1–2 mA/m² (Figure 8.5) that is usually sufficient to dissipate most of the driving voltage ΔE. Therefore, with $r \approx 1$, the macrocell is under cathodic control and produces negligible or only modest corrosion effects.

A less favorable but rarer case exists when the ratio r tends to zero, namely the anodic surface is very small compared to the cathodic surface. In this case, since

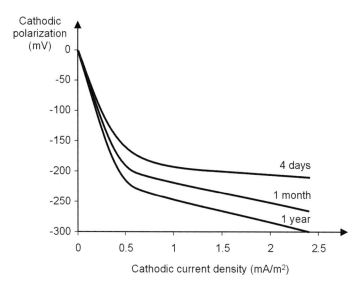

Figure 8.5 Cathodic polarization of a passive rebar measured after varying times of application of different current densities [9].

the cathodic surface is large, its current density is small and thus the cathodic polarization is negligible, so that:

$$\Delta E = \psi_a(i) + R \cdot i \cdot A_a \tag{8.8}$$

Only anodic polarization and ohmic drop then dissipate the driving voltage. If the resistivity of the concrete is high, the ohmic drop contribution tends to be high around the small corroding site and consequently the macrocell does not produce a significant increase of the corrosion rate. Conversely, if the resistivity is low, the driving voltage may lead to a remarkable anodic polarization and consequently to a significant increase of the corrosion rate at the anodic site. It should then be realized that in concrete with low resistivity, macrocell effects may be large because much larger cathodic areas than usual contribute to the corrosion process.

8.4
Modeling of Macrocells

Well-documented practical examples of macrocell action in concrete are rare. Nevertheless, because the consequences may be serious, this problem has attracted the attention of researchers using new techniques, in particular (numerical) modeling. One early approach involved experimental work and boundary element modeling [10]. Based on both anodic and cathodic polarization curves and macrocell experiments, the general validity of numerical modeling of macrocell effects in concrete was demonstrated and relevant details such as the influence of temperature and the geometry were described.

In a further collective effort, a Task Group of European Federation of Corrosion's Working Party 11 "Modeling of reinforcement corrosion in concrete" has initiated a collective study of numerical modeling of corrosion in concrete in order to investigate, among others, macrocell effects [11]. The idea was to set up propagation models for concrete based on the electrochemical nature of the processes involved. The input for such models includes the polarization behavior of active and passive steel and the resistivity of concrete. In general, a geometrical model of a specimen or a concrete element or structure is set up and current densities and potentials are calculated at every point in the modeled geometry by solving Laplace's equation. The steel/concrete interface is modeled using a Butler–Volmer-type expression. Due to lack of sufficient experimental data, this involves some simplifications. A special issue of Materials and Corrosion was devoted to several solution methods for three case studies: a simple cylindrical concrete/steel specimen, a bridge deck with a crack down to the reinforcement exposed to external chloride and a bored tunnel lining element with chloride-containing ground water on the outside and semidry air inside. The latter case is treated here briefly, as it represents a typical macrocell/hollow-leg situation. For more details the reader is referred to the papers [12–16]. Using finite element modeling and starting from practical dimensions for the concrete and the reinforcement layout, realistic resistivities for saturated and semidry concrete, limited oxygen transport on the outside and realistic steel polarization behavior, the effect of cathodic coupling of the inner reinforcement increases the corrosion rate of the outer (depassivated) reinforcement by a factor of four to five. Essentially similar results were obtained with boundary element modeling. All authors state that more research and in particular experimental data are needed to improve the models and the reliability of the results. However, the present results suggest that macrocell effects can be significant, increasing the corrosion rate.

Macrocells not only increase corrosion rate of active steel in the propagation phase, but may also promote corrosion initiation on passive steel during the initial phases in the life of a structure. Considering, again, the case of immersed and hollow marine structures, the macrocell between passive steel in dry and chloride-free concrete and (still) passive steel in wet and chloride-contaminated concrete may promote corrosion initiation on the latter due to anodic polarization [17]. In principle, if the kinetics of chloride and water penetration are also considered, and if reliable correlations are available in terms, for instance, of pitting potential versus chloride content (see Pedeferri's diagram, Section 7.4.1), numerical modeling could be useful in estimating the effects of a macrocell on the time of initiation of corrosion. Also, possible beneficial effects of preventative or protective measures (e.g., a cathodic prevention or protection system) can be evaluated through numerical models. This is an example where the difficulties in achieving direct and reliable data on the evolution of corrosion conditions of steel (due to the limited accessibility and to the different behavior of monitoring probes compared with structures exposed to the atmosphere [18]) can be at least partially mitigated by a numerical approach, provided reliable input data are available.

References

1 Andrade, C., Maribona, I.R., Feliu, S., González, J.A., and Feliu, S., Jr. (1992) The effect of macrocells between active and passive areas of steel reinforcements. *Corrosion Science*, **33** (2), 237–249.

2 Schiegg, Y., Audergon, L., Elsener, B., and Böhni, H. (2007) Online monitoring of the corrosion in reinforced concrete structures, in *Corrosion of Reinforcement in Concrete. Mechanisms, Monitoring, Inhibitors and Rehabilitation Techniques*, European Federation of Corrosion Publication number 38 (eds M. Raupach, B. Elsener, R. Polder, and J. Mietz), Woodhead Publishing Limited, Cambridge, pp. 133–145.

3 Sagues, A.A., Lau, K., Powers, R.G., and Kessler, R.J. (2010) Corrosion of epoxy-coated rebar in marine bridges – Part 1: a 30-year perspective. *Corrosion*, **66** (6), 0650011–06500113.

4 Sergi, G. and Page, C.L. (2000) Sacrificial anodes for cathodic prevention of reinforcing steel around patch repairs applied to chloride-contaminated concrete, in *Corrosion of Reinforcement in Concrete. Corrosion Mechanisms and Corrosion Protection*, The European Federation of Corrosion Publication number 31 (eds J. Mietz, R. Polder, and B. Elsener), The Institute of Materials, London, pp. 93–100.

5 Bertolini, L., Pedeferri, P., Pastore, T., Bazzoni, B., and Lazzari, L. (1996) Macrocell effect on potential measurements in concrete cathodic protection systems. *Corrosion*, **52**, 552–557.

6 Bertolini, L., Gastaldi, M., Pedeferri, M.P., and Redaelli, E. (2002) Prevention of steel corrosion in concrete exposed to seawater with submerged sacrificial anodes. *Corrosion Science*, **44** (7), 1497–1513.

7 Laase, H. and Stichel, W. (1983) Rehabilitation of retaining walls in Berlin, special aspects of corrosion of the back side. *Die Bautechnik*, **4**, 124–129 (in German).

8 Polder, R.B. and Larbi, J.A. (1995) Investigation of concrete exposed to North Sea water submersion for 16 years. *Heron*, **40** (1), 31–56.

9 Bertolini, L., Bolzoni, F., Pastore, T., and Pedeferri, P. (1996) New experiences on cathodic prevention of reinforced concrete structures, in *Corrosion of Reinforcement in Concrete Construction* (eds C.L. Page, P.B. Bamforth, and J.W. Figg), Society of Chemical Industry, pp. 389–398.

10 Jaeggi, S., Böhni, H., and Elsener, B. (2007) Macrocell corrosion of steel in concrete – experiments and numerical modelling, in *Corrosion of Reinforcement in Concrete. Mechanisms, Monitoring, Inhibitors and Rehabilitation Techniques*, European Federation of Corrosion Publication number 38 (eds M. Raupach, B. Elsener, R. Polder, and J. Mietz), Woodhead Publishing Limited, Cambridge, pp. 75–88.

11 Gulikers, J. and Raupach, M. (2006) Preface. Modelling of reinforcement corrosion in concrete. *Materials and Corrosion*, **57** (8), 603–604.

12 Raupach, M. (2006) Models for the propagation phase of reinforcement corrosion – an overview. *Materials and Corrosion*, **57** (8), 605–613.

13 Warkus, J., Raupach, M., and Gulikers, J. (2006) Numerical modelling of corrosion – Theoretical backgrounds. *Materials and Corrosion*, **57** (8), 614–617.

14 Gulikers, J. and Raupach, M. (2006) Numerical models for the propagation period of reinforcement corrosion – Comparison of a case study calculated by different researchers. *Materials and Corrosion*, **57** (8), 618–627.

15 Redaelli, E., Bertolini, L., Peelen, W., and Polder, R. (2006) FEM-models for the propagation of chloride induced reinforcement corrosion. *Materials and Corrosion*, **57** (8), 628–635.

16 Warkus, J., Brem, M., and Raupach, M. (2006) BEM-models for the propagation period of chloride induced reinforcement corrosion. *Materials and Corrosion*, **57** (8), 636–641.

17 Della Pergola, A., Lollini, F., Redaelli, E., and Bertolini, L. (2013) Numerical modelling of initiation and propagation of corrosion in hollow submerged marine concrete structures, submitted.

18 Raupach, M., Polder, R., Frolund, T., and Nygaard, P. (2007) Corrosion monitoring at submerged concrete structures – Macrocell corrosion due to contact with aerated areas? EUROCORR 2007, Freiburg im Breisgau, Germany, 9–13 September 2007 (CD-ROM).

9
Stray-Current-Induced Corrosion

Stray current, arising for instance from railways, cathodic protection systems, or high-voltage power lines, often induces corrosion on buried metal structures, leading to severe localized attack [1]. The current may be DC or AC, depending upon the source. Stray current deviates from its intended path because it finds a parallel and alternative route. It may also find a low-resistance path by flowing through metallic structures buried in the soil (pipelines, tanks, industrial and marine structures). For instance, underground pipelines can pick up current strayed from a railway system at some point remote from the traction power substation and discharge the current to the soil and then back to the rail near to the substation.

In the case of direct current (DC) interference, a cathodic reaction (e.g., oxygen reduction or hydrogen evolution) takes place where the current enters the buried structure, while an anodic reaction (e.g., metal dissolution) occurs where the current returns to the original path, through the soil (Figure 9.1). Metal loss results in the anodic points, where the current leaves the structure; usually, the attack is localized and can have serious consequences, especially on pipelines. The effects of AC stray current are more complex; however, alternating current interference is known to be much less dangerous than direct current.

Stray current can also flow through reinforced or prestressed concrete and produce an alteration of the potential distribution inside the concrete, which can influence corrosion of embedded steel [2, 3]. Several types of structures may be subjected to stray current, such as bridges and tunnels of railway networks or structures located in the neighborhood of railways. Here, the concrete, like the soil surrounding buried structures, is the electrolyte and the reinforcing bars or prestressing tendons can pick up the stray current. Laboratory studies have shown that stray DC current rarely has corrosive consequences on steel in concrete, in contrast to their effect on metallic structures in the soil [2–7]. In fact, passive steel in alkaline and chloride-free concrete has a high intrinsic resistance to stray current. Nevertheless, under particular circumstances corrosion can be induced on the passive reinforcement, especially if chlorides contaminate the concrete, even at levels in themselves too low to initiate pitting corrosion. A few cases have been documented [8, 9].

Corrosion of Steel in Concrete: Prevention, Diagnosis, Repair, Second Edition. Luca Bertolini,
Bernhard Elsener, Pietro Pedeferri, Elena Redaelli, and Rob Polder.
© 2013 Wiley-VCH Verlag GmbH & Co. KGaA. Published 2013 by Wiley-VCH Verlag GmbH & Co. KGaA.

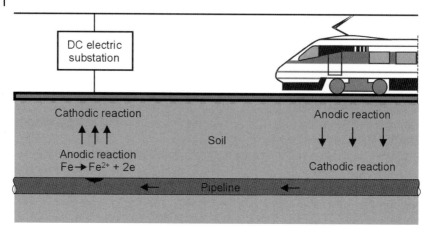

Figure 9.1 Example of stray current from a DC railway line picked up by a buried pipeline.

9.1
DC Stray Current

Consequences of DC stray current in reinforced concrete vary, depending on the properties of the concrete (alkaline, carbonated or contaminated by chlorides), the duration of the current circulation and the current density. It is therefore necessary to distinguish concrete structures noncontaminated by chlorides and noncarbonated from those contaminated by chlorides in quantities insufficient to initiate corrosion and, finally, from those that already have corroding rebars due to chlorides or carbonation.

9.1.1
Alkaline and Chloride-Free Concrete

Passivity of steel in alkaline and chloride-free concrete also provides resistance to stray current. In fact, before the stray current is picked up by the reinforcement, a significant driving voltage (ΔE) has to be present between the point where the current enters the reinforcement (cathodic site, A of surface A_c in Figure 9.2) and the point where the current returns to the concrete (anodic site, B of surface A_a) [2]. ΔE equals the sum of the dissipative contributions due to the cathodic (ψ_c) and anodic (ψ_a) polarizations and the ohmic drop through the reinforcement (ψ_Ω). The latter is negligible and, thus, the sum of the anodic and cathodic polarizations, that both depend on the circulating current, must equal the driving voltage ΔE between points A and B.

For passive reinforcement in noncarbonated and chloride-free concrete, current can flow only if there is a great enough increase in the potential of the anodic area to exceed the threshold for oxygen evolution (Figure 9.3). It was shown in Chapter

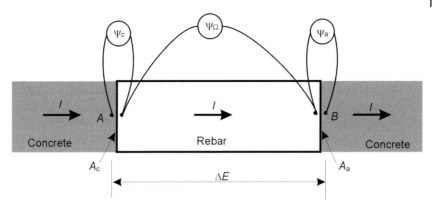

Figure 9.2 Schematic representation of the electrical interference on reinforcement in concrete.

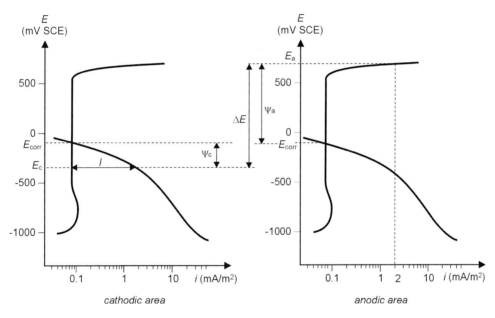

Figure 9.3 Schematic representation of electrochemical conditions in the cathodic and anodic zones of reinforcement in noncarbonated and chloride-free concrete that is subject to stray current.

7 that at potentials below +600 mV SCE no iron dissolution or any other anodic process takes place, and thus it is impossible for the current to leave the metal.

Furthermore, even if such a condition is reached and current circulates through the reinforcement, this does not automatically lead to corrosive attack, since the anodic process that takes place at potentials higher than about +600 mV SCE is oxygen evolution, instead of iron dissolution. Nevertheless, attack may occur if the current flows for sufficiently long periods of time [5, 7]. Initiation of corrosion can be ascribed to the depletion of the alkalinity in the vicinity of the anodic area promoted by the anodic reaction of oxygen evolution ($2H_2O \rightarrow O_2 + 4H^+ + 4e$).

Consequently, for corrosion to initiate, two preconditions are necessary: a) the electric field must be strong enough to create conditions for the circulation of current through the reinforcement, and b) it must persist for a time long enough to lead to acidification and the destruction of passivity.

First Precondition Passive reinforcement in noncarbonated and chloride-free concrete offers a high intrinsic resistance to stray current, since the driving force required to produce the circulation of an appreciable current density in the anodic areas (i.e., $>2\,mA/m^2$) is at least 600 mV [2].

Figure 9.3 shows that even for the circulation of a small current density of $2\,mA/m^2$ through the reinforcement, the cathodic polarization (ψ_c) is of the order of 200–300 mV and the anodic polarization (ψ_a) is at least 500–600 mV.

Therefore, the driving voltage required to force the circulation of current through the reinforcement is:

$$\Delta E = \psi_c + \psi_a \approx 700\text{–}800 \text{ mV} \tag{9.1}$$

Figure 9.4 shows that the potential difference in concrete between point A (where current enters the steel) and B (where current leaves the steel) in order to cause a considerable current density, should be even greater.

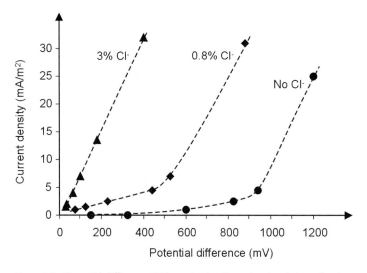

Figure 9.4 Potential difference (ΔE) required to force the circulation of a given current density through the reinforcement, as a function of the chloride content in concrete (% by mass of cement) [2].

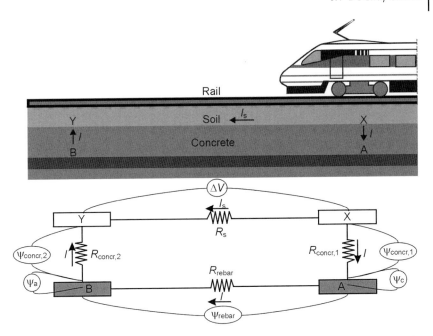

Figure 9.5 Schematic representation of the alternative routes for the stray current through the soil and through the reinforcement.

The anodic and cathodic areas may not have the same size, as has been assumed so far. The driving voltage ΔE may decrease when the cathodic area is significantly larger than the anodic area, such that the current density on the cathodic site is negligible compared to that of the anodic site, and thus $\psi_c \rightarrow 0$, so that $\Delta E \rightarrow \psi_a$. Therefore, under particular circumstances, ΔE may be as low as 500–600 mV.

Values of driving voltage ΔE considered so far are measured in points A and B in the vicinity of the steel surface. To evaluate the conditions that can lead to the circulation of current through the reinforcement, also the dissipation due to the ohmic drop through the concrete cover has to be considered (Figure 9.5). The stray current in the soil will divide into two contributions: I_s that will flow through the soil and I that will flow through the reinforcement. Current I_s will generate a voltage difference $\Delta V = I_s \cdot R_s$ between points X and Y, assuming that R_s is the equivalent resistance of the soil between the two points. R_s depends on the resistivity of the soil and geometrical parameters.

ΔV is the driving voltage that allows the current I to flow through the reinforcement, following the path X→A→B→Y. This path presents, along with ohmic drops in the concrete cover ($\psi_{concr,1} + \psi_{concr,2}$) and in the reinforcement (ψ_{rebar}), the polarization contributions on the anodic and cathodic surfaces ($\psi_a + \psi_c$). Therefore the current I is determined by the relation:

$$\Delta V = \psi_{concr,1} + \psi_a + \psi_{rebar} + |\psi_c| + \psi_{concr,2} \tag{9.2}$$

that is:

$$I_s \cdot R_s = R_{concr,1} \cdot I + |\psi_c| + R_{rebar} \cdot I + \psi_a + R_{concr,2} \cdot I \qquad (9.3)$$

where: R_{rebar}, $R_{concr,1}$, and $R_{concr,2}$ are the electrical resistances of the reinforcement and the concrete cover, and both ψ_a and ψ_c are a function of current I.

We have already seen that the interference current is negligible if the potential difference between points A and B (i.e., $\Delta E = \psi_a + \psi_{rebar} + |\psi_c|$) is lower than at least 600 mV. Therefore, ΔV must be even higher, since it also includes ohmic drops in the concrete cover. This means that current will flow through the reinforcement only if the stray current is high enough and the distance between points X and Y is long enough to generate a significant ohmic drop ($I_s \cdot R_s$) in the path via the soil. Only in special cases can this occur, such as when concrete elements are located right next to the sources of stray current.

Furthermore, the current I will be reduced if the concrete has a high resistivity, so that $\psi_{concr,1}$ and $\psi_{concr,2}$ increase, or the reinforcement is disconnected at certain lengths, so that the distance between points X and Y where current enters and leaves the structure is limited and, consequently the driving voltage ΔV is reduced.

Second Precondition Dissolution of iron will take place only when the current circulates for sufficiently long periods and in high enough intensity to produce acidity that can destroy the conditions of passivity. For instance, Figure 9.6 shows the results obtained with laboratory tests on specimens of cement paste subjected to the circulation of current densities of 1 and 10 A/m² [5, 6]. Corrosion initiated on steel embedded in cement paste without chloride only after more than 200 h

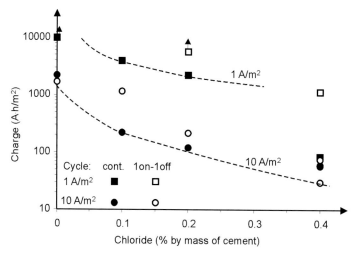

Figure 9.6 Charge required for initiation of corrosion on steel plates embedded in cement pastes with different chloride contents, which were subjected to anodic current densities of 1 A/m² or 10 A/m², applied continuously or at alternate hours [5, 6].

of application of an anodic current density of $10\,A/m^2$. A current density of $1\,A/m^2$ could not initiate corrosion even after 14 months (i.e., more than $10\,000\,h$) of continuous application, although the charge that circulated ($>10\,000\,A\,h/m^2$) was much higher than the charge that could initiate corrosion during the test with a current density of $10\,mA/m^2$ ($2200\,A\,h/m^2$). These results show that, even if some current can circulate through the passive reinforcement (i.e., the first precondition is fulfilled), the risk of corrosion induced by stray current on steel in alkaline and chloride-free concrete is extremely low. In fact, only high current density circulating for a very long time can induce corrosion at anodic sites. Since the reinforcement is not coated (and thus current is not forced to concentrate in small areas of defects of the coating, as occurs on coated steel structures), it is rare that a high current density can be induced by stray current in concrete.

The experimental results and interpretation mentioned allow us to comment on accelerated corrosion tests involving direct current polarization of steel in concrete specimens. In some cases, high current or voltage is applied until cracking of the concrete occurs and the results are interpreted as indicating service life. This is incorrect, because the strong polarization increases the potential into the oxygen evolution range, thus altering the mechanism substantially and rendering the results not representative for concrete under normal service conditions.

9.1.2
Passive Steel in Chloride-Contaminated Concrete

DC stray currents may have more serious consequences in chloride-contaminated concrete. On passive reinforcement in concrete containing chloride in a quantity below the critical content and thus in itself insufficient to initiate localized corrosion, the driving voltage ΔE required for current to flow through the reinforcement is lower than in chloride-free concrete; and it decreases as the chloride content increases (Figure 9.7). This is a consequence of less-perfect passivity, and in particular a lower pitting potential.

Furthermore, initiation of corrosion is also favored in the presence of chloride. In fact, localized breakdown of the passive film can take place at anodic sites where the pitting potential (E_{pit}) is exceeded. The experimental results illustrated in Figure 9.6 show a remarkable decrease in the charge required for the onset of corrosion on steel embedded in cement pastes when the chloride content was increased up to 0.4% by mass of cement. Even a current density of $1\,A/m^2$ can initiate corrosion in the presence of small amounts of 0.1 and 0.2% chloride by mass of cement (i.e., not dangerous for pitting corrosion in the absence of stray current) [6]. Figure 9.6 also shows the higher risks connected with higher anodic current densities: the charge required for corrosion initiation with a current density of $10\,A/m^2$ was more than one order of magnitude lower than that due to $1\,A/m^2$, that is, times for initiation of corrosion were more than 100 times lower.

Once initiated, the attack may proceed even if the causes that made it possible are no longer present, for example, if the stray current decreases so that the potential at the anodic site falls to values below E_{pit}.

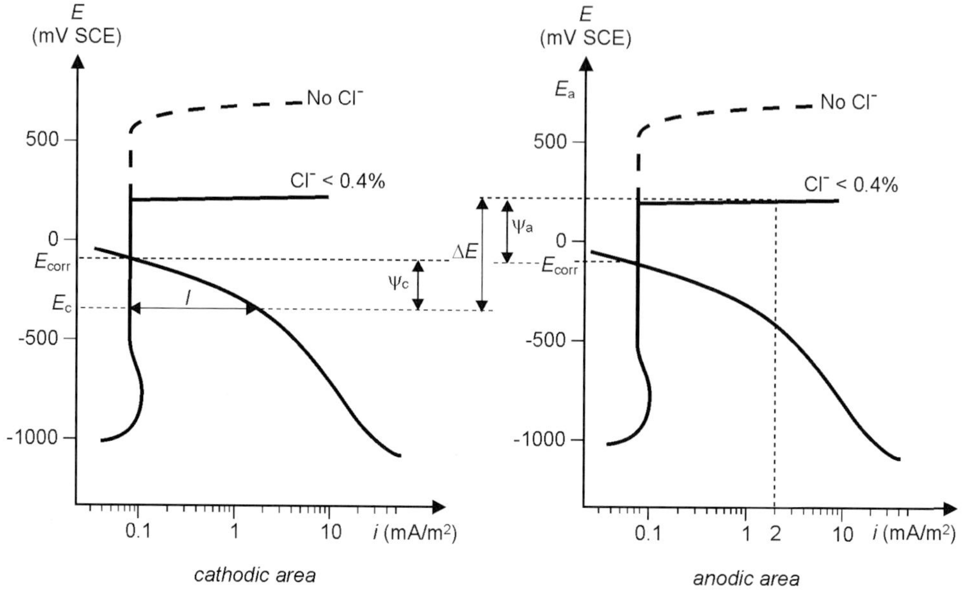

Figure 9.7 Schematic representation of electrochemical conditions in the cathodic and anodic zones of passive reinforcement in chloride-contaminated concrete that is subject to stray current.

Interruptions in the Stray Current Stray currents produced by rail traction systems are nonstationary, and thus the effect of interruptions of the current should be taken into consideration. In fact, gradients of ionic concentration in the pore solution near the steel surface, produced by the depletion of alkalinity due to the anodic reaction and increase in chloride concentration due to migration, can be mitigated during the interruption of current. Therefore, interruptions in the stray current may have a beneficial effect, as shown in Figure 9.6 where results of tests with continuous application of the current are compared with tests with circulation of current at alternated hours (*1on-1off*). The periodical interruption of current had a beneficial effect, since it increased the charge required for initiation of corrosion. This effect was remarkable in cement pastes with chloride contents lower than 0.4% by mass of cement.

9.1.3
Corroding Steel

The protection that concrete offers to steel against stray current ceases when corrosion of the reinforcement has initiated, for example, due to carbonation, chloride contamination, or the stray current itself. In this case, any current flowing through the steel will increase the corrosion rate at the anodic site, similarly as in buried

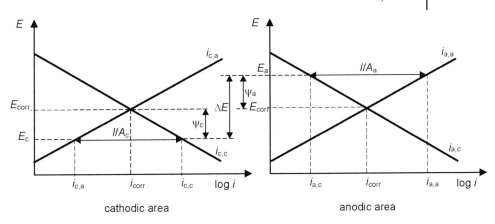

Figure 9.8 Schematic representation of electrochemical conditions in the cathodic and anodic zones of corroding reinforcement that is subject to stray current.

steel structures. Figure 9.8 shows that even small driving voltages can lead to an increase in the corrosion rate on the anodic area (from i_{corr} to $i_{a,a}$). Furthermore, it has been observed that if steel is subjected to pitting corrosion in chloride-contaminated concrete, the anodic current increases the size of the attacked area [5, 6].

9.2
AC Stray Current

Alternating stray currents are known to be much less dangerous than direct stray currents [10]. Even in neutral environments, such as soil, it was shown that AC with current below 100 A/m² will not give rise to noticeable effects on steel, even over long periods. Effects become significant only at anodic current densities above 1000 A/m² for active materials and at even higher values for passive materials. Some authors suggest a conservative threshold value of 30 A/m² [11].

For steel embedded in concrete, it was observed that current densities up to 50 A/m² applied for 5 months to passive steel in concrete with up to 0.4% chlorides did not lead to corrosion initiation [6]. Since steel in reinforced concrete structures is not coated, it is not actually possible to reach such high current densities. It can be assumed, therefore, that interference from AC current cannot induce corrosion on passive steel in concrete. AC stray current may affect the corrosion rate or macrocell effect in the presence of previously depassivated steel. For instance, Figure 9.9 shows that the superposition of 50 A/m² AC for 5 min led to a temporary remarkable increase in the macrocell current between corroding steel in concrete with 0.8% chloride and passive steel in concrete with 0.4% chloride [7]. It was also shown that extremely high AC current densities of 100 and 500 A/m² could also promote corrosion initiation in chloride-contaminated concrete [12].

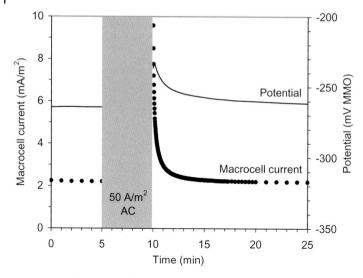

Figure 9.9 Effect of 50 A/m² AC on the macrocell current density between corroding steel in concrete with 0.8% chlorides by mass of cement and passive steel in concrete with 0.4% chlorides by mass of cement. Potential of anodic steel bar is also shown [7].

Although there is no experience, interaction between AC and DC stray currents cannot be excluded, since AC can influence the anodic behavior of steel [10]. Therefore, attention should be dedicated to possible synergistic effects of AC and DC stray currents that might, under specific circumstances, be able to stimulate the corrosion rate of depassivated steel or promote corrosion on passive steel.

9.3
High-Strength Steel

High-strength steel used for pretensioned or post-tensioned structures may be more susceptible to stray currents than ordinary reinforcing steel. In the case of DC current, failures may be induced not only at the anodic site but also at the cathodic site, where very low potentials may be reached. High-strength steel is susceptible to hydrogen embrittlement, leading to failure without warning, if it is cathodically polarized at sufficiently negative potentials to allow hydrogen evolution. This phenomenon will be dealt with in Chapter 10.

There are no known cases in which any possible negative effects can be traced to hydrogen embrittlement in the presence of AC current. Nevertheless, considering that AC current reduces electrodic overvoltages, in prestressing tendons that are polarized cathodically by DC current interference, a superposition with alternating current could increase hydrogen evolution (due to reduction of the overvoltages) and thus increase the risk of hydrogen embrittlement.

9.4
Fiber-Reinforced Concrete

It has been argued that steel-fiber-reinforced concrete (SFRC) is susceptible to the effects of stray current. Steel fibers can pick up stray current as can normal steel reinforcement. Nevertheless, due to the short length of a single fiber, it is unlikely that stray current may flow through it. On passive steel fibers in concrete free of chloride it was shown that stray current may flow only if a potential drop higher than 1 V is produced between the opposite ends of the fiber [13]. Therefore, if fibers are electrically disconnected, a large amount of the driving voltage is dissipated each time the current leaves or enters a fiber. Similarly to steel reinforcement, steel fibers may become more susceptible to stray current if corrosion has initiated, for example, in the presence of chloride contamination; however, it should be observed that a higher threshold for corrosion initiation was shown for steel fibers [14] and this may help in also increasing the resistance of SFRC to stray current in chloride-bearing environments.

9.5
Inspection

Inspection of structures subjected to stray current can be based on potential measurements. For instance, fluctuation of potential on passing of trains strongly suggests the presence of stray current in concrete. Interpretation of the potential values of the reinforcement is different from that used for structures not subjected to stray current (Chapter 16). In the case of stray currents, the values of potential that indicate the presence or absence of corrosion are outlined in Figure 9.10. The zone of corrosion is that polarized anodically, and corrosion phenomena are produced at higher potentials. In order to correctly interpret potential measurements, it is necessary to consider that the value of potential measured contains a contribution of ohmic drop, caused by the circulation of current between the reference electrode and the surface of the reinforcement. Such a contribution should be eliminated, but in the presence of variable electrical fields this is not easy, since these currents are usually beyond the inspector's control. Only in the case of stray current from cathodic protection systems can the interfering current be momentarily interrupted, by turning off the source of the current at the time of measurement. This is known as the "on–off" technique. In the presence of stray current caused by DC traction systems the on–off technique cannot be applied. Among other methods used to evaluate interference in buried metal structures, a promising one for concrete structures is the method of transversal gradients [15]. Other measurements that can help in detecting the presence of stray current in reinforced concrete are based on the potential difference present between different parts of the structure, due to the ohmic drop produced by the stray current.

Monitoring of stray current may also be carried out by embedding permanent reference electrodes in concrete at positions identified as critical, in order to track

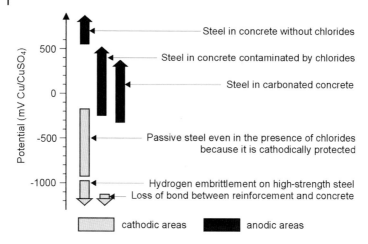

Figure 9.10 Correlation between the potential and the state of corrosion of carbon steel reinforcement affected by stray currents [4].

the presence of stray currents, determine the direction of flow, locate anodic and cathodic areas, etc. [3, 16, 17]. In the case of prestressed concrete structures, expected cathodic areas should also be monitored in order to investigate the risk of hydrogen embrittlement.

9.6
Protection from Stray Current

It was shown that stray current can hardly induce corrosion on passive steel in noncarbonated and chloride-free concrete. However, the potential adverse effects of stray current on concrete structures may become increasingly important with the increased use of underground concrete construction. Stray-current effects are rarely recognized as such. The importance increases further due to the increase of the required service lives (i.e., 100 years or more).

The best protection against stray current is, therefore, provided by concrete. Those methods that can improve the resistance of concrete to carbonation or chloride contamination, which are illustrated in Chapters 11 and 12, are also beneficial with regard to stray-current-induced corrosion. It should be observed that this may not apply for preventative techniques, since conditions leading to corrosion initiation due to stray current are different, in terms of potential, from those leading to corrosion initiation due to carbonation or chloride contamination. For instance, the use of stainless steel or galvanized steel bars, which improves the resistance to pitting corrosion in chloride-contaminated concrete (Chapter 15), does not substantially improve the resistance to stray current in chloride-free and noncarbonated concrete [5]. In any case, a high concrete resistivity will reduce the current flow due to stray current.

Specific measures should be adopted to limit, as far as possible, the stray current from the interfering structure. These may be based on three different approaches [3]:

- prevent the stray current from developing at the source;
- prevent the stray current from reaching the structure;
- design the structure in such a way that the harmful effects are reduced.

In the case of railway lines, the current returning towards the power unit must be forced to flow through the rail. To achieve this, it is possible to reduce the longitudinal electrical resistance of the rails (to reduce leakage of current) and increase their resistance, that is, insulation, with respect to the ground (e.g., improving efficiency of water drainage and maintenance and cleaning of the ballast; using insulating membranes between the rails and soil; isolating the rails from other metallic structures and in particular, from the grounding network). Application of waterproof coatings to buried reinforced concrete structures may reduce the stray current flowing through the concrete.

Electrical discontinuity in reinforcement also contributes to mitigate the circulation of current. In fact, it forces the absorbed current to enter and leave the reinforcement, each time losing part of the driving voltage to overcome cathodic or anodic overvoltages. In tunnels, for instance, electrical longitudinal sectioning may allow to reduce the driving voltage below the critical value that allows the stray current to flow through the reinforcement.

References

1 Schütze, M. (ed.) (2000) *Corrosion and Environmental Degradation*, Wiley-VCH Verlag GmbH, Weinheim.

2 Bertolini, L., Bolzoni, F., Pastore, T., and Pedeferri, P. (1993) Stray-current induced corrosion in reinforced concrete structures, in *Progress in the Understanding and Prevention of Corrosion* (eds J.M. Costa and A.D. Mercer), Institute of Materials, London, p. 568.

3 NACE (2010) *Stray-Current-Induced Corrosion in Reinforced and Prestressed Structures*, NACE International, Houston, p. 34.

4 Bertolini, L., Lazzari, L., and Pedeferri, P. (1996) Factors influencing stray current induced corrosion in reinforced concrete structures. *L'Industria Italiana del Cemento*, **709**, 268–279.

5 Bertolini, L., Bolzoni, F., Brunella, M.F., Pastore, T., and Pedeferri, P. (1996) Stray current induced corrosion in reinforced concrete structures: resistance of rebars in carbon, galvanized and stainless steels. *La metallurgia italiana*, **88**, 345–351 (in Italian).

6 Bertolini, L., Carsana, M., and Pedeferri, P. (2007) Influence of stray currents on corrosion of steel in concrete, in *Corrosion of Reinforcement in Concrete. Mechanisms, Monitoring, Inhibitors and Rehabilitation Techniques*, European Federation of Corrosion Publication number 38 (eds M. Raupach, B. Elsener, R. Polder, and J. Mietz), Woodhead Publishing Limited, Cambridge, pp. 105–119.

7 Bertolini, L., Carsana, M., and Pedeferri, P. (2007) Corrosion behaviour of steel in

concrete in the presence of stray current. *Corrosion Science*, **49** (3), 1056–1068.

8 Buhr, B., Nielsen, P.V., Bajernaru, F., and McLeish, A. (1999) Bucharest metro: dealing with stray current corrosion. Tunnel Construction and Piling Conference '99, London, September 8–10, 1999.

9 Bazzoni, B., Briglia, M.C., Cavallero, G., Melodia, D., and Panaro, F. (1999) Monitoring of stray current interference in the reinforced concrete structures of the Turin underground railway loop, *54th NACE Corrosion/99*.

10 Jones, D.A. (1978) Effect of alternating current on corrosion of low alloy and carbon steels. *Corrosion*, **34**, 428–433.

11 Heim, G., Heim, T., Heinzen, H., and Schwenk, W. (1993) Research on corrosion of steel under cathodic protection due to alternate current. *3R International*, **32**, 246–249 (in German).

12 Ormellese, M., Lazzari, L., and Brenna, A. (2010) AC-induced corrosion on passive metals. NACE International Corrosion/10 Conference, Paper 10109.

13 Solgard, A., Carsana, M., Geiker, M.R., Küter, A., and Bertolini, L. (2013) Experimental observations of stray current effect on steel fibres embedded in mortar, submitted.

14 Mangat, P.S. and Gurusamy, K. (1988) Corrosion resistance of steel fibres in concrete under marine exposure. *Cement and Concrete Research*, **18** (1), 44–54.

15 Bertolini, L., Pedeferri, P., Bazzoni, B., and Lazzari, L. (1998) Measurements of ohmic drop free potential in the presence of stray current. NACE International Corrosion/98 Conference, paper 98560.

16 Peelen, W.H.A. and Polder, R.B. (2004) Durability assessment of the concrete sheet piling for the new Dutch heavy-duty "De Betuweroute" railway line. *Corrosion Prevention & Control*, 11–16 March.

17 Peelen, W.H.A. and Polder, R.B. (2004) Durability assessment of concrete sheet piling in the "De Betuweroute" railway line. *The Structural Engineer*, **82** (3), 25–27.

10
Hydrogen-Induced Stress Corrosion Cracking

Failures of prestressing steel induced by corrosion are well known in the civil engineering field. They are mainly related to prestressed structures (buildings, bridges, pipelines, tanks), anchorages in rocks or soils, and tendons exposed to the atmosphere. The analysis of these failures shows that they are mainly induced by stress corrosion cracking. In this regard, it should be pointed out that failures of prestressing steel can only occur if the steel has some defects at the time of construction or if the temporary protection (to which it should be subjected from the manufacturing plant until the time of grouting) is poor or if the final protection is insufficient. A review of selected failures in concrete structures in which prestressing steels break in a brittle way due to stress corrosion cracking shows that these failures are often due to an accumulation of causes such as poor design, errors during construction, careless detailing and, in some cases, use of unsuitable materials [1]. Even though the number of failures is modest compared with the worldwide number of prestressed structures, their consequences may be extremely serious or even catastrophic since they affect the stability of the structure.

Of course, corrosion of prestressing steel is not only related to hydrogen-induced stress corrosion cracking but it may also occur due to the usual penetration of chloride or carbonation. In fact, if tendons are not adequately protected due to poor detailing or poor workmanship and inadequate grouting, aggressive species (e.g., water and deicing salts) can penetrate especially through the most vulnerable parts, such as anchorages, joints, or cracks. Consequently, properly grouted tendons contribute decisively to the serviceability, safety and durability of prestressed concrete bridges. Sudden collapses of bridges in Belgium and the United Kingdom led to a temporary ban of grouted post-tensioning tendons in the United Kingdom from 1992 to 1996. This has initiated a review of all aspects related to durability of post-tensioning [2].

Recently cases of corrosion of post-tensioning tendons as a consequence of segregation of the cement grout have been reported [3, 4]. Severe localized attacks leading to in-service failure of external post-tensioning cables were associated to the presence of a whitish and incoherent paste produced by the segregation of the cement grout. The segregated grout had a remarkably high content of alkalis and sulfate ions and negligible content of chloride ions. Thus, the usual mechanisms of steel corrosion in concrete could not explain this phenomenon [4]. These cases

Corrosion of Steel in Concrete: Prevention, Diagnosis, Repair, Second Edition. Luca Bertolini,
Bernhard Elsener, Pietro Pedeferri, Elena Redaelli, and Rob Polder.
© 2013 Wiley-VCH Verlag GmbH & Co. KGaA. Published 2013 by Wiley-VCH Verlag GmbH & Co. KGaA.

have shown the importance of preventing the segregation of the grout both during and after the injection. Specific tests, for example, the so-called inclined-tube test [5], have been developed for testing the stability of injection grouts.

10.1
Stress Corrosion Cracking (SCC)

Metals, under specific environmental, metallurgical, and loading conditions, may be subjected to corrosion phenomena that lead to the initiation and propagation of sharp cracks, which can lead to failure without warning. This phenomenon is known as *stress corrosion cracking* (*SCC*). Under certain conditions, cracks propagate owing to the dissolution at their tip, that is, due to the anodic reaction of the corrosion process; this is known as anodic stress corrosion cracking. Under other conditions, cracks propagate because atomic hydrogen, usually produced by the cathodic reaction of a corrosion process, penetrates the metal lattice and accumulates near the crack tip.

Anodic Stress Corrosion Cracking This mechanism may occur only for specific combinations of metal and environment. For example, with regard to low-alloyed steels, it is possible only in environments with certain chemical compounds (nitrates, chromates, carbonates/bicarbonates, etc.) and in specific ranges of pH; in austenitic stainless steels only in the presence of chlorides and at temperature above 50 °C or even lower in acidic environment. These are conditions that can hardly ever be reached in concrete structures. Nürnberger [6], for example, states that "in prestressed concrete constructions in fertilizer storage and in stable ceilings, the condition for *SCC* can be reached. In stables brickwork saltpeter $Ca(NO_3)_2$ may be formed by urea. In the presence of moisture the nitrate may diffuse into the concrete and may cause *SCC* in the case of pretensioned concrete components affecting the tendons if the concrete cover is carbonated due to an inferior quality of the concrete". In any case, failures of this type are rare in the field of reinforced and prestressed concrete structures. In some cases failure took place in the presence of defect or outside the concrete. Among these, the collapse of a concrete ceiling in a swimming pool near Zurich in 1985 should be mentioned, which caused the death of 12 people and originated from chloride-induced anodic stress corrosion of stainless steel (type AISI 304) hangers just outside the concrete that supported the ceiling.

Hydrogen-Induced Stress Corrosion Cracking (HI-SCC) High-strength steel used in prestressed concrete, under certain metallurgical and environmental conditions that will be specified below, may be subject to cracking caused by atomic hydrogen produced on its surface. This type of failure belongs to the larger family of phenomena known as *hydrogen-induced stress corrosion cracking* (*HI-SCC*).

Failure of prestressing steel is usually induced by atomic hydrogen that penetrates the metal lattice. The conditions required for cracking are: a sensitive material, a tensile stress and an environment that produces atomic hydrogen on the steel surface.

Evolution of atomic hydrogen on the steel surface may induce the initiation and propagation of cracks starting from the metal surface, especially in the presence of notches or localized corrosion attack. Even in the absence of flaws on the surface, atomic hydrogen may penetrate the steel lattice, accumulate in the areas subjected to the highest tensile stress, above all at points corresponding to lattice defects, and lead to brittle failure beginning at one of these sites.

This type of attack does not require any specific environment to take place, since it can take place simply in neutral or acidic wet environments. Failure due to hydrogen is named *hydrogen embrittlement* since it leads to a brittle-like fracture surface. Indeed, the ductility of the bulk metal does not change, but the propagation of the crack is due to the mechanical stresses induced in the lattice by hydrogen accumulated near the crack tip. If hydrogen is present in the metal lattice before the application of loads a *delayed fracture* may occur, that is, the steel does not fail when the load is applied, but after a certain time.

10.2
Failure under Service of High-Strength Steel

Fracture of high-strength steel due to SCC normally takes place in three stages (Figure 10.1): (i) a first stage of incubation or crack initiation; (ii) a second stage of slow (subcritical) propagation of cracks; (iii) a third stage of fast propagation,

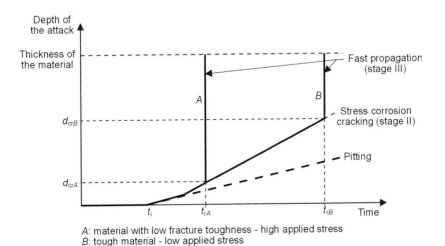

A: material with low fracture toughness - high applied stress
B: tough material - low applied stress

Figure 10.1 Sequence of phenomena that lead to the initiation and propagation of hydrogen-induced cracks and subsequent failure in two materials (A and B) with different fracture toughness (d_{cr} = critical flaw size; t_i = incubation time, t_r = time of failure) [7].

which occurs when some critical conditions are reached and suddenly leads to failure.

10.2.1
Crack Initiation

Crack initiation is promoted by corrosion and, particularly, by pitting or crevice attack, even of low depth, which can cause local conditions of acidity and thus the development of atomic hydrogen (see Chapter 6). The duration of this stage depends both on the characteristics of the steel and the environment (e.g., the surface finishing of the steel bars, the pH and chemical composition of the environment, etc.) and the time required to initiate the preliminary pitting or crevice attack. However, it does not depend on the stress applied to the steel.

The duration of the first stage can be nil if:

- hydrogen is already present in the metal lattice, for instance because the steel has been subjected to treatments of pickling or galvanizing or it has been precorroded due to lack of protection during transportation or storage at the construction site;

- hydrogen is produced on the surface of the steel due to a cathodic polarization (e.g., produced by stray current, cathodic protection or macrocells);

- steel has sharp flaws on its surface.

If steel is passive and is embedded in noncarbonated and chloride-free concrete, the first stage does not take place and, consequently, also the others do not. Conversely, it can take place if concrete is carbonated or chloride contaminated. Similarly, it can occur if part of the steel surface is not embedded in concrete and is in contact with soil, aqueous solutions with neutral or acid pH or simply with moist atmosphere.

10.2.2
Crack Propagation

Once the crack has initiated, hydrogen concentrates in the zones of the metal lattice where the tensile stress is highest. Specifically, the hydrogen content increases where flaws or notches are present on the surface of the steel. The metal in these zones becomes locally brittle, since hydrogen atoms hinder the movement of dislocations in the metal lattice. A triaxial stress builds up and leads to a dilatation of the lattice that allows other hydrogen to come from neighboring (less-stressed) parts. When the amount of hydrogen and the tensile stress are high enough to lead to a microfracture, a real crack forms. This crack propagates according to the mechanism just described, since hydrogen accumulates at the tip of the crack. This is named *subcritical propagation*. Its rate can vary from negligible values to several millimeters per year, depending on the characteristics of the material and the environment and also on the stress at the crack tip.

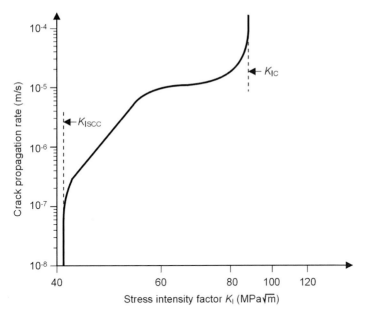

Figure 10.2 Example of crack propagation rate as a function of the stress intensity factor K_I.

σ_s and K_{ISCC} To study the propagation of cracks due to SCC, a parameter called the *stress intensity factor* (K_I) should be considered:

$$K_I = \beta \cdot \sigma \sqrt{\pi w} \qquad (10.1)$$

where: β is a factor that varies between 0.5 and 2 according to the bar and crack geometry, σ is the nominal stress, w is the size of the crack on the surface of the material. The rate of propagation of the crack is determined by K_I. If K_I is below a critical value called K_{ISCC}, the crack does not propagate due to stress corrosion. For $K_I > K_{ISCC}$ the crack propagates with a rate that increases as the size of the crack increases (and thus also K_I increases). Figure 10.2 shows a typical relationship between the stress intensity factor and the rate of crack propagation.

K_{ISCC} depends on metallurgical factors (it usually decreases as the strength of the steel increases, even though it also depends on the microstructure of the material, for example, it is lower in quenched and tempered steel than in cold-worked steels) and on environmental factors (for instance, in alkaline environments and in the absence of chlorides, K_{ISCC} is so high that normal mechanical failure takes place before stress corrosion cracks can develop).

10.2.3
Fast Propagation

It is worth underlining that the rate of crack propagation decreases as K_I decreases (Figure 10.2). If K_I is lower than K_{ISCC}, for example, due to low loading conditions or the small size of the crack or flaws, the crack does not initiate or does not propagate even if it has already initiated. This means that the second stage either does not take place or does not finish. Therefore, corrosion attack that occurs on the surface of high-strength steel, for example, due to pitting or carbonation, does not necessarily lead to stress corrosion cracking.

As the size of the crack increases due to its propagation, K_I may reach the fracture toughness of the material (i.e., a threshold value K_{IC}) and fast propagation takes place (the propagation rate is of the same order as the sound rate through the material), which leads to brittle fracture. From Eq. (10.1) the critical size of the crack can be calculated as:

$$w_{cr} = \frac{K_{IC}^2}{\beta^2 \sigma^2 \pi} \tag{10.2}$$

For a material of given fracture toughness (K_{IC}) and for a given nominal stress (σ), a critical crack size can be calculated above which failure occurs. For instance, for bars made of steels that have fracture toughness values of 60 and 30 MPa√m, respectively, in the presence of cracks with the typical geometry of those produced by stress corrosion cracking (for which a β value of about 0.6 can be assumed) and a stress $\sigma = 800$ MPa, the maximum allowable size of defects is, respectively, 5 and 1.2 mm. Figure 10.3 depicts the fracture surfaces of bars with different fracture toughness.

10.2.4
Critical Conditions

Figure 10.4 summarizes the critical conditions that lead to crack propagation due to mechanical stress (third stage) and those that lead to propagation due to stress corrosion cracking (second stage). In the absence of flaws or cracks on the surface of the steel, critical conditions leading to failure are defined by maximum allowable

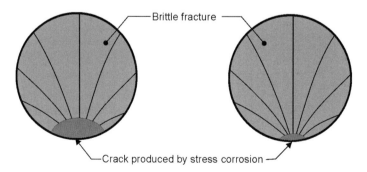

Figure 10.3 Comparison of the fracture surfaces of two bars of high-strength steel with higher K_{IC} (left) and lower K_{IC} (right) under the same applied tensile stress.

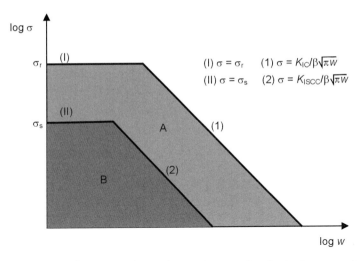

Figure 10.4 Safe regions with regard to crack propagation due to stress corrosion cracking (B) or mechanical failure (A) in a log σ–log w diagram [8].

values of the applied stress σ; conversely, in the presence of flaws or cracks they are defined by critical values of the stress intensity factor K_I. More precisely, the critical conditions leading to fracture due to either pure mechanical stress or stress corrosion cracking are defined by pairs of parameters: for pure mechanical fracture these are the ultimate tensile stress (σ_r) and the fracture toughness (K_{IC}), while for stress corrosion cracking they are a critical stress σ_s and a critical value of the stress intensity factor K_{ISCC}. Hence, mechanical failure cannot occur if $\sigma < \sigma_r$ and $K_I < K_{IC}$ and, similarly, failure due SCC cannot occur if $\sigma < \sigma_s$ and $K_I < K_{ISCC}$. Conditions under which cracks do not propagate are shown in the log σ–log w diagram of Figure 10.4, where w is the crack size and σ is the applied stress.

The critical conditions with regard to the stress are shown by two horizontal lines $\sigma = \sigma_r$ and $\sigma = \sigma_s$, those with regard to K_I can be obtained from Eq. (10.1) by replacing K_I with K_{ISCC} and K_{IC}, respectively:

$$\sigma = K_{IC}/(\beta\sqrt{\pi w}) \quad \sigma = K_{ISCC}/(\beta\sqrt{\pi w}) \tag{10.3}$$

Each pair of lines in Figure 10.4 defines a field where cracks cannot propagate due to brittle mechanical failure (A) and stress corrosion cracking (B). In aggressive environments and for steels susceptible to SCC, field B may be smaller than field A and thus stress corrosion cracking may occur for combinations of stress and flaw size that would not be sufficient to produce pure mechanical failure. Conversely, in mild environments or in the presence of steel with low susceptibility to SCC, σ_s and K_{ISCC} approach σ_r and K_{IC}, respectively, and thus subcritical propagation of the crack does not take place (field B approaches field A in Figure 10.4).

Figure 10.5 Fracture surface of a prestressing bar that failed due to hydrogen embrittlement.

10.2.5
Fracture Surface

The fracture surface of high-strength steel subject to hydrogen-induced stress corrosion cracking has a brittle morphology. In the case of prestressing bars, the fracture is perpendicular to the applied stress and no necking can be observed. The brittle fracture initiates on the surface of the steel where a crack produced by stress corrosion with an elliptical shape can be observed (Figure 10.5). The final brittle fracture starts from the tip of the *SCC* crack and, in the case of bars, it shows the typical marks of brittle failure (chevrons). The size of the flaw that initiates the brittle fracture (or the area of the crack initially produced by stress corrosion cracking) depends on the fracture toughness of the steel and the applied stress (Section 10.2.3). Usually, smaller cracks can be observed around the main crack that initiated the brittle fracture (Figure 10.6).

The microscopic aspect of the surface where the crack propagates due to stress corrosion and that where the final brittle fracture occurs are rather different. Stress corrosion cracks are intercrystalline, while crack surfaces of the final brittle fracture are transcrystalline (cleavage). For cold-worked wires the stress corrosion crack initially propagates perpendicularly to the crystalline grains of the steel, which are strongly elongated, and is transcrystalline; subsequently, it propagates longitudinally (Figure 10.7).

10.3
Metallurgical, Mechanical and Load Conditions

Hydrogen-induced *SCC* can affect, above all, high-strength steels used in prestressed concrete (Section 15.1.2). It does not affect steels with tensile strength

Figure 10.6 Secondary cracks observed around the main crack that initiated the brittle fracture [9].

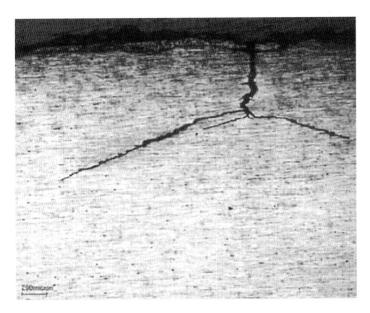

Figure 10.7 Stress corrosion cracking initially propagates perpendicularly to the crystalline grains of the steel; subsequently it propagates longitudinally [9].

lower than 700–900 MPa and thus it does not occur in ordinary reinforcing bars. In general, the amount of hydrogen necessary to induce cracking decreases as the strength of steel increases. However, tensile strength being equal, the amount of hydrogen needed to induce cracking will vary with the microstructure of the steel and in particular with the degree of residual stress. This amount will thus vary with the thermal and mechanical treatments used to give the steel the mechanical properties required and in short with the type of process used to produce it. In practice, production of steel may take place: by cold drawing followed by stretching; hot rolling followed by cold deformation and stretching; and by quenching and tempering (Section 15.1.2).

Quenched and tempered steels, nowadays produced in very small quantity in Germany and in Japan, are undoubtedly the most susceptible to hydrogen embrittlement. The martensitic microstructure produced by quenching generates residual stresses that are not always completely removed by tempering and a relatively low number of dislocations remains in the metal lattice, so that a small amount of hydrogen can interfere with their movement and cause hydrogen embrittlement. Hot-rolled steels are less susceptible in that, while they have a similarly low dislocation density, they contain almost no internal stresses. The most common cold-drawn steels, finally, are even less susceptible, since the drawing process introduces a large number of dislocations and thus of sites to which hydrogen can migrate. Hydrogen is thus more widely distributed and rarely reaches the critical levels required for crack propagation.

The most critical conditions are reached in those materials that are subjected to tensile stress that vary in time, such as to induce a slow strain rate. In fact, slow changes in the applied stress, even of a modest level, for example, increases of a few % in the applied stress during one day or one week, may considerably increase the susceptibility to stress corrosion cracking.

10.3.1
Susceptibility of Steel to HI-SCC

Quantitative tests have been developed to evaluate the susceptibility to hydrogen embrittlement of different types of prestressing steel (Table 10.1). The most widely used is that known as the FIP standard test that consists of immersion of the prestressing steel in a solution of ammonium thiocyanate (NH_4SCN) at 50 °C under the application of a stress equal to 80% of its tensile strength. This test results in a brittle fracture of the steel after hydrogen charging and general embrittlement of the whole cross section. The time to rupture is measured. Prestressing steel is classified as extremely susceptible if time to failure remains below the following values: 2–3 h for cold-drawn wires; 10–15 h for quenched and tempered wires; 30–50 h for hot-rolled bars. However, because of the high aggressiveness of the solution used in the tests, results cannot be simply transferred to the practical behavior of prestressing steel. Furthermore, this test is not suitable for comparison of different prestressing steels [10].

Table 10.1 Parameters and criteria for testing susceptibility to stress corrosion cracking of prestressing steel.

Test	Test solution	Temperature	Load	Life time required (h)
FIP	200 g NH_4SCN + 800 g H_2O	50 °C	80% of tensile strength	Cold-drawn wire: >2–3 Quenched and tempered wire: >10–15 Hot-rolled bar: >30–50
DIB	0.5 g/l Cl^- + 5 g/l $SO_4^=$ + 1 g/l SCN^-	50 °C	80% of tensile strength	>2000

Another test (the DIB test) utilizes a solution containing 5 g/l of $SO_4^=$, 0.5 g/l of Cl^- and 1 g/l of SCN (added as potassium salt). This test has the advantage that the test solution is less aggressive and the amount of hydrogen that is developed during the test is comparable with that produced in practice on the surface of prestressing tendons within the ducts in defective conditions. The time to fracture required is 2000 h. All types of steel that caused damage in the field, failed in this DIB standard test within 2000 h [11].

10.4 Environmental Conditions

A necessary condition for hydrogen-induced stress corrosion cracking to take place is that the surface of the steel is covered with a layer of adsorbed atomic hydrogen. Atomic hydrogen may be formed through the reduction of hydrogen ions:

$$H^+ + e \rightarrow H_{ad} \tag{10.6}$$

or through the hydrolysis of water:

$$H_2O + e \rightarrow H_{ad} + OH^- \tag{10.7}$$

In practice, these processes take place when steel is subject to acid corrosion, is cathodically polarized, is coupled with less-noble materials, or is interfered by stray current or has undergone special production or finishing processes (such as pickling or galvanizing).

On the other hand, atomic hydrogen usually is consumed by recombination:

$$2H_{ad} \rightarrow H_2 \tag{10.8}$$

If there are no particular species in the environment (such as sulfur compounds, thiocyanates, cyanates, or compounds of arsenic, or antimony), recombination of atomic hydrogen occurs rapidly, so that the amount of hydrogen adsorbed in the metal lattice is modest. On the contrary, if such substances are present, recombination slows down, and hydrogen accumulates on the surface. Also, in the presence of oxygen the concentration of atomic hydrogen on the metallic surface may be reduced because of the reaction:

$$2H_{ad} + \frac{1}{2}O_2 \rightarrow H_2O \tag{10.9}$$

In conclusion, the environmental conditions that can promote hydrogen-induced stress corrosion cracking are: an acidic medium, cathodic polarization of the steel, and the presence of promoters such as sulfur compounds, thiocyanates, cyanates, or compounds of arsenic, antimony.

Critical Intervals of Potential and pH The reduction of hydrogen ions to atomic hydrogen takes place when the potential (E) of steel is more negative than the equilibrium potential for that reaction ($E_{eq,H}$). The latter decreases linearly with the pH, according to Nernst's law:

$$E_{eq,H} = E^\circ - 0.059 \cdot pH \tag{10.10}$$

The reaction can therefore only take place in the lower-left field of the potential versus pH plot depicted in Figure 10.8.

10.5
Hydrogen Generated during Operation

In general, steel in concrete operates in the interval of potential and pH outside the critical ranges for hydrogen evolution. Under particular conditions, however, the situation may be different. Situations that make it possible for hydrogen to

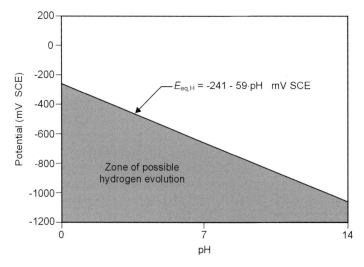

Figure 10.8 Potential–pH field in which hydrogen evolution is possible.

develop are: localized corrosion on the reinforcement that leads to oxygen depletion (and thus depresses the potential), acidity production at the anodic zones, and external cathodic polarization applied to the steel (due to, for example, excessive cathodic protection or stray currents).

Noncarbonated and Chloride-Free Concrete In concrete that is not carbonated and does not contain chlorides, and in the absence of external cathodic polarization, hydrogen evolution, and thus consequent embrittlement, cannot take place. In this type of concrete, characterized by a pH above 12, hydrogen evolution can only occur at potentials below about −900 mV SCE. Passive steel under free corrosion conditions has much less negative potentials (Chapter 7); in the case of atmospherically exposed structures, the potential is between 0 and −200 mV SCE (zone A of Figure 10.9).

Carbonated Concrete The pH of carbonated concrete falls to values just slightly above 8 and so the critical threshold for hydrogen evolution rises to −700 mV SCE. In aerated carbonated concrete the potential of reinforcement reaches values between −300/−500 mV SCE and thus hydrogen evolution cannot take place (zone B of Figure 10.9). In nonaerated and carbonated concrete (i.e., water-saturated concrete), instead, the potential can fall, under very special circumstances, even below −700 mV SCE and hydrogen evolution may take place (zone C of Figure 10.9).

Concrete Containing Chlorides Inside corrosion pits that can develop in chloride-contaminated concrete, there are large reductions in pH and low potentials are reached (for example, inside a pit, values below pH 2.5 and −600 mV SCE,

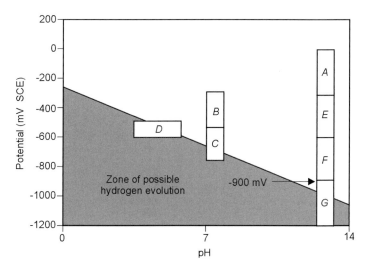

Figure 10.9 Diagram indicating the field of potential and pH for reinforcement: passive (A), in aerated (B) and nonaerated (C) carbonated concrete, subject to pitting (D), under cathodic prevention (E), under cathodic protection (F), and under overprotection (G) [12].

respectively, have sometimes been measured). Under these conditions, hydrogen may develop (zone D of Figure 10.9).

Cathodically Protected Structures Cathodic protection (Chapter 20) interferes with the process of hydrogen embrittlement in a complex way. On the one hand, it favors embrittlement when it brings the potential below that of equilibrium for hydrogen evolution ($E_{eq,H}$), but on the other hand it opposes it because it brings conditions of high pH (above 12.5) to the entire surface of the reinforcement (if it is already corroding) or maintains such conditions (if it is not yet corroding). It thus brings the "safe limit" for hydrogen evolution ($E_{eq,H}$) to very negative values (for pH 12.5: $E_{eq,H} = -978$ mV SCE). In practice, as shown in Figure 10.10, obtained under the most critical conditions (notched samples and slow strain rate), phenomena of hydrogen embrittlement are not produced at potentials above −900 mV SCE, and in order to produce considerable effects, the potential must fall below −1100 mV SCE. For this reason, if the reinforcement is maintained at potential values less negative than −900 mV SCE there will be no real risks of hydrogen embrittlement.

In Figure 10.9 zones E and F represent the operational conditions of reinforcement in structures to which cathodic prevention and cathodic protection, respectively, have been applied (Chapter 20). Values at which cathodic protection normally operates are not sufficiently negative to induce hydrogen evolution. Even if it should operate in conditions of overprotection (potential below −900 mV SCE), the same diagram shows that up to potentials of −1100 mV SCE on the protected steel, the situation will nevertheless be less critical than on that unprotected reinforcement where pitting attack occurs.

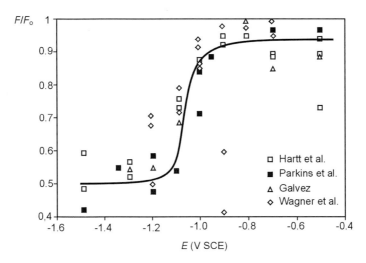

Figure 10.10 Normalized tensile strength (F/F_o) on precracked samples with varying potential, according to different authors (F_o = tensile strength in air; F = tensile strength at varying potentials in concrete) carried out on cracked samples to which a slow strain is applied [13].

Because of the low values of pH and potential that are created inside the pits, the driving voltage available for the process of hydrogen evolution ($\Delta V = E_{eq,H} - E$) on steel that is unprotected and affected by pitting corrosion corresponds to what would exist in conditions of extreme overprotection. For example, this driving voltage inside a pit where the pH is below 4 and the potential is below –600 mV SCE (zone D in Figure 10.9) corresponds to what is found when steel is brought to potentials below –1100 mV SCE at pH > 12.5 (zone G).

Considerations analogous to those used for cathodic protection, and also chloride extraction and realkalization (see Chapter 20), are also valid for the action of stray currents (Chapter 9), in zones where the cathodic process occurs.

10.6
Hydrogen Generated before Ducts Are Filled

Collapses of prestressed concrete structures due to hydrogen embrittlement, of high-strength strands in post-tensioned structures built in the 1960s and 1970s, have occurred generally without warning. The collapses occurred after service lives varying from a few months to 30 years and presented several common characteristics [14]. They were found only on certain types of quenched and tempered steels or cold drawn steels, all having a tensile strength higher, respectively, than 1600 MPa and 1800 MPa. Nevertheless, some episodes occurred even though none of the environmental conditions required to produce hydrogen existed during service of the structures. In particular, in the areas where failure took place, the grout was found to be perfectly injected in the ducts, noncarbonated and containing insignificant levels of chlorides. The fracture began at a point of the wires affected by a slight corrosion attack that must have initiated before injection of the grout. Finally, all these fractures occurred at the lowest part of the duct, where water tends to accumulate and stagnate before grout is injected.

These observations would attribute the cause of anomalous failure to the presence of an aggressive solution that came into contact with the steel before the ducts were filled with grout. A solution with these characteristics is the aqueous phase that is produced in concrete in the first hours following its mixing [6, 14].

Today, the recommended procedure is to: conserve properly the wires or strands before they are utilized, so that localized attack cannot occur; make sure before injection that aggressive solutions cannot penetrate the duct; inject the mortar immediately after prestressing. To avoid corrosion during longer periods between tensioning and grouting specific temporary protection methods should be used.

10.7
Protection of Prestressing Steel

Corrosion prevention has to be applied to prestressing steel from the moment the strands are delivered until they are embedded in concrete or in grouted ducts. The essential elements to be considered are: prestressing steel, the ducts containing the tendons, the anchorage system, the protective system overall. The protection of prestressed tendons from external corrosive agents, in particular from the infiltration of deicing salts, requires that they are completely surrounded with a protective barrier.

A laboratory and field study investigated the performance of different temporary corrosion protection methods for prestressing steel [15]. For post-tensioning systems a multibarrier approach consisting in concrete cover, duct and alkaline grout is used. New systems developed by industry use polymer ducts and electrically isolated tendons [16, 17].

References

1 *fib*, International Federation for Structural Concrete (2003) Influence of material and processing on stress corrosion cracking of prestressing steel – case studies, Bulletin 26, Lausanne.
2 The Concrete Society Technical Report 47 (1996) Durable post-tensioned concrete bridges, The Concrete Society, Crowthorne, UK.
3 Godart, B. (2001) Status of durability of post-tensioned tendons in France, *fib* Workshop Durability of post-tensioning tendons, Ghent, 15–16 November, 25–42.
4 Bertolini, L. and Carsana, M. (2011) High pH corrosion of prestressing steel in segregated grout, in *Modelling of Corroding Concrete Structures* (eds C. Andrade and G. Mancini), Springer, RILEM Bookseries, pp. 147–158.
5 CEN (2007) EN 445. Grout for Prestressing Tendons – Test Methods, European Committee for Standardization.
6 Nürnberger, U. (2002) Corrosion induced failures of prestressing steel. *Otto Graf Journal*, **13**, 9–23.
7 Brown, B.F. (1968) The application of fracture mechanics to stress corrosion cracking. *Metallurgical Review*, **13**, 171.
8 Sinigaglia, D., Re, G., and Pedeferri, P. (1971) *Cedimento per fatica e ambientale dei materiali metallici*, Ed. CLUP, Milano.
9 Bertolini, L., Brunella, M.F., and Pedeferri, P. (1997) Cedimento in servizio degli acciai da compressione. *L'Edilizia*, **XI** (9), 28–36.
10 Fédération Internationale de la Précontrainte (FIP) (1980) Report on prestressing steel. Stress Corrosion Cracking resistance test for prestressing tendon.
11 Nürnberger, U. (2001) Corrosion induced failures mechanism, *fib* Workshop Durability of post-tensioning tendons, Ghent, 15–16 November.
12 Pedeferri, P. and Bertolini, L. (2000) *Durabilità del calcestruzzo armato*, McGrawHill Italia, Milano.
13 Klisowski, S. and Hartt, W.H. (1996) Qualification of cathodic protection for corrosion control of pretensioned tendons in concrete, in *Corrosion of Reinforcement in Concrete Construction* (eds C.L. Page, P.B. Bamforth, and J.W. Figg), Society of Chemical Industry, pp. 354–368.
14 Nürnberger, U. (1993) Special problems in post-tensioned structures, COST 509 Workshop Corrosion and protection of metals in concrete, La Rochelle.
15 Marti, P., Ullner, R., Faller, M., Czaderski, C., and Motavalli, M. (2008) Temporary corrosion protection and bond of prestressing steel. *ACI Structural Journal*, **105** (1), 51–59.
16 Della Vedova, M. and Elsener, B. (2006) Quality control and monitoring of electrically isolated tendons. in Proc. of the 2nd *fib* Congress, Naples, 5–8 June 2006.
17 COST 534 (2009) New materials, systems, methods and concepts for prestressed concrete structures, Final Report (eds R. Polder, M.C. Alonso, D. Cleland, B. Elsener, E. Proverbio, Ø. Vennesland, and A. Raharinaivo), European Commission, Cost office.

11
Design for Durability

Prevention of reinforcement corrosion and other types of deterioration begins in the design phase, when a structure is conceived and structural calculations are made, details are designed, materials and their proportions as well as possible additional preventative measures are selected. Prevention is further materialized as the concrete is prepared, placed, compacted and cured. It will continue throughout the entire service life of the structure, with programmed inspections, monitoring and maintenance. Civil engineers have become increasingly aware of the importance of these matters in the past decades and various international organizations have given guidance on this topic [1–8].

This chapter deals with the characteristics required of concrete so that corrosion of reinforcement and its consequences are not likely throughout the service life. The techniques of additional protection that may be needed for particular conditions of aggressiveness are also outlined in this chapter and will be illustrated in the following ones. After a description of the factors affecting the durability of reinforced concrete structures, prescriptive approaches to durability as well as performance-based approaches with regard to carbonation and chloride-induced corrosion will be considered. Measures against chemical and physical attack of concrete have been addressed in Chapter 3.

Experience has shown that a few design details are frequently the cause of failures, which will be further discussed in Chapter 12. Let it suffice here that the origin of corrosion can often be traced to simple errors that could have been avoided without any appreciable increase in cost.

In fact, the cost of adequate prevention carried out during the stages of design and execution are minimal compared to the savings they make possible during the service life and even more so, compared to the costs of rehabilitation that might be required at later dates. This is depicted by the so-called De Sitter's "law of five" that states as follows: one dollar spent in getting the structure designed and built correctly is as effective as spending 5$ when the structure has been constructed but corrosion has yet to start, 25$ when corrosion has started at some points, and 125$ when corrosion has become widespread [9].

This concept of a sequence of events with increasing levels of costs implies that the structure should be accessible to inspection and maintenance. If accessibility is limited, such as in underground structures, even more emphasis should be

Corrosion of Steel in Concrete: Prevention, Diagnosis, Repair, Second Edition. Luca Bertolini,
Bernhard Elsener, Pietro Pedeferri, Elena Redaelli, and Rob Polder.
© 2013 Wiley-VCH Verlag GmbH & Co. KGaA. Published 2013 by Wiley-VCH Verlag GmbH & Co. KGaA.

placed on service-life design. On the other hand, if planned and regular inspection, monitoring and maintenance of the structure are taken into account in the design stage, the initial requirements could be relaxed. In such cases, programmed maintenance should be included in the design. For relatively short-lived components or materials this is done in practice. For example, in some cases precast sidewalks on bridges and vehicle barriers in tunnels are used that are designed to be replaced after some time, for example, after 50 years in a structure with a design life of 100 years. Prescheduled reapplication of a hydrophobic treatment may be another example.

11.1
Factors Affecting Durability

11.1.1
Conditions of Aggressiveness

Environmental aggressiveness is a function of numerous factors that are not always independent of each other. They have, in fact, enormous and complex synergistic effects connected to both the macroclimate and local microclimatic conditions that the structure itself helps create, such as: humidity of the environment and its variability in time and place, the presence of chlorides and oxygen and the temperature. The following outline summarizes the environmental aggressiveness under the most frequent conditions of exposure.

The environment is not aggressive if it is sufficiently dry. In the case of carbonated concrete that does not contain chlorides, the relative atmospheric humidity (R.H.) below which the corrosion rate becomes negligible is about 70% and 60% R.H. in temperate and in tropical climates, respectively. If, on the other hand, the concrete contains chlorides, this is reduced to 60% R.H. or, if the chloride level is high, to even less than 50% R.H.

The environment is not aggressive, even in the presence of chlorides, if it maintains the concrete in conditions of total and permanent saturation with water, because under these conditions oxygen cannot effectively reach the surface of the reinforcement. Obvious exceptions may be gross defects in the concrete cover such as honeycombs and wide cracks. Submerged concrete without such gross defects can be experiencing aggressive conditions when parts of the cross section are aerated (macrocell mechanism, in hollow structures such as tunnels or offshore platform legs, Chapter 8), or in the presence of stray current (Chapter 9).

In the absence of chlorides, for R.H. > 70% that remains constant or shows only modest changes that do not lead to condensation, the environment is moderately aggressive in temperate climates and aggressive in tropical or equatorial climates. For R.H. > 70% with widespread and frequent variations, or if condensation takes place on the surface of the concrete or wetting–drying cycles occur, the environment is aggressive in temperate climates and very aggressive in hot climates. For humidities between 70% and 95% R.H., the corrosion rate depends considerably on the quality of the concrete (Figure 5.9), which on the other hand has only a modest influence for R.H. < 70% or R.H. > 95%.

In the presence of chlorides, the environment may be aggressive if the R.H. remains above 50% (or even 40% if the chloride content is very high and hygroscopic chlorides such as magnesium or ammonium chloride are present). Aggressiveness increases with humidity (until it reaches a maximum at R.H. of about 90–95% for dense concrete and 95–98% for more porous concrete), with chloride content and with temperature.

Conditions of exposure to marine atmosphere, even if not in direct contact with seawater, are aggressive.

Conditions of contact with seawater and subsequent drying, as in the splash zone of marine structures or those found on viaducts where deicing salts are used, are very aggressive.

Finally, exposure conditions of horizontal surfaces or surfaces subject to water stagnation in the splash zone of marine structures and parts of structures that come in contact with deicing salts after prolonged drying out, such as below expansion joints that are leaking after some time, are extremely aggressive.

11.1.2
Concrete Quality

The resistance of a structure against corrosion depends strongly on a wide range of properties that are generally taken together in the term "quality of concrete", which includes its composition and the care with which it is executed. Important concrete quality items are [10]:

- water/cement ratio (w/c);
- cement content;
- cement type;
- mixing, placing, compaction, and curing;
- cracking, both on the macroscopic and microscopic scale;
- other aspects, such as air content.

Some compositional features also have a strong influence on the mechanical strength of the concrete, in particular the w/c ratio. However, in particular in chloride-contaminated environments, the cement type is even more important. In other chapters, the microstructure of the hardened cement paste and the beneficial role of blast furnace slag and pozzolana such as fly ash have been outlined. The other most important factor is of course the thickness of the concrete cover, which will be discussed in Section 11.1.4.

11.1.3
Cracking

The risk of corrosion of reinforcement has often been correlated to the width of cracks. In the past, standards provided only a maximum limit for this parameter which varied, from 0.3 mm for internal exposure in a nonaggressive environment, to 0.1 mm for exposure in an aggressive environment. Later, other more detailed standards were issued to take into account the type of reinforcement and the fact that the crack width may fluctuate in time (for example, higher limits can be used for crack widths that are expected only for relatively short times with respect to the service life of the structure). It should be realized that cracks in concrete structures may have many different origins and characteristics. Cracks due to bending of concrete members tend to run perpendicular to the bars and to be tapered, closing near the depth of the steel bars. Cracks due to plastic shrinkage or restrained shrinkage tend to run through large parts of the cross section and could be parallel to the steel bars.

Indeed, a vast number of experiments show that there is no precise correlation between the crack width (as long as it remains below 0.5 mm) and the risk of corrosion. This risk will depend on factors such as environmental conditions (in particular the humidity) and the properties of the concrete (permeability and thickness of the concrete cover) that will determine the corrosion behavior of the reinforcement even in the absence of cracks. For example, in dry concrete (i.e., with high resistivity) and, conversely, in water-saturated concrete (i.e., with low oxygen availability), even when there are cracks of considerable width, the corrosion rate remains negligible. This does not happen, on the other hand, in the presence of wetting–drying cycles or in conditions of moisture content just below saturation. In any case, cracks may reduce the corrosion initiation time in that they provide a preferential path for the penetration of carbonation or chlorides (Figure 11.1). Experiments with sectioned steel bars in intentionally cracked concrete beams have shown that the depassivation time decreases as the crack width increases; however, there is no relationship between crack width and corrosion rate; actually the corrosion rate decreases with increasing cover of the uncracked concrete (between cracks) due to the influence of the cathodic process [11].

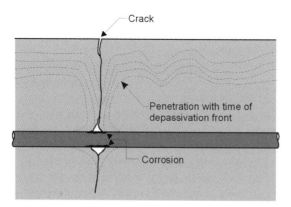

Figure 11.1 Illustration of the penetration of the depassivation front (as a consequence of carbonation or chloride ingress) in time in cracked concrete [1].

Generally, if the crack width is modest (e.g., it is below 0.3–0.5 mm), after the initiation of corrosion on the steel surface, the corrosion rate is low. Chemical processes in the cement paste and formation of corrosion products may seal the crack near the reinforcement and allow the protective oxide film to form again. For carbonation-induced corrosion, repassivation can take place when the migration of alkalinity from the surrounding concrete brings the pH of the pore solution in contact with the corrosion products to values above 11.5. Repassivation may have trouble taking place or may not take place at all in the following situations:

- for concrete cover of very low thickness (below 20 mm) or high porosity;
- when variations in load make the width of the cracks vary cyclically;
- when water flows through the crack (e.g., because it is exposed to water only on one side) and tends to remove the corrosion products and reduce the local alkalinity;
- in the combined presence of both carbonation and chlorides, even of low levels.

Parallel cracks pose a more serious threat to reinforcement than transverse cracks, as they sustain higher corrosion rates [12]. In any case, understanding of the corrosion mechanisms in relation to cracks is poor. In general, crack widths should be limited in particular in aggressive environments.

Apart from structural design, other factors related to concrete technology and execution influence the occurrence of cracks. Cracking due to restrained shrinkage as a result of temperature gradients can be restricted by limiting the dosage of cement, using low-heat cement, delaying the onset of cooling of concrete after placing and extending the curing period [13].

11.1.4
Thickness of the Concrete Cover

Besides concrete quality, a minimum value of the concrete cover also has to be specified. An increase in the thickness of the concrete cover brings about different beneficial effects and only in extreme cases some adverse effects. First, increasing the cover increases the barrier to the various aggressive species moving towards the reinforcement and increases the time for corrosion initiation, even though different transport laws apply depending on the characteristics of the concrete and the cause of corrosion (carbonation or chlorides).

It may be remembered that the carbonation depth in time assumes values equal to or (for long periods) below those expressed by the square-root law: $s = k \cdot t^{1/2}$. Therefore, if the thickness of the concrete cover in some areas of the structure is halved with respect to its nominal value, in these areas the initiation time is reduced to less than one fourth of that predicted (Figure 11.2). Analogous considerations are valid if concrete is exposed to chlorides. Therefore, to prevent steel corrosion adequate measures shall be taken to guarantee the fulfillment of the design thickness of the concrete cover.

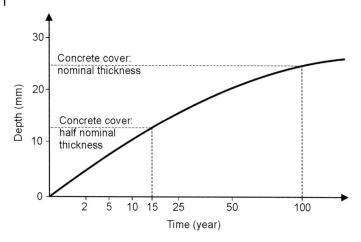

Figure 11.2 Reduction of the initiation time of corrosion due to local reductions in the thickness of the concrete cover [1].

As the environmental aggressiveness increases, it is theoretically possible to maintain a constant level of durability by increasing the thickness of the concrete cover. In reality, however, the cover thickness cannot exceed certain limits, for mechanical and practical reasons. In particular, a very high cover may have less-favorable barrier properties than expected. In extreme cases or in the case of inadequate curing, a thick unreinforced layer of concrete cover may form (micro) cracks due to tensile forces exerted by drying shrinkage of the outer layer, while the wetter core does not shrink. In practice, having cover depths above 70 to 90 mm is not considered realistic.

11.1.5
Inspection and Maintenance

Regular inspection of reinforced concrete structures allows detection of signs of deterioration and, if necessary, application of remedial measures in an early stage of damage. As regular inspections usually are visual only, deterioration can be detected only when it manifests at the surface (e.g., by cracking or spalling). Chloride-induced localized corrosion might be detected in a very late state only. Planning of inspection activities should be considered in the design stage, on the basis of assumptions on the expected behavior of the structure and the aggressiveness of the environment. This approach is referred to as *proactive* conservation since it enables a structure to maintain its condition and to meet its design service life by avoiding or minimizing signs of deterioration [14, 15]. This is also an example of the application of the previously mentioned De Sitter's rule, since the extra costs connected with planned inspection allow reduction of future (much higher) costs connected with repair interventions. According to environmental conditions and to the reliability required to the structure, different levels of inspection measures can be considered, ranging from visual inspection to more sophisticated measurements, as described in Chapter 16. Permanent monitoring systems, such as for instance internal probes that allow monitoring of the initiation or propagation of corrosion of steel reinforcement or other durability-related parameters, can also be used (Chapter 17).

Regular maintenance of concrete structures, based for instance on the replacement of temporary components (e.g., drainpipes) or on the periodic reapplication of protective measures (e.g., a surface coating), is a further tool that helps achieve durability requirements.

Inspection and maintenance activities allow a record to be kept of the actual conditions of a structure, from its construction on. In this regard, it was proposed to introduce a "birth certificate", that is, a document that records the initial "as built" conditions of a structure (in terms, for instance, of concrete quality and thickness of concrete cover) for future reference [4, 16]. This document should contain all data relevant to durability from the structural design and the construction phase, and it is expected to be updated in the course of time with data from inspection and (if applicable) information of interventions, in order to help the future management of the structure.

11.2
Approaches to Service-Life Modeling

The appearance of premature and unexpected corrosion damage in reinforced concrete structures, which at the time of construction were considered of almost unlimited duration, led to the introduction of the concepts of *durability* and *service life* in the 1970s. The service life of a structure can be defined as the period of time in which it is able to comply with the given requirements of safety, stability and serviceability, without requiring extraordinary costs of maintenance and repair [1, 3, 4]. Any approach to the durability design of a reinforced concrete structure should start from the definition of the target service life, which should be clearly specified. The target service life should be concerted between the owner and the designer; its definition stems from considerations such as, among others, the type and relevance of the structure and its function. Qualitative and fuzzy definitions of service life can only lead to a vague consideration of durability issues.

Together with the definition of the target service life goes the definition of the limit state, that is, an event in the life of the structure that separates the desired from the undesired behavior and marks the end of the service life. Depassivation of steel reinforcement and cracking of the concrete cover are examples of limit states.

Approaches to the design for durability of reinforced concrete structures can be divided into two categories, *prescriptive* approaches and *performance-based* approaches. The former consist of simple rules usually presented in the form of threshold values of different parameters, often derived from experience, that should be fulfilled in order to guarantee an expected service life of the order of 50 years, using ordinary materials and technologies. The latter consist in design procedures aimed at determining (or verifying) the service life as a function of the desired performance of the structure, allowing the quantification of the effect of all the parameters involved, including the use of innovative materials and technologies or additional protection measures.

11.2.1
Prescriptive Approaches

Prescriptive approaches are also referred to as *standard* approaches since they are proposed by normative organizations as standardized methods to deal with durability. These methods start from the identification of the environmental aggressiveness through an exposure class, according to which limiting values of durability-related parameters are specified. For instance, in the case of reinforcement corrosion, requirements are usually given in terms of a maximum w/c ratio and a minimum thickness of concrete cover, together with execution-related indications, such as curing times. The fulfillment of requirements specified in prescriptive approaches is expected to guarantee a certain target service life, usually of the order of 50 years. Prescriptive methods are usually easy to apply even by those who may not be experts in design procedures or deterioration phenomena of reinforced concrete.

Even though the introduction of the standard approach has been a step forward towards the consideration and improvement of the durability of reinforced concrete structures, it should be recognized that this approach is not exhaustive of all the aspects related to durability. Exposure classes simply refer to average conditions, and not to actual microclimatic conditions that, as already mentioned, may differ significantly from the average. Experience showed that the recommendations are adequate for structures exposed to "ordinary" aggressive environments: for instance, it is generally accepted that in the case of carbonation-induced corrosion the target service life (usually 50 years) can be guaranteed. Conversely, in exposure conditions that are harsher compared to carbonation, as in the case of chloride-bearing environments (both marine and from deicing salts), the standard recommendations are not enough to avoid corrosion initiation on steel reinforcement during the target service life. In this regard, as will be seen later with reference to the European standard (Section 11.3), it should be recalled that standard approaches often do not consider (and, so, do not allow evaluation of the impact of) the use of innovative technologies, such as for instance the use of blended cements, whose effects in chloride environments are well known, or corrosion-resistant reinforcement.

Requirements given by standard approaches only provide *deemed-to-satisfy* rules. Such rules cannot be used to quantify the performance of the structure in general, the effects of additional measures (for instance increasing the cover to the steel), or the consequences of substandard practice (for example, using a higher w/c) on the resulting service life or its reliability.

11.2.2
Performance-Based Approaches

Performance-based approaches are procedures that allow design of each structural element so that it will withstand the actual local conditions of exposure during the design service life. Such approaches are based on the modeling of deterioration mechanisms due to a certain type of attack as a function of time, in order to estimate the evolution of damage as a function of the different factors affecting it. Such methods should draw a picture of the characteristics that the concrete and the structure should possess to protect the reinforcement for the requested service life from a predictive model of the corrosion attack. These "refined" methods (as opposed to "standard" methods) may be based on long-term experience with local practices in local environments, on data from an established performance test method for the relevant mechanism, and on the use of proven predictive models.

Calculating the service life is based on the modeling of degradation mechanisms due to attack by a particular aggressive agent and/or on empirical formulas that estimate the evolution of deterioration depending on the environmental conditions and the properties of concrete. Presently, this mainly concerns modeling initiation or propagation of reinforcement corrosion because for this type of deterioration there are reliable kinetic models, while for other forms of attack these models are lacking or only rudimentary. As a function of the environment and the required service life, these methods allow an evaluation of the thickness of the concrete cover, the type of cement and the w/c ratio. Maintaining the same service life, it is possible, for example, to reduce the concrete cover by using a higher quality of concrete or a different type of cement.

In the early years, service-life models for reinforcement corrosion basically followed the square-root-of-time approach or slight modifications, such as the example illustrated in Figure 11.2. Based on empirical data on the rate of carbonation or chloride ingress, a minimum cover depth was determined that was expected to delay the onset of corrosion for a required period. Following this approach, more data were collected from existing structures and exposure sites, and the influence of various factors such as cement type and local environment became clearer [1, 17].

More recently, several proposals have been presented for verifying the durability of concrete with respect to various types of deterioration. One of the first internationally recognized methods was proposed by *DuraCrete* ("Probabilistic Performance based Durability Design of Concrete Structures") that was finalized in the late 1990s and available to the public at large in the next decade [18]. Today, one of the most authoritative approaches is the *fib* Model Code for Service-Life Design [16], which was issued in 2006 by the International Federation for Structural Concrete and is expected to be included in the framework of standardization of performance-based design approaches.

A common feature of performance-based methods is the need for reliable input parameters, particularly with regard to the correlation between short-term performance (typically obtained from laboratory tests on "young" concrete specimens) and long-term performance in real exposure conditions. In the rest of this section, after reporting basic equations for the service life in the case of carbonation- and chloride-induced corrosion, the general procedure of design will be outlined.

For carbonation-induced corrosion, the service life (t_{SL}) is expressed as the sum of the initiation (t_i) and propagation (t_p) periods up to the limit state at which deterioration becomes unacceptable: $t_{SL} = t_i + t_p$ (Figure 4.1). The initiation time (t_i) may be calculated as a function of the properties of concrete, the environment and the thickness of the concrete cover (x) (Chapter 5). The propagation time (t_p) can be estimated if the corrosion rate is known, once the maximum acceptable penetration of corrosion has been fixed. Often, a maximum penetration for corrosion attack (e.g., 100 µm) is considered. In more refined approaches modeling of expansive stresses induced by corrosion products is considered [19].

When corrosion is caused by chloride ingress, the service life is usually assumed to be equal to the initiation time: $t_{SL} = t_i$. The period of propagation, which may be of short duration, is traditionally not taken into account because of the uncertainty with regard to the consequences of localized corrosion.

The initiation time (t_i) may be calculated as a function of the chloride transport properties of concrete (usually the apparent diffusion coefficient), the surface chloride content dictated by the environment, the thickness of the concrete cover and the critical chloride content determining the onset of corrosion (a parameter that might be difficult to establish, see Section 6.2). The arrival of the critical chloride content at the steel depth x at time t is normally calculated using equations formally derived from Fick's second law of diffusion (Chapter 6). Using this type of calculation, it is possible to find values for D_{app} (assumed constant, but see warnings in later sections) that can be used to obtain a particular service life as a function of the thickness of the concrete cover and the critical chloride content C_{th}, assuming a fixed chloride surface content C_s, as seen in Table 11.1 [20].

Table 11.1 Maximum acceptable values of D_{app} (10^{-12} m²/s, assumed constant) as a function of the concrete cover thickness, the service life and the chloride threshold (C_{th}) for a constant surface content of 4% chloride by mass of cement [20].

Thickness of concrete cover (mm)	Maximum acceptable D_{app}			
	C_{th} = 0.4%	C_{th} = 0.4%	C_{th} = 1%	C_{th} = 1%
	t = 50 years	t = 120 years	t = 50 years	t = 120 years
30	0.106	0.0442	0.217	0.0906
40	0.189	0.0786	0.387	0.161
50	0.295	0.123	0.604	0.252
65	0.486	0.207	1.02	0.425
75	0.663	0.276	1.36	0.566
90	0.954	0.398	1.96	0.816
100	1.18	0.491	2.42	1.01

Limit States and Design Equation   Performance-based approaches are based on the definition of the limit state, that is, an event in the life of the structure that marks the boundary between the desired and the adverse behavior, or, in other words, a state "beyond which the structure no longer fulfils the relevant design criteria", as defined by EN 1990 [21]. *Serviceability* limit states (*SLS*) correspond to conditions beyond which specified requirements are no longer met; the need for repair due to cracking or spalling of concrete cover is an example of *SLS*. *Ultimate* limit states (*ULS*) are associated with collapse or with other forms of structural failure, usually involving possible loss of human lives or high economic losses.

The condition of fulfillment of the limit state is expressed in the *design equation* g (or limit state equation), which, in its simplest form, is presented as:

$$g = R(t) - S(t) > 0 \tag{11.1}$$

where: $R(t)$ is a "resistance" variable and $S(t)$ is a "load" variable, both taken as time dependent. A limit state equation is positive only if the structure considered is fully capable to show the desired behavior, that is, to deliver the required performance. Similarly to design equations for structural design, loads and resistances are considered; however, due to the nature of the involved phenomena, the time variable is also considered, in order to take proper account of the evolution of deterioration.

In structural design, loads and resistances are generally thought to be independent of time and reversible. The *loads* are actions such as traffic, snow and wind. The *resistances* are material parameters such as the yield strength of steel and the compressive strength of concrete. Service-life design requires time-dependent formulation of the resistances and the loads also taking into account irreversible accumulation of aggressive substances. Material variables in general will be resistance variables and variables describing the environment will be load variables, for instance the presence of chloride from deicing salts. The environmental load variables may increase in time (e.g., due to chloride accumulation). The time dependency of the resistance considers the degradation of the materials properties, for instance the loss of cross section due to corrosion of a reinforcing bar.

Variability   Variables and parameters considered in the design Eq. (11.1) are dependent on a number of factors that are difficult to predict and, in addition, are often characterized by an intrinsic variability. For instance, in the case of the simple square-root law for carbonation in concrete (Section 5.2.1), local variations in the properties of concrete or in the conditions of exposure may result in variability of the carbonation depth; a similar variability may characterize the concrete cover. Often, average values of the variables (or nominal values) are considered to account for their variability. This approach is referred to as *deterministic*, since it does not consider explicitly the variability, but only a determined value (usually the average) of the variable or parameter. As a consequence, in a deterministic approach the resulting service life will also be an average value, that is, it will have a 50% probability to be achieved, which may be unacceptable.

The consideration of the variability of parameters and variables in Eq. (11.1) and, as a consequence, of the uncertainty in the resulting service life, led to *probabilistic* approaches and to the introduction of the concept of *reliability*. In probabilistic approaches the design equation is formulated as:

$$P_f = P\{g < 0\} < P_0 \tag{11.2}$$

where P_f is the failure probability, that is, the probability that the limit state occurs, and P_0 is the maximum acceptable failure probability. The reliability is expressed through a *reliability index* β, which is defined as:

$$\beta = \Phi^{-1}(P_0) \tag{11.3}$$

where Φ^{-1} is the inverse of the normal distribution. The maximum failure probability is selected taking into account, among others, the possible consequences of failure in terms of risk to life, injury and potential economical losses [21].

In contrast to the deterministic approach and depending on the severity of the limit state, the failure probability can be set to be (much) lower than 50%. Serviceability limit states should have a low failure probability of the order of 1:10 to 1:100. Ultimate limit states involve safety (human lives) or loss of the structure (high economic damage) and should have a very low failure probability of the order of 1:10 000. Failure probabilities like this are defined by for example, EN 1990, Annex B and C [21]. As initiation of corrosion does not immediately have extreme consequences, a probability of failure has been proposed for this event of 1:10 [16, 22].

The probabilistic approach to durability design can be depicted in terms of evolution of service life as a function of time [23, 24], as is shown in Figure 11.3. The

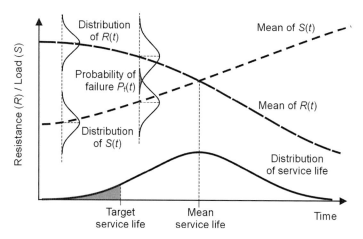

Figure 11.3 Graph showing load and resistance as a function of time, their mean and distribution and their overlap demonstrating the failure probability.

load on a structure ($S(t)$) will remain the same over time or may increase, for instance due to increasing traffic. On the other hand, the resistance of the structure ($R(t)$) will decrease due to degradation processes. Both functions $S(t)$ and $R(t)$ are stochastic and should be described by distributions around a mean value at every moment in time. The service-life distribution can be found through the convolution of distributions of $S(t)$ and $R(t)$, that is, the total probability of the situation that the load will be higher than the resistance of the structure (which is the failure probability P_f).

Due to the complexity of full probabilistic approaches (i.e., approaches based on Eq. (11.2) with all variables and parameters expressed in stochastic terms), "simplified" performance-based approaches have been proposed. Semiprobabilistic methods are examples of simplified approaches, since the procedure for the evaluation of service life is similar to that of deterministic approaches, in which variability is considered through the introduction of corrective coefficients (partial factors) that are applied to each variable. Other examples are some guidelines that prescribe concrete properties in terms of durability-related parameters (e.g., maximum allowed values of the chloride diffusion coefficient measured according to a standardized procedure) given as a function of the design service life, concrete cover, composition and environmental conditions. Examples of these simplified approaches will be briefly discussed in Section 11.5.

11.3
The Approach of the European Standards

Since the 1990s, the European standards EN 206-1 [2] and Eurocode 2 [3] have dealt with the problem of durability of concrete structures in a much wider sense than previous regulations. For instance, Eurocode 2 makes reference to durability in Section 4, stating that: "A durable structure shall meet the requirements of serviceability, strength and stability throughout its intended working life, without significant loss of utility or excessive unforeseen maintenance". European standards face durability of reinforced concrete through a standardized method based on the definition of an exposure class and the subsequent requirements for the w/c ratio, the cement content and the thickness of the concrete cover.

Table 11.2 shows the exposure classes defined by European standard EN 206-1 on the basis of environmental conditions. The environment is considered as the sum of chemical and physical actions to which the concrete is exposed and that result in effects on the concrete or the reinforcement or embedded metal. Only corrosion of reinforcement will be considered here (degradation of concrete was treated in Chapter 3). The following classes are available: (i) no risk of corrosion or other forms of attack, (ii) carbonation-induced corrosion, (iii) corrosion induced by chlorides from seawater, and (iv) corrosion induced by chlorides other than from seawater. Each class is divided into subgroups to distinguish different exposure conditions for each type of attack. So, in the first step, the designer is required

Table 11.2 Exposure classes according to EN 206-1 [2].

Class designation	Description of the environment	Informative examples where exposure classes may occur
1 – No risk of corrosion or attack		
X0	For concrete without reinforcement or embedded metal: all exposures except where there is freeze–thaw, abrasion or chemical attack. For concrete with reinforcement or embedded metal: very dry	Concrete inside buildings with very low air humidity.
2 – Corrosion induced by carbonation[a] Where concrete containing reinforcement or other embedded metal is exposed to air and moisture, the exposure shall be classified as follows:		
XC1	Dry or permanently wet	Concrete inside buildings with low air humidity. Concrete permanently submerged in water.
XC2	Wet, rarely dry	Concrete surfaces subject to long-term water contact. Many foundations.
XC3	Moderate humidity	Concrete inside buildings with moderate or high air humidity. External concrete sheltered from rain.
XC4	Cyclic wet and dry	Concrete surfaces subject to water contact, not within exposure class XC2.
3 – Corrosion induced by chlorides other than from seawater[a] Where concrete containing reinforcement or other embedded metal is subject to contact with water containing chlorides, including deicing salts, from sources other than from seawater, the exposure class shall be classified as follows:		
XD1	Moderate humidity	Concrete surfaces exposed to airborne chlorides.
XD2	Wet, rarely dry	Swimming pools. Concrete exposed to industrial waters containing chlorides.
XD3	Cyclic wet and dry	Parts of bridges exposed to spray containing chlorides. Pavements. Car park slabs.
4 – Corrosion induced by chlorides from seawater Where concrete containing reinforcement or other embedded metal is subject to contact with chlorides from seawater or air carrying salt originating from seawater, the exposure shall be classified as follows:		
XS1	Exposed to airborne salt but not in direct contact with seawater	Structures near to or on the coast.
XS2	Permanently submerged	Parts of marine structures.
XS3	Tidal, splash, and spray zones.	Parts of marine structures.
5 – Freeze–thaw attack with or without deicing agents see Chapter 3		
6 – Chemical attack see Chapter 3		

a) The moisture condition relates to that in concrete cover to reinforcement or other embedded metal but, in many cases, conditions in the concrete cover can be taken as reflecting that in the surrounding environment. In these cases classification of the surrounding environment may be adequate. This may not be the case if there is a barrier between the concrete and its environment.

Table 11.3 Recommendations (informative) for the choice of the limiting values of concrete composition and properties in relation to exposure classes according to EN 206-1 for the exposure classes shown in Table 11.2 [2].

Exposure class		Maximum w/c	Minimum strength class	Minimum cement content (kg/m³)
No risk of corrosion or attack	X0	–	C12/15	–
Carbonation-induced corrosion	XC1	0.65	C20/25	260
	XC2	0.60	C25/30	280
	XC3	0.55	C30/37	280
	XC4	0.50	C30/37	300
Chloride-induced corrosion – seawater	XS1	0.50	C30/37	300
	XS2	0.45	C35/45	320
	XS3	0.45	C35/45	340
Chloride-induced corrosion – Cl⁻ other than from seawater	XD1	0.55	C30/37	300
	XD2	0.55	C30/37	300
	XD3	0.45	C35/45	320

The values in this table refer to the use of cement type CEM I conforming to EN 197-1 and aggregates with nominal maximum size in the range of 20 to 32 mm. The minimum strength classes were determined from the relationship between water/cement ratio and the strength class of concrete made with cement of strength class 32.5. The limiting values for the maximum w/c ratio and the minimum cement content apply in all cases, while the requirements for concrete strength class may be additionally specified.

to identify the conditions of environmental exposure and the expected cause of deterioration.

Depending on the aggressiveness of the environment, that is, for each exposure class, recommended (informative) values are given in terms of maximum w/c ratio and minimum cement content, as indicated in Table 11.3. Informative values are only given with regard to the use of portland cement (CEM I, Table 1.3). Also, a minimum concrete compressive strength should be required, as is also shown in Table 11.3 (the minimum compressive strengths are obtained considering a cement of strength class 32.5). In addition, EN 206-1 specifies the maximum allowed content of mixed-in chlorides (due to contamination of raw materials), as indicated in Table 11.4. Other special requirements are given for cases where, in addition to reinforcement corrosion, degradation of concrete is also expected, as, for example, a minimum content of entrained air in concrete exposed to freeze–thaw attack (as indicated in Table 3.2) or specific types of cement for concrete exposed to sulfate attack.

The requirements for concrete composition provided by EN 206-1 are recommended (informative, that is non-normative) values and they were evaluated assuming an intended service life of 50 years, the use of portland cement (CEM I) and maximum aggregate size between 20 and 32 mm. National documents further specify these values (this gives rise to variations of national provisions, as highlighted by a recent survey [25]) as well as additional requirements as national

Table 11.4 Maximum chloride content of concrete according to EN 206-1 [2].

Concrete use	Chloride content class[a]	Maximum Cl⁻ by mass of cement[b]
Not containing steel reinforcement or other embedded metal with the exception of corrosion-resisting lifting devices	Cl 1.0	1.0%
Containing steel reinforcement or other embedded metal	Cl 0.2	0.20%
	Cl 0.4	0.40%
Containing prestressing steel reinforcement[b]	Cl 0.10	0.10%
	Cl 0.20	0.20%

a) For a specific concrete use, the class to be applied depends upon the provisions valid in the place of use of the concrete.
b) Where type II additions (e.g., fly ash) are used and are taken into account for the cement content, the chloride content is expressed as the percentage chloride ion by mass of cement plus total mass of additions that are taken into account.

normative values.[1] Execution aspects that should be followed in addition to previous requirements are reported in EN 13670 [26] that gives indications on mixing, placing and curing of concrete.

Besides concrete quality, a minimum value of the concrete cover also has to be specified. Eurocode 2 [3] fixes minimum values ranging from 10 mm for a dry environment up to 55 mm for prestressing steel in chloride-bearing environments, as shown in Table 11.5. It should be kept in mind that these values are minimum values that should be increased to obtain nominal values by 10 mm, to also take into consideration construction variability. Besides the protection of steel to corrosion, further requirements of minimum cover depth are fixed to ensure adequate transmission of mechanical forces and fire resistance.

Following the limiting values indicated by Table 11.3, the concrete is deemed to satisfy the durability requirements for the intended use, provided that:

- it is placed, compacted and cured properly according to EN 13670 [26], which deals with execution-related factors;
- the minimum cover to the reinforcement is respected according to EN 1992-1-1 [3];

1) It should be noted that National Annexes to EN 206-1 show a strong variation with regard to lower and upper limits for w/c and cement content for various exposure classes [25]. For example, across CEN countries the maximum w/c for XS3 may vary from 0.35 to 0.50 and the minimum cement content from 150 to 400 kg/m³. Such extreme differences can be explained, at least partially, by local experience, both with regard to cement types used in particular climates and the severity in different countries.

Table 11.5 Minimum values for concrete cover depth, simplified from Eurocode 2 with regard to corrosion protection of steel for structure class 4, service life of 50 years, and exposure classes defined by EN 206-1 (Table 11.2). These values should be increased by 10 mm to obtain the nominal cover depth.

Action	Exposure class	Minimum cover thickness (mm) to	
		reinforcing steel	prestressing steel
No risk	X0	10	10
Carbonation-induced corrosion	XC1	15	25
	XC2, XC3	25	35
	XC4	30	40
Chloride-induced corrosion	XS1, XD1	35	45
	XS2, XD2	40	50
	XS3, XD3	45	55

- the appropriate exposure class was selected; and
- the anticipated maintenance is applied (although no detailed indication on this point is given in EN 206-1).

In most structures exposed to the atmosphere, the compliance with normative recommendations will provide a service life of about 50 years. Therefore, by simply following these standards it would be possible to avoid the vast majority of forms of deterioration, including corrosion, which are found today and that are connected to incorrect design, material composition or construction practice.

Nevertheless, in relatively few but very important conditions of environmental exposure, associated above all with the presence of chlorides, the requirements are not adequate. For example, Table 11.6 shows the initiation time for corrosion in the most critical parts of a structure operating in a marine environment (i.e., tidal/splash zone), assuming a critical chloride content that initiates corrosion of 1% by mass of cement and a surface content of 4%. Times-to-initiation were evaluated for different concrete cover depths on the basis of apparent diffusion coefficients for chlorides (D_{app}) determined from specimens made with two qualities of concrete (w/c 0.40 and 0.54) using portland cement, and submerged for 16 years in the North Sea [27]. It can be seen that an initiation period of 30 years is predicted with a w/c ratio of 0.40 and a cover depth of 70 mm. The limits recommended by EN 206-1 and Eurocode 2, which for this environment require a maximum w/c of 0.45, and a minimum concrete cover of 45 mm, are insufficient for this case. Table 11.6 also shows that the use of a slag cement would considerably increase the time to depassivation, due to the higher imperviousness to chlorides compared to ordinary portland cement. In fact, the addition of pozzolanic materials or ground granulated blast furnace slag to portland cement (see Section 12.6.1) as well as other additional preventive measures (Section 15.2) can bring about notable improvements.

Table 11.6 Initiation time of corrosion as a function of thickness of concrete cover and concrete composition, calculated on the basis of a critical chloride content of 1% by mass of cement, a surface content of 4%, and constant diffusion coefficients for chlorides (D_{app}) determined on two OPC concretes (with 420 kg/m³ of portland cement and w/c of 0.4 or 300 kg/m³ and w/c 0.54) and a GGBS concrete (with 420 kg/m³ of slag cement and w/b of 0.4) after submersion in the North Sea for 16 years. OPC is comparable to present day CEM I 32.5, GGBS cement (with 70% slag) is comparable to modern CEM III/B 42.5 [27].

Type of cement/w/b	OPC/0.40	OPC/0.54	GGBS/0.4
D_{app} (10^{-12} m²/s)	2	3	0.3
Cover thickness (mm)	Time to depassivation (y)		
30	5	4	36
50	15	10	100
70	30	20	200

Another possible limitation is the classification of the environmental aggressiveness reported in Table 11.2. Exposure classes refer to average conditions and not to local microclimatic conditions, including those created by the structure itself, where aggressiveness may strongly differ from the average. For example, on the beams of a viaduct where deicing salts are used, the situation is more aggressive where water tends to stagnate (and thus chlorides accumulate) or is diverted, thus in correspondence to joints, in zones near the intrados of a curve, in places where drainages function poorly, etc. Inside a building, where the carbonation front may reach the reinforcement in a relatively short time, the corrosion rate is usually negligible because the relative humidity is low; but wherever there is water leakage or frequent and abundant condensation, the corrosive attack may occur at a rate that is certainly not negligible. Yet another example is given by the external parts of a building, where the aggressiveness will change in passing from areas shielded from rain to those exposed to it. Furthermore, in the course of time, accumulation of aggressive loads may occur, especially when the use of the structure is changed or when its maintenance is neglected. Finally, it should be noted that situations of great aggressiveness can be caused by the simultaneous presence of environmental factors that, taken individually, would not lead to corrosion.

Even the maximum allowed chloride contents indicated by EN 206-1 may not be adequate. From the point of view of availability of raw materials (sea-dredged aggregate, brackish mixing water), allowing a maximum of 0.4% of chloride for reinforced concrete seems understandable. However, in view of corrosion initiation by chloride contents from 0.4% to 0.5% in many cases, this also consumes a large part of the concrete's ability to delay corrosion in chloride-contaminated environments. In particular, in marine or deicing salt environment, it seems more appropriate to fix much lower chloride contents in fresh concrete, like class Cl 0.10

in Table 11.4. As a matter of facts, this is a typical example of the rule of fives: short-term benefits (chloride contaminated sea-dredged aggregate is cheaper) may cause higher costs in the future (repair). Here, a preventative strategy by specifying a lower chloride content (to be obtained by washing the aggregate) could save large amounts of money on the time scale of the service life.

Finally, it should be reminded that the standard approach only gives deemed-to-satisfy rules and does not allow quantification of the possible benefits that could be achieved, for instance, using a blended cement, increasing the concrete cover with respect to prescribed values, or resorting to additional protection measures.

11.4
The *fib* Model Code for Service-Life Design for Chloride-Induced Corrosion

For important concrete structures it has become increasingly required that a service life of 80, 100 years or more is demonstrated by the designer or the contractor, including its reliability, as illustrated by at least some of the examples given in Table 11.7. Using the standard prescriptive method, this is not possible. The standard method gives deemed-to-satisfy rules, without specifying the resulting service life or its reliability. Moreover, innovative designs, materials or preventative measures cannot be judged properly.

The issue of the design for durability in the case of long required service lives (in particular in aggressive environments) can be faced through performance-based methods, that allow evaluation of the performance of a structure in terms of probability of failure of a given limit state. Based on previous work in *DuraCrete* [17], the International Federation for Structural Concrete (*fib*) has issued a Model Code for Service-Life Design [16] which is expected to prepare the framework for standardization of performance-based design approaches. The Model Code includes several approaches to the design; in this section, as an example, the full

Table 11.7 Service life (year) required for some reinforced concrete structures.

Offshore platforms	40
King Fahd Causeway (Saudi Arabia–Bahrain)	75
Tejo River Bridge (Portugal)	99
Great Belt Link (Denmark)	100
Sidney Harbor Tunnel (Australia)	100
Øresund link (Denmark–Sweden)	100
Western Scheldt Tunnel (The Netherlands)	100
Green Heart Tunnel (The Netherlands)	100
Channel Tunnel (France–England)	120
Alexandria Library (Egypt)	200
Eastern Scheldt Storm Barrier (The Netherlands)	200
National Library (London)	250

probabilistic approach with regard to corrosion initiation due to chlorides will be considered.

The design equation is expressed as the difference between the critical chloride concentration and the chloride concentration at the depth of the reinforcing steel at a time t, and is expressed in probabilistic term as:

$$P\{\ \} = P\{C_{crit} - C(a, t_{SL}) < 0\} < P_0 \tag{11.4}$$

where:

$P\{\}$	=	probability that depassivation occurs,
C_{crit}	=	critical chloride content (% by mass of cement),
$C(a,t_{SL})$	=	chloride content at depth a and time t (% by mass of cement),
a	=	concrete cover (mm),
t_{SL}	=	design service life (year),
P_0	=	target failure probability.

The variables a, C_{crit} and $C(a,t_{SL})$ need to be quantified in a full probabilistic approach.

The limit-state condition will be reached when $g=0$, that is, when $C_{crit} = C(x=a,t_{SL})$. The evolution of chloride concentration is modeled through the equation:

$$C(x = a, t) = C_0 + (C_{S,\Delta x} - C_0) \cdot \left[1 - \mathrm{erf}\frac{a - \Delta x}{2 \cdot \sqrt{D_{app,C} \cdot t}}\right] \tag{11.5}$$

where:

$C(x,t)$	=	content of chlorides in the concrete at a depth x (structure surface: $x = 0$ m) and at time t (% by mass of cement),
C_0	=	initial chloride content of the concrete (% by mass of cement),
$C_{S,\Delta x}$	=	chloride content at a depth Δx and at a certain time t (% by mass of cement),
Δx	=	depth of the convection zone (concrete layer, up to which the process of chloride penetration differs from Fick's 2nd law of diffusion) (mm),
$D_{app,C}$	=	apparent coefficient of chloride diffusion through concrete (mm²/year),
t	=	time (year),
erf	=	error function (see Section 2.2.2).

The apparent diffusion coefficient of chloride diffusion can be determined by:

$$D_{app,C} = k_e \cdot D_{RCM,0} \cdot k_t \cdot A(t) \tag{11.6}$$

where:

k_e	=	environmental transfer variable (–) defined as:

$$k_e = \exp\left(b_e\left(\frac{1}{T_{ref}} - \frac{1}{T_{real}}\right)\right) \tag{11.7}$$

11.4 The fib Model Code for Service-Life Design for Chloride-Induced Corrosion

where:

b_e	=	regression variable (K),
T_{ref}	=	standard test temperature (K),
T_{real}	=	temperature of the structural element or the ambient air (K),
$D_{RCM,0}$	=	chloride migration coefficient of water-saturated concrete, measured at the reference period t_0 on specimens that have been cast and stored under defined conditions (mm²/year),
k_t	=	transfer parameter, which converts the chloride migration coefficient measured under accelerated conditions into a chloride diffusion coefficient, for specimens that have been exposed to natural conditions (–),

$A(t)$ = subfunction considering the "aging", which can be evaluated as:

$$A(t) = \left(\frac{t_0}{t}\right)^a \tag{11.8}$$

a	=	aging exponent representing the time-dependent behavior of the apparent diffusion coefficient (–),
t_0	=	reference point of time, usually 28 days (0.0767 year),
t	=	time of exposure (year).

Table 11.8 summarizes the variables involved and reports the types of distribution indicated by the Model Code, their mean value and standard deviation. Some variables should be obtained from direct measurements, such as for instance the environmental temperature T_{real}, which can be obtained from data from the nearest weather station, or the initial chloride content in concrete C_0, which can be obtained from laboratory analyses. For other variables, values are provided by the Model Code, such as for instance in the case of the critical chloride content that was determined (for ordinary steel) through measurements on specimens exposed to different environmental conditions and different concrete qualities, or in the case of environmental and age coefficients. The chloride migration coefficient $D_{RCM,0}$ should be obtained from laboratory data or, less favorable, the literature [28–30]; values of $D_{RCM,0}$ for concrete mixes made with different binders and w/c ratios are also given for orientation purposes, as reported in Table 11.9. Other variables should be determined from less-immediate procedures and considerations, as for $C_{S,\Delta x}$, which will not be reported here.

The full probabilistic approach provided by the fib Model Code can be used to evaluate different design solutions, for instance evaluating the concrete cover that allows guarantee, with a certain P_f or reliability level (i.e., β index), of a given target service life. It also allows evaluation of design solutions involving different materials, for instance concrete mixes made with blended cements. Figures 11.4 and 11.5 show, as an example, results obtained applying the *fib* Model Code to the design of a reinforced concrete structural element exposed to the splash zone in a marine environment. Figure 11.4 reports results in terms of the correlation between the

Table 11.8 Type of distribution, mean value (m), and standard deviation (s) for variables in Eqs. (11.5–11.8) [16].

Variable	Unit	Distribution	Mean value	Standard deviation
C_0	% by cement mass	Constant	Laboratory analysis	–
$C_{S,\Delta x}$	% by cement mass	Lognormal	To be determined	To be determined
Δx	mm	Beta, Normal $0 \leq \Delta x \leq 50$	Splash condition: 8.9 Submerged: 0 Tidal condition: to be determined	5.6 – –
T_{real}	K	Normal	Weather station data	Weather station data
T_{ref}	K	Constant	293	–
b_e	K	Normal	4800	700
$D_{RCM,0}$	mm²/year	Normal	Laboratory test	$0.2\,m$
k_t	–	Constant	1	–
a	–	Beta $0 \leq a \leq 1$	OPC: 0.3 FA: 0.6 GGBS: 0.45	0.12 0.15 0.20
t_{SL}	Year	Constant	Concerted owner/designer	–
C_{crit}	% by cement mass	Beta $0.2 \leq C_{crit} \leq 2$	0.6	0.15
a	mm	Lognormal, Beta, Weibull, Neville	Construction documents	8–10[a] 6[b]

a) Without particular execution requirements.
b) With additional execution requirements targeted.

Table 11.9 Values of $D_{RCM,0}$ ($10^{-12}\,m^2/s$) for different concrete mixes [16].

	w/c_{eqv} [a]					
Cement type	0.35	0.40	0.45	0.50	0.55	0.60
CEM I 42.5R	–	8.9	10.0	15.8	19.7	25.0
CEM I 42.5R + FA (k = 0.5)	–	5.6	6.9	9.0	10.9	14.9
CEM I 42.5R + SF (k = 2)	4.4	4.8	–	–	5.3	–
CEM III/B 42.5	–	1.4	1.9	2.8	3.0	3.4

a) equivalent w/c ratio, considering FA (fly ash) or SF (silica fume) with the respective k value (efficiency factor). The considered contents were: FA: 22% by cement mass; SF: 5% by cement mass.

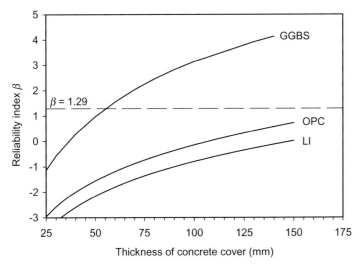

Figure 11.4 Mean thickness of concrete cover of a reinforced concrete element to guarantee a service life of 50 years as a function of the reliability index β (exposure condition: splash zone, mean annual temperature 20°C) for concretes made with ordinary portland cement (OPC), 30% limestone cement (LI) and 70% ground granulated blast furnace slag cement (GGBS), all with w/c ratio of 0.45 [31].

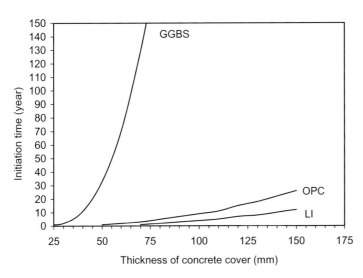

Figure 11.5 Initiation time of reinforcement corrosion as a function of the mean thickness of concrete cover for a given reliability index β of 1.29 ($P_f = 10^{-1}$) for different types of concrete (same as in Figure 11.4) [31].

reliability index β and the concrete cover, considering a target service life of 50 years (defined through a limit state corresponding to steel depassivation), for three different types of concrete: an ordinary portland cement concrete (OPC), a 30% limestone cement concrete (LI), and a 70% ground granulated blast furnace slag cement concrete (GGBS), all with w/c ratio of 0.45 (details on other parameters are reported in [31]). If a probability of failure of 10^{-1} is selected (which corresponds to $\beta = 1.29$), as is usually done for serviceability limit states involving depassivation, with GGBS concrete a minimum thickness of concrete cover of 55 mm should be adopted, while for OPC and LI concretes even a very high concrete cover of 150 mm would not be enough to guarantee the chosen probability of failure in 50 years.

In Figure 11.5 the results for the same case are reported in terms of initiation time of corrosion (i.e., service life) as a function of the concrete cover for a given maximum probability of failure (again 10^{-1}). Here, it can be observed that OPC and LI concretes with w/c of 0.45 would lead to service lives lower than 30 and 10 years, respectively, even with a concrete cover of 150 mm, while much higher service lives would be achieved with a GGBS concrete with the same w/c ratio, which, for instance, would guarantee a service life higher than 150 years with a concrete cover of about 75 mm.

Obviously, the application of any performance-based approach, such as that presented here, requires reliable data for all the input parameters of the design equation, including the parameters selected by the user (for instance, in the previous examples the resistance to chloride penetration of the different concretes as well as the critical chloride threshold, which had been evaluated through experimental tests) and those given by the Code (for instance, the environmental factor). In this regard, it should be reminded that the use of full probabilistic models, including that proposed by fib, is still relatively limited as well as long-term evaluation of their actual reliability. Comparison with the actual performance of existing structures (evaluated by means of a condition assessment) may be useful to evaluate the output of modeling and, possibly, to adjust design parameters [32–34].

11.5
Other Methods

A number of approaches to durability design of reinforced concrete structures exist, which may be prescriptive, performance-based and hybrid, other than those considered in the previous sections. For instance, a prescriptive approach based on the definition of exposure classes is available in North America (which differ from those considered by European standards, reflecting different traditions and experiences), and for each exposure class maximum values of water/binder ratio and minimum values of compressive strength are given [6].

Many performance-based approaches have been proposed either by single researchers or institutions and they are available in the scientific literature.

The *DuraCrete* method [18] deserves a special mention, since it was one of the first to be available to a wide public of researches and users. It was the outcome of a European BRITE EuRam research project carried out in the late 1990s called "Probabilistic Performance based Durability Design of Concrete Structures". The project team, consisting of designers and researchers, both civil engineers and materials scientists, aimed at formulating a service-life design approach that was scientifically correct and that followed the design philosophy for structural performance and safety. The base was laid in the 1980s [35] and several publications give more details [23, 36, 37]. The full-probabilistic framework for service-life design developed in *DuraCrete* is similar to that for structural design [36]. A simplified service-life design method was also derived [18], using characteristic values and safety factors, which is similar to the conventional engineering approach of structural design using the so-called *Load and Resistance Factor Design* (*LRFD*) method. The *DuraCrete* method has been applied to the design of a number of relevant structures (mainly tunnels and bridges) exposed to harsh environments and a relatively large documenting literature exists [37–41].

Simplified methods are often proposed to help the user in the design procedure. Full probabilistic approaches are mainly considered in the case of exceptional structures [16], while for more "ordinary" structures semiprobabilistic approaches may be suitable. Both the *fib* Model Code and the *DuraCrete* propose alternative methods based on partial factors [16, 18]. It should be noted that the introduction of simplified methods often stemmed from the intrinsic complexity of full probabilistic methods. However, it is evident that the same considerations and limitations as for full probabilistic approaches apply, like for instance those regarding the reliability of input parameters and of prediction outcomes.

Finally, a semiprobabilistic simplification of the *DuraCrete* methodology has been recently proposed [42, 43]. Here, indications are given in the form of limiting values of performance-related parameters in design tables, as a function of concrete cover, environmental conditions and binder type. The performance-based nature of the approach is represented by considering the chloride migration coefficient D_{RCM} as the dominant performance parameter directly related to durability, and clearly specifying the target service life and probability of failure. An example of the outcomes of this simplified approach, provided by a Dutch Guideline, is reported in Table 11.10, that refers to a design service life of 100 years and target failure probabilities of 10% and 5%, respectively, for reinforcing steel and prestressing steel. Analysis of more than five hundred D_{RCM} test results for different binder type, w/b and age has provided a range of "obtainable" values, in particular as a function of w/b (Figure 11.6). These values are reflected in the Table 11.10. These data were obtained from various laboratories, on concrete made with various cements (for type, strength class and manufacturer). It should be realized that in particular CEM I produces widely varying D_{RCM} values for nominally identical cements that are produced in different plants.

Table 11.10 Maximum $D_{RCM,28}$ for various cover depths as a function of binder type and environmental class for a design service life of 100 years. Note: boldface values are practically achievable by present-day concrete technology with currently used w/b; italic values are not achievable (lower values) or not recommended (higher values) [42, 43].

Mean cover (mm)		Maximum value $D_{RCM,28}$ (10^{-12} m²/s)							
Reinforcing steel	Prestressing steel	CEM I		CEM I + III 25–50% S		CEM III 50–80% S		CEM II/B-V, CEM I + 20–30% V	
		XD1/2/3 XS1	XS2 XS3	XD1/2/3 XS1	XS2 XS3	XD1/2/3 XS1	XS2 XS3	XD1/2/3 XS1	XS2 XS3
35	45	*3.0*	*1.5*	*2.0*	*1.0*	*2.0*	*1.0*	6.5	5.5
40	50	**5.5**	*2.0*	*4.0*	*1.5*	*4.0*	*1.5*	12	10
45	55	**8.5**	**3.5**	**6.0**	*2.5*	**6.0**	*2.5*	18	15
50	60	12	**5.0**	**9.0**	**3.5**	**8.5**	**3.5**	26	22
55	65	17	**7.0**	*12*	**5.0**	12	**5.0**	36	30
60	70	22	**9.0**	*16*	**6.5**	15	**6.5**	47	39

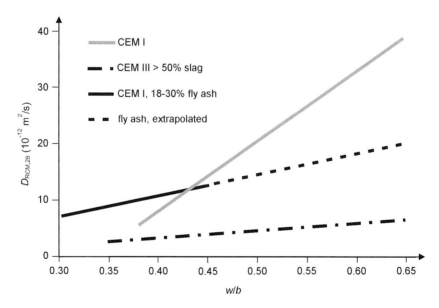

Figure 11.6 $D_{RCM,28}$ as a function of w/b for various binders.

11.6
Additional Protection Measures

In particular situations or conditions characterized by:

- very strong environmental aggressiveness;
- the impossibility of having adequate thickness of the concrete cover;
- the unavailability of good quality concrete;
- the necessity to guarantee a very long service life (Table 11.7); or
- the inaccessibility for maintenance,

it may be opportune or necessary to increase the durability of the structure, with respect to what would be achieved by following EN 206-1. This can be done by resorting to specific preventative measures that modify the characteristics of the concrete, the reinforcement, the external environment, or the structure itself (Figure 11.7). For various reasons, an option is to apply these protection measures only to critical parts of the structure (joints, supports, anchors, or any area where aggressiveness is higher) or only to the outer mat of reinforcement ("skin reinforcement").

Preventative measures, often referred to as additional protection measures, are employed as shown in Figure 11.8. They may operate by hindering aggressive species from reaching the reinforcement, or by controlling the corrosion process through inhibition of the anodic process or the corrosion current flow in the concrete. It should be noted that it is not possible to prevent the cathodic reaction from taking place. No technique available today can inhibit oxygen supply to the reinforcement, unless there is a way to keep the structure totally and permanently saturated with water. Preventative techniques will be described in the following chapters.

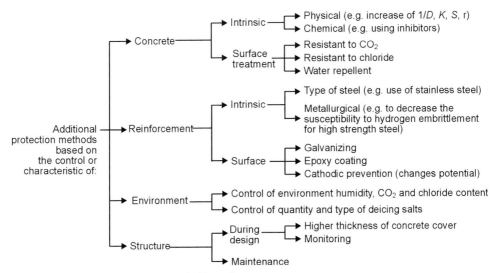

Figure 11.7 Classification of methods of additional protection.

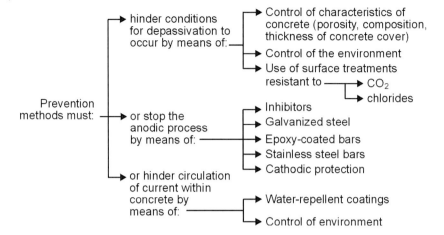

Figure 11.8 Mechanisms of additional protection measures.

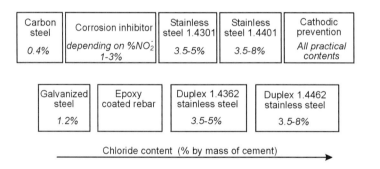

Figure 11.9 Indicative values of the maximum chloride content (% by mass of cement) that can be reached at the surface of reinforcement before corrosion initiates with some additional protection measures.

11.7
Costs

In the selection process on design options that allow fulfilling the required service life with an adequate reliability, economic considerations cannot be disregarded. Any action in favor of increasing the durability of the structure, for example, the increase of the concrete cover thickness, the reduction of w/c ratio, the use of preventative techniques, has an extra cost that should be justified by future savings.

The first step is the assessment of the advantage of any action in terms of increase of service life or reliability. For instance, various preventative measures have different limits with regard to their application. Figure 11.9 depicts indicative values of the content of chlorides that can be tolerated by the main available preventative techniques.

The cost of various techniques can only be given very roughly, and any estimate will be incomplete, since the actual cost will vary from one application to another. Furthermore, different types of prevention mechanisms are not directly comparable. Beyond this, it can be said that with respect to normal carbon steel reinforcement, use of galvanized and epoxy-coated bars costs about twice as much, and the cost of stainless steel reinforcement is about 5 to 10 times higher (see Chapter 15). The use of nitrite inhibitors (that will be described in Chapter 13) in high doses costs approximately 30 €/m^3 of concrete. Coatings may vary from 7 to 50 €/m^2 of concrete surface, hydrophobic treatment costs about 10 €/m^2 (see Chapter 14). Cathodic prevention varies from 50 to 100 €/m^2 (see Chapter 20).

In any case, it is often not possible to find well-consolidated criteria that allow a decision to be made in favor of one method or another on the basis of technical and economic considerations. Ideally, prevention options should be compared on the basis of the cost over the complete life cycle of the structure (*Life-Cycle Costing*). The calculations should take into account the initial cost, the expected cost of inspection and maintenance (repair) and the cost of demolition as influenced by the various prevention options, including the stochastic nature of many variables involved. Another important parameter is the (real) rate of interest. In this regard, performance-based approaches presented in Section 11.2.2 allow comparison of the behavior with respect to durability of different solutions, for instance the performance of corrosion-resistance reinforcement. As an example, Figure 11.10 shows the reliability index as a function of the concrete cover, for different types

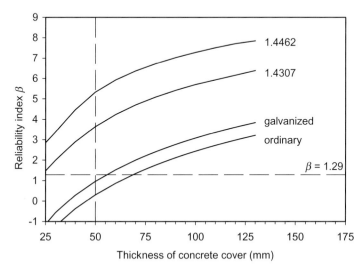

Figure 11.10 Mean thickness of concrete cover of a reinforced concrete element to guarantee a service life of 50 years as a function of the reliability index β (exposure condition: splash zone, mean annual temperature 30°C) for different types of reinforcement in concrete made with GGBS and w/c ratio of 0.45 [31].

of reinforcement (characterized by different values of critical chloride threshold) in a particular slag cement concrete [31]. Such correlations may quantitatively support the selection of the type of reinforcement that optimizes the required performance.

References

1 CEB (1989) *Durable concrete structures*, CEB Design Guide, Bulletin d'information 182.
2 CEN (2000) EN 206-1. Concrete – Part 1. Specification, Performance, Production and Conformity, European Committee for Standardization.
3 CEN (2004) EN 1992-1-1. Eurocode 2: Design of Concrete Structures – Part 1-1: General Rules and Rules for Buildings, European Committee for Standardization.
4 Rostam, S. (1999) Durability, in *Structural Concrete – Textbook on Behaviour, Design and Performance*, fib Bulletin 3, Vol. 3, 1–54.
5 Rostam, S. (1999) Assessment, maintenance and repair, in *Structural Concrete – Textbook on Behaviour, Design and Performance*, fib Bulletin 3, Vol. 3, 205–236.
6 ACI (2005) ACI 318-05. Building Code Requirements for Structural Concrete and Commentary, ACI Committee 318, American Concrete Institute.
7 ACI (2000) ACI 365.1R-00. Service-Life Prediction – State-of-the-Art Report, American Concrete Institute.
8 ISO (2012) ISO/FDIS 16204. Durability – Service Life Design of Concrete Structures, International Organization for Standardization.
9 De Sitter, R.W. (1984) Costs for service life optimization "The Law of Fives", CEB-RILEM Workshop Durability of concrete structures, 18–20 May 1983, Workshop Report, 131–134, Copenhagen.
10 Neville, A.M. (1995) *Properties of Concrete*, 4th edn, Longman, Harlow.
11 Schiessl, P. and Raupach, M. (1997) Laboratory study and calculations on the influence of crack width on chloride-induced corrosion of steel in concrete. *ACI Materials Journal*, **94**, 56–62.
12 Wilkins, N.J.M. and Lawrence, P.F. (1983) The corrosion of steel reinforcements in concrete immersed in seawater, in *Proc. Corrosion of Reinforcement in Concrete* (ed. A.P. Crane), Society of Chemical Industry, 1983, pp. 119–141.
13 ACI (2007) ACI 224.1R-07. Causes, Evaluation, and Repair of Cracks in Concrete Structures, ACI Committee 224, American Concrete Institute.
14 fib, International Federation for Structural Concrete (2010) *Model code 2010 – First complete draft – Volume 1*, Bulletin 55, Lausanne.
15 fib, International Federation for Structural Concrete (2010) *Model code 2010 – First complete draft – Volume 2*, Bulletin 56, Lausanne.
16 fib, International Federation for Structural Concrete (2006) *Model code for service life design*, Bulletin 34, Lausanne.
17 Vesikari, E. (1988) *Service life of concrete structures with regard to corrosion of reinforcement*, Technical Research Centre of Finland, Research Reports 553.
18 DuraCrete R17, DuraCrete Final Technical Report (2000) The European Union – Brite EuRam III, DuraCrete – Probabilistic performance based durability design of concrete structures, Document BE95-1347/R17, May 2000; CUR, Gouda, The Netherlands.
19 Angst, U., Elsener, B., Jamali, A., and Adey, B. (2012) Concrete cover cracking owing to reinforcement corrosion – theoretical considerations and practical experience. *Materials and Corrosion*, **63** (12), 1069–1077.
20 Bamforth, P.B. (1994) Specification and design of concrete for the protection of reinforcement in chloride contaminated environments, in Proc. of UK Corrosion

and Eurocorr 94, Bournemouth, November, 1994.
21 CEN (2002) EN 1990, Eurocode – Basis of Structural Design, European Committee for Standardization.
22 Fluge, F. (2001) Marine chlorides – A probabilistic approach to derive provisions for EN 206-1, Third DuraNet Workshop on Service life design of concrete structures, from theory to standardisation, Tromsø, 10–12 June, 2001.
23 Siemes, A.J.M. and Rostam, S. (1996) Durable safety and serviceability – A performance based design format, IABSE report 74, Proceedings IABSE colloquium "Basis of design and Actions on Structures – Background and Application of Eurocode 1", Delft, 41–50.
24 Helland, S. (2001) Basis of design – Structural and service life design, a common approach, Third DuraNet Workshop on Service life design of concrete structures, from theory to standardisation, Tromsø, 10–12 June 2001.
25 CEN/TC 104/SC1 *Survey of national provisions for EN 206-1:2000*, in preparation.
26 CEN (2009) EN 13670. Execution of Concrete Structures, European Committee for Standardization.
27 Polder, R.B. and Larbi, J.A. (1995) Investigation of concrete exposed to North Sea water submersion for 16 years. *Heron*, **40** (1), 31–56.
28 NT BUILD 492 (1999) *Concrete, mortar and cement-based repair materials: chloride migration coefficient from non-steady-state migration experiments*, Nordtest.
29 Tang, L. (1996) Electrically accelerated methods for determining chloride diffusivity in concrete. *Magazine of Concrete Research*, **48**, 173–179.
30 Tang, L. (1996) *Chloride transport in concrete – measurement and prediction*, PhD Thesis, Dept. of Building Materials, Chalmers University of Technology, Gothenburg, Sweden.
31 Bertolini, L., Carsana, M., Gastaldi, M., Lollini, F., and Redaelli, E. (2010) Performance-based approaches to durability. Design of reinforced concrete structures. *Structural – L'Edilizia*, **165**, year XVIII, 68–75 (in Italian).
32 Gehlen, C. and Sodeikat, C. (2002) Maintenance planning of reinforced concrete structures: redesign in a probabilistic environment inspection update and derived decision making, in Proc. 9th Conference on Durability of Building Materials and Components, Brisbane, Australia, 17–21 March 2002, pp. 1–10.
33 Bertolini, L., Lollini, F., and Redaelli, E. (2011) Durability design of reinforced concrete structures. *Proceedings of ICE – Construction Materials*, **164** (6), 273–282.
34 Lollini, F., Redaelli, E., and Bertolini, L. (2012) Analysis of the parameters affecting probabilistic predictions of initiation time for carbonation-induced corrosion of reinforced concrete structures. *Materials and Corrosion*, **63** (12), 1059–1068.
35 Siemes, A.J.M., Vrouwenvelder, A.C.W.M., and van den Beukel, A. (1985) Durability of buildings: a reliability analysis. *Heron*, **30** (3), 2–48.
36 Vrouwenvelder, A. and Schiessl, P. (1999) Durability aspects of probabilistic ultimate limit state design. *Heron*, **44** (1), 19–30.
37 Siemes, A.J.M., Polder, R.B., and de Vries, J. (1998) Design of concrete structures for durability – Example: chloride penetration in the lining of a bored tunnel. *Heron*, **43** (4), 227–244.
38 DuraCrete R9 (2000) Statistical quantifications of the variables in the limit state functions, Document BE95-1347/R9, January 2000, The European Union – Brite EuRam III, DuraCrete – Probabilistic Performance based Durability Design of Concrete Structures.
39 de Rooij, M., Polder, R.B., and van Oosten, H. (2007) Validation of durability of cast in situ concrete of the Groene Hart railway tunnel. *Heron*, **52** (4), 225–238.
40 Gehlen, C. (2000) Probabilistische Lebensdauerbemessung von Stahlbetonbauwerken, Deutscher Ausschuss für Stahlbeton 510, Berlin.

41 Breitenbücher, R., Gehlen, C., van den Hoonaard, J., and Siemes, T. (1999) Service Life Design for the Western Scheldt Tunnel, in Proc. 8th Conf. Durability of Building Materials and Components, Vancouver, Canada, 1999, pp. 3–15.

42 Polder, R.B., van der Wegen, G., and van Breugel, K. (2011) Guideline for service life design of structural concrete–a performance based approach with regard to chloride induced corrosion, in Proc. *fib* Workshop Performance-Based Specifications for Concrete, F. Dehn, H. Beushausen (eds.), Leipzig, 14–15 June 2011, pp. 25–34

43 CUR Leidraad 1 (2009) Guideline: Durability of structural concrete with respect to chloride induced corrosion, Duurzaamheid van constructief beton met betrekking tot chloride-geïnitieerde wapeningscorrosie, CUR Gouda (in Dutch).

12
Concrete Technology for Corrosion Prevention

Prevention of reinforcement corrosion is strongly related to the properties of concrete. Even if additional preventative measures are adopted, a suitable quality has to be assured for the concrete. This chapter outlines aspects of concrete technology that are related to the protection of the embedded reinforcement.

12.1
Constituents of Concrete

Concrete is formed by mixing cement, coarse and fine aggregate, water, and often by the incorporation of admixtures [1–4]. Selection of the constituents is the first step for obtaining durable concrete, as their properties influence the behavior of fresh and hardened concrete.

12.1.1
Cement

The presence of many types of cement [5], as illustrated in Chapter 1, shows that there is no single cement that is the best choice under all circumstances. In the past, pure portland cement was most favored; nowadays advantages related to the use of other types of cements have been clearly demonstrated, especially regarding durability. Environmental issues, and especially the reduction of carbon dioxide emissions during manufacturing, also contribute to the development of new binders. The type of cement may influence various properties of the hardened concrete or its production technology, such as high early strength, heat of hydration, resistance to environmental aggressiveness, potential reactivity with reactive aggregates, curing conditions. Proper selection and use of cement can contribute to finding the most economical balance of properties desired for a particular concrete mixture [1, 4].

For general use, that is, when the environment has a mild aggressiveness and the strength requirements for concrete are in the usual range, most cements can be suitable. In that case the selection is made on the basis of cost or local availability. Portland and portland-composite cements with strength class 32.5 or 42.5 are

Corrosion of Steel in Concrete: Prevention, Diagnosis, Repair, Second Edition. Luca Bertolini,
Bernhard Elsener, Pietro Pedeferri, Elena Redaelli, and Rob Polder.
© 2013 Wiley-VCH Verlag GmbH & Co. KGaA. Published 2013 by Wiley-VCH Verlag GmbH & Co. KGaA.

nowadays available as general-purpose cements. Although in several European countries portland-limestone cements can be found (Table 1.3), cements blended with pozzolanic addition or blast furnace slag should be preferred, because these types of addition contribute to hydration reactions and thus to the reduction in porosity and permeability of the cement paste, while limestone has a negligible effect (Figure 1.10).

When a high early strength is required, such as for post-tensioned structures or pretensioned precast elements, portland cement of strength class 52.5 may be used. Blended cements are usually not suitable because of the slow rate of hydration, with the exception of portland-silica fume cement and special (fast) slag cements.

Special requirements on the chemical composition of the cement may be necessary for certain applications, such as those requiring higher resistance to sulfate attack, reduced heat evolution or where using aggregates susceptible to alkali aggregate reaction cannot be avoided [4].

Pozzolanic and blast furnace cements may be the most suitable choice for many structures that are critical from a durability point of view. In fact, they reduce the rate of development of heat of hydration, they lead to a lower content of alkalis and lime in the cement paste, and they can produce a more dense cement paste. They should be preferred, for instance, for massive structures (to reduce the rate of development of heat of hydration), or in sulfate-contaminated environments (Section 3.3), when there is risk of ASR (Section 3.4), or in chloride-contaminated environments (Section 12.5.1).

Besides the use of blended cement, mineral additions may also be added to portland cement at the construction site; in this case suitability of the addition should be checked; for instance, EN 450-1 [6] provides conformity criteria for the use of fly ash.

12.1.2
Aggregates

Aggregates are natural or artificial materials with particles of size and shape suitable for the production of concrete. Normally, aggregates occupy a high fraction of the volume of concrete (60–85%). They allow a reduction in the amount of cement paste (and thus a reduction in both the cost of concrete and the consequences of heat of hydration, drying shrinkage and creep) and they contribute to the mechanical properties of concrete (compressive strength, elastic modulus, wear resistance, etc.)

To be suitable for making concrete, an aggregate has to comply with several requirements, which are dealt with by the European standard EN 12620 [7]. Regarding geometrical requirements, aggregate particles should have a shape (e.g., round, angular, flake), grading and maximum size such as to favor casting of fresh concrete. For instance, although increasing the maximum size of aggregate particles may lead to an increase in workability, the maximum size is limited by the size of the structural element, the distance between the rebars, and the thickness of the concrete cover. The practical range for ordinary concrete is 16–32 mm.

The porosity of the mineral that constitutes the aggregate influences the physical properties (especially density and water absorption), strength and elastic modulus of the concrete. These and other properties of aggregates are dealt with in specialized literature [1, 4, 7].

Under some circumstances, aggregates can have characteristics that negatively affect the durability of reinforced concrete. For instance, they can be susceptible to freeze–thaw attack, they can contain harmful ions such as sulfates or chlorides, or they can be potentially reactive with alkalis in the cement paste (see Chapter 3).

12.1.3
Mixing Water

Water used for concreting must be free of salts or impurities that can interfere with setting and hardening of the cement paste or negatively affect concrete properties. As far as durability of reinforced concrete is concerned, the content of sulfates and chlorides is of major importance. Seawater shall not be used for mixing water for reinforced concrete structures, because chloride and sulfate concentrations are high (respectively, around 19 g/l and 4 g/l).

The European standard for mixing water is EN 1008 [8]. It states that potable water is suitable and requires no testing. Water from underground sources, surface water and industrial waste water shall be tested; sewage water is not suitable. Tests include analyses for chloride, sulfate, alkali, oils, and fats, detergents, acids, humic matter, dissolved solids, color and odor. The chloride content has specific limits for prestressed (500 mg/l), reinforced (1000 mg/l), and plain concrete (4500 mg/l). Sulfate is limited to 2000 mg/l. Setting times and compressive strength shall be tested and shall comply to limits referring to concrete mixed with demineralized or distilled water, or chemical analysis for harmful substances (sugars, phosphates, nitrates, lead, and zinc) shall be carried out. Seawater and brackish water may be used for concrete without reinforcement or other embedded metal, but is in general not suitable for reinforced or prestressed concrete. Water recycled from processes in the concrete industry may be used, but specific limits are set for the amount of dissolved solid matter. A maximum density of 1.15 kg/l is given, in general corresponding to 0.286 kg of solid matter per liter. Such water may stem from cleaning concreting equipment, surplus concrete, sawing or grinding and water blasting of hardened concrete.

12.1.4
Admixtures

Admixtures are substances that are added during the mixing process in small quantities related to the mass of cement, in order to improve the properties of fresh or hardened concrete [9, 10]. The most utilized admixtures are: *water reducers* and *superplasticizers* that may be added to improve the workability of concrete or reduce the amount of mix water; *viscosity-modifying agents* that help in reducing the tendency to bleeding and segregation of mixes; *hardening accelerators* that are used to increase the rate of development of early strength of concrete; *set retarders* that reduce the setting time of concrete; *air-entraining agents* that increase the freeze–thaw resistance of concrete (Section 3.1). Admixtures with several combined effects are also available. *Corrosion inhibitors* have been developed in order to increase the corrosion protection of embedded steel; these will be treated in Chapter 13.

Admixtures for concrete should not have any adverse side effect; for instance they shall not contain appreciable amounts of chloride ions that may be harmful with regard to steel corrosion (in the past, the most utilized accelerating admixture was calcium chloride) [11].

Water Reducers and Superplasticizers These admixtures permit a reduction in the water content of a given concrete mix without affecting the consistence, or, alternatively, increase the workability without affecting the water content. They are chemical substances that improve the mobility of the particles in the fresh concrete mix. They contain surface-active agents that are adsorbed to the surface of solid particles and modify their electrical surface charge, so that they do not tend to flocculate, and thus are more dispersed, and the water can better lubricate the mix.

Superplasticizers are important in order to obtain durable reinforced concrete structures. They allow a high workability, which is essential to achieve proper compaction of fresh concrete especially in slender elements or in the presence of dense reinforcement, without increasing the amount of mixing water. The increase in workability is then achieved without changing the w/c ratio and thus without affecting the strength and permeability of concrete.

According to the European standard EN 934-2 [12], they are divided into water reducers (or plasticizers) and superplasticizers (or high-range water reducers), depending on the degree of their effect. The effect depends on the chemical composition and the dosage [9].

Superplasticizers can be used with the purpose of increasing the durability, by decreasing the w/c ratio. For instance, they can guarantee the same workability with a remarkable reduction of the water content; hence, if the dosage of cement is not changed, the w/c ratio decreases.

12.2
Properties of Fresh and Hardened Concrete

Concrete in the fresh state should have good workability in order to be properly placed and compacted in the mold so that it can achieve the expected properties after hardening. The most relevant properties of hardened concrete are strength, deformation due to loading or thermal and moisture variations, resistance to cracking, and durability.

12.2.1
Workability

Fresh concrete must be mixed, transported, placed in the mold and compacted without compromising the homogeneity of the mix. Voids that remain in the mix after compaction, due both to entrapped air and to separation of coarse and fine aggregate from cement paste (*segregation*), are deleterious for strength and durability.

In order to favor the placing of concrete and the removal of entrapped air, concrete should be fluid and thus able to easily flow; however, the mix should have a good cohesion in order to prevent segregation during handling, transport and placing. After placing, part of the mixing water tends to rise towards the upper surface (*bleeding*). This water can reach the surface or can be trapped below coarse aggregates or steel bars, causing a local increase in the w/c ratio that will lead to a weaker and more permeable cement paste in the interfacial transition zone.

The term *workability* (or *consistence*) is used to indicate the properties of fresh concrete that are related to its ability to flow during placing, to be compacted (i.e., vibrated in order to remove entrapped air), and to be sufficiently cohesive.

Fluidity primarily increases if the water content is increased or a superplasticizing admixture is used; nevertheless an excess of water or superplasticizer may also reduce the cohesion of concrete and thus favor segregation. Cohesion of the mix may be improved by increasing the amount of fines (cement, filler or fine sand) or using viscosity-modifying admixtures. The properties of aggregates, such as the shape of particles, the grading and maximum size, also play an important role with regard to both fluidity and cohesion of fresh concrete. Table 12.1 shows an example of recommendations by the American Concrete Institute for the amount of mixing water required to achieve a certain slump.

Measurement of Workability Since the workability of concrete is a combination of properties (fluidity, cohesion, and compactability), there are several standardized tests that measure one or more of these properties. These tests are based on measuring the behavior of fresh concrete subjected to a standardized procedure.

Table 12.1 Approximate amount of mixing water (kg/m^3) required to obtain different slumps for various nominal maximum sizes of aggregates and typical contents of entrapped air that may result [4]. Data refer to nonair-entrained concrete with angular aggregates.

Slump	Nominal maximum size of aggregate (mm)							
(mm)	9.5	12.5	19	25	37.5	50	75	150
25–50	208	199	187	178	163	154	130	113
75–100	228	216	201	193	178	169	145	124
150–175	243	228	213	202	189	178	160	–
Typical entrapped air (%)	3	2.5	2	1.5	1	0.5	0.3	0.2

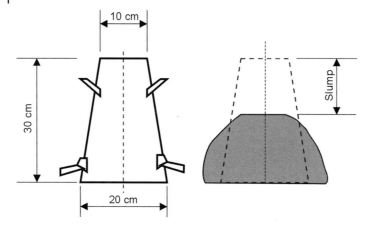

Figure 12.1 Slump test for measurement of workability of fresh concrete.

The most commonly used is the *slump test*. Fresh concrete is placed in the apparatus shown in Figure 12.1, following a standardized procedure; then, the mold is removed, and the concrete will slump under its own weight. The decrease in height measured with respect to the initial height is taken as a measure of the workability of concrete.

Other tests normally used for measuring concrete workability are [3]:

- *flow tests*, which measures the spread of concrete subjected to standardized vibration or jolting;
- *Vebe test*, which measures the time required to compact the concrete on a vibrating table;
- *compacting factor test*, which measures the degree of compaction of concrete after it falls from a standardized height.

In general, results of different types of tests are not correlated. Compared with the slump test, flow tests are more suitable for flowable concrete, while the compacting factor and Vebe tests are suitable for concrete with low workability. According to the European standard EN 206, the workability of concrete should be classified as shown in Table 12.2.

The workability of fresh concrete decreases in time (slump loss), especially in hot conditions [1]; the consistence should be tested at the time of use of concrete or, in the case of ready-mixed concrete that is transported from a plant to the site, at the time of delivery.

12.2.2
Strength

In the structural design of reinforced concrete structures the compressive strength of concrete is of primary importance [13]. In Chapter 1 it was shown that the

Table 12.2 Classes of consistence of concrete measured with slump test, Vebe test, degree of compactability, and flow table test [3].

Slump class	Slump (mm)	Vebe class	Vebe time (s)	Compaction class	Degree of compactability	Flow class	Flow diameter (mm)
S1	10–40	V0	≥31	C0	≥1.46	F1	≤340
S2	50–90	V1	21–30	C1	1.26–1.45	F2	350–410
S3	100–150	V2	11–20	C2	1.11–1.25	F3	420–480
S4	160–210	V3	6–10	C3	1.04–1.10	F4	490–550
S5	≥210	V4	3–5			F5	560–620
						F6	≥630

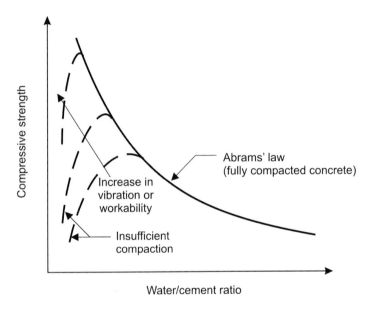

Figure 12.2 Schematic representation of the relationship between compressive strength of concrete and water/cement ratio.

strength of cement paste depends on its porosity. When concrete is considered, the strength is also influenced by the interfacial transition zone between the aggregates and the cement matrix, which is often a weaker zone, being constituted by cement paste of lower strength and higher permeability.

Nevertheless, the water/cement ratio and curing remain the main factors that influence the strength of concrete. Figure 12.2 outlines the relationship between the compressive strength of concrete and w/c ratio, for a given curing (Abrams' law). The strength increases by decreasing the w/c, due to the reduction in the porosity of the cement paste both in the bulk and at the interface with aggregate particles.

In practice, since the cement content cannot be increased too much, the water content has to be reduced in order to achieve low w/c ratios; as a consequence, the workability of concrete decreases. Below a certain w/c ratio, which depends on the technique used for compaction, proper consolidation of fresh concrete cannot be achieved and thus the strength of hardened concrete decreases (as shown by the dashed lines in Figure 12.2). The use of a superplasticizing admixture is therefore necessary to obtain a suitable workability in particular for a low water/cement ratio.

Other factors that affect the strength of concrete are the type of cement, some properties of aggregates (e.g., shape and grading, by influencing workability of fresh concrete, can have an indirect influence on the strength of hardened concrete) and admixtures (for instance, the use of air-entraining agents leads to a reduction in strength, while accelerators increase the early strength, but in general lead to a reduction in the strength measured at later ages).

Compressive Strength and Strength Class The compressive strength of concrete can be measured with compressive tests on 150 mm cubes or cylinders (150 mm in diameter and 300 mm in height). The shape of the specimen influences the result of the test; the strength measured with the cube test is about 1.25 times that measured with the cylinder test.

Since significant scatter is present in the strength of concrete, a statistical approach is followed to define a *characteristic strength* utilized in the structural design [13]. Assuming that the compressive strength has a normal distribution with mean value f_{cm} and standard deviation σ, the characteristic strength can be calculated as:

$$f_{ck} = f_{cm} - K \cdot \sigma \quad (12.1)$$

where K is a constant that depends on the percentage of strength values that are statistically allowed to be lower than f_{ck}. According to European standards, a percentage of 5% is considered and the characteristic strength is measured on specimens cured for 28 days under wet conditions [3]. Figure 12.3 shows the relationship between mean strength and characteristic strength.

The standard deviation measures the variability of concrete strength and thus defines its quality; in fact, the variability increases if poor quality control is carried out on raw materials, moisture content of aggregates, proportioning, mixing, transport, and placing.

EN 206-1 defines the compressive strength classes for normal-weight concrete shown in Table 12.3; the numbers in each class indicate the minimum characteristic compressive strength (measured at 28 days) on cylinders ($f_{ck,cyl}$) and cubes ($f_{ck,cube}$), respectively.

Tensile Strength The tensile strength of concrete controls its resistance to cracking. It can be measured indirectly by means of the flexure test or the splitting test.

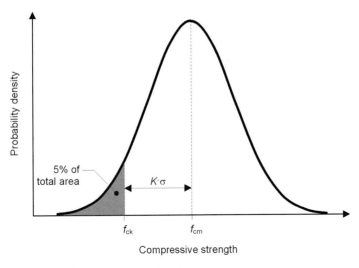

Figure 12.3 Characteristic strength (f_{ck}) and mean strength (f_{cm}) of concrete.

Table 12.3 Classes of compressive strength of normal-weight concrete [3].

Compressive strength class	Minimum characteristic cylinder strength, $f_{ck,cyl}$ (MPa)	Minimum characteristic cube strength, $f_{ck,cube}$ (MPa)
C8/10	8	10
C12/15	12	15
C16/20	16	20
C20/25	20	25
C25/30	25	30
C30/37	30	37
C35/45	35	45
C40/50	40	50
C45/55	45	55
C50/60	50	60
C55/67	55	67
C60/75	60	75
C70/85	70	85
C80/95	80	95
C90/105	90	105
C100/115	100	115

Often, it is estimated with empirical formulae that correlate the tensile strength to the compressive strength; they normally are of the type: $f_t = k \cdot f_c^n$ [1, 13]. These correlations can only give an approximate evaluation of the tensile strength, since the correlation between f_t and f_c is affected by many factors, such as the composition of the mix, the moisture condition, the type of cement and curing.

12.2.3
Deformation

Concrete has a viscous behavior: when it is loaded with a constant stress it shows a strain that increases with time. Conventionally an *elastic* deformation is considered when it occurs immediately after application of the load, while subsequent deformation is attributed to *creep*. It is possible to define a modulus of elasticity for concrete that can be evaluated with short-term tests [13]. Similarly as for the tensile strength, empirical formulae are available that give an approximate correlation of the modulus of elasticity with the compressive strength [1, 13]. A dynamic modulus can also be estimated with nondestructive tests that measure the rate of propagation of ultrasonic vibrations through concrete [1].

12.2.4
Shrinkage and Cracking

Deformation of concrete can occur due to temperature or humidity changes. When the deformation is restrained, stresses are induced in the concrete. Shrinkage produced by drying out or cooling is of particular interest with regard to the durability of concrete. During the first days after mixing, concrete has a high moisture content and the heat of hydration may increase its temperature. Subsequently, loss of water, for example, due to evaporation, or a decrease in temperature, induce shrinkage in the concrete. Depending on the cause, shrinkage is called plastic, autogeneous, drying, or thermal shrinkage [1].

For instance, drying shrinkage occurs when concrete is allowed to dry out after curing. It depends on the w/c ratio, the aggregate to cement ratio, the elastic modulus of aggregates, the slenderness of the element, and the humidity of the environment.

When the shrinkage cannot take place due to internal (e.g., the reinforcement) or external restraints, tensile stresses arise in the concrete that can lead to cracking if they exceed the tensile strength of the concrete. Shrinkage is more harmful when it takes place at early ages, that is, when the tensile strength of concrete is low; thus, proper curing is essential to avoid early cracking due to evaporation of water.

12.3
Requirements for Concrete and Mix Design

In addition to the durability requirements illustrated in Chapter 11, concrete has to fulfill requirements of compressive strength and workability. According to EN 206-1, the basic requirements of concrete can be quantitatively expressed in terms of exposure class (Tables 3.1 and 11.2), compressive strength class (Table 12.3), class of consistence (Table 12.2) and maximum nominal aggregate size. Further requirements may apply under particular circumstances (e.g., low heat of hydration or high early strength). Normally, the designer of a structure should only

12.3 Requirements for Concrete and Mix Design

specify the required properties to the producer, who is responsible for providing a suitable concrete (this is called *designed concrete*, to be distinguished from *prescribed concrete* in which the composition of the concrete and the constituent materials to be used are specified [3]). Of course, in the case of designed concrete additional specifications may be added to fulfill the durability requirement. For instance, if a performance-based durability procedure is followed in the design stage, the type of binder or a maximum value of the apparent chloride diffusion coefficient measured by a specific test should be specified.

The concrete composition and the constituent materials for designed or prescribed concrete are chosen to satisfy the requirements specified for fresh and hardened concrete, taking into account the production process and the intended method of execution of concrete works. A single mix has to be defined that is able to satisfy all the design requirements. The procedure used for the selection of a concrete mix (type and proportion of cement, water, aggregate, and admixtures) is called the *mix design*. Several procedures for the mix design can be found in the technical literature, which are based on empirical and experimental correlations that associate the requirements for concrete to different parameters of its composition, such as the w/c ratio, the water content, etc. [4, 14–16]. Some of these parameters are associated with more than one of the requirements. This is the case for the w/c ratio.

Figure 12.4 shows that both strength and durability requirements concur with the assessment of a suitable w/c ratio. In Chapter 11 it was shown that to guarantee

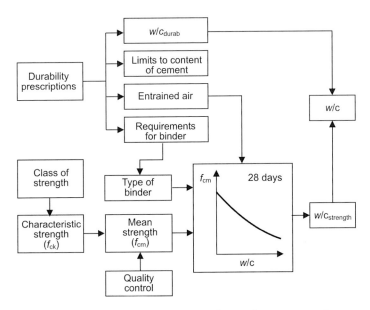

Figure 12.4 Interaction between durability and strength requirements for concrete in the procedure for mix design.

the required service life of a reinforced concrete structure, a widely used approach is to fix an upper limit for the w/c ratio.

To fulfill the strength class, the concrete has to reach a characteristic strength (f_{ck}). Depending on the quality control, a minimum value of the average strength measured at 28 days of curing (f_{cm}) can be calculated from the required f_{ck}. For a given type of cement, a maximum value for the w/c ratio can then be defined in order to satisfy the required mean compressive strength.

The actual w/c ratio will be the lower of the two values deriving from compressive strength requirements and from the need for durability. Therefore, if it is necessary to decrease the w/c ratio because of the aggressiveness of the environment or the particular service life required, the strength will automatically increase. To avoid ambiguity in the specifications for concrete, EN 206-1 suggests that the durability requirements that are expressed in terms of w/c ratio should also be converted in an equivalent value of minimum strength. For instance, Table 11.3 shows the minimum strength classes evaluated from the w/c ratios for concrete made with cements of strength class 32.5 (the correlation between w/c and strength depends on the type of cement). This approach has the advantage that the structural design of a reinforced concrete structure can be optimized according to the minimum strength of the concrete that is required for durability reasons (in some countries structural designers are still used to specify concrete with low strength, for example, with strength class C20/25, which cannot guarantee a reasonable service life even in mildly aggressive environments). A second advantage is related to quality control: while it is difficult to evaluate the w/c in hardened concrete, it is much simpler to measure the strength of concrete, and this is already done for structural reasons.

Interaction between durability and strength requirements is not limited to the w/c ratio. For instance, Figure 12.4 shows that the air entrained to increase the resistance to freeze–thaw cycles reduces the strength of concrete and thus a lower w/c ratio is required to obtain the required strength. Once the w/c ratio has been found, the dosage of cement, water, and aggregates has to be calculated. The water content (kg/m^3 of concrete) can be estimated on the basis of the consistence class, the properties of the aggregate, and the use of a plasticizer. For instance, Table 12.1 shows the water content suggested by the ACI Manual of concrete practice [4].

The cement content (kg/m^3) is equal to the water content divided by w/c, provided it is suitable according to limits fixed for durability (Figure 12.4). Although a lower limit of cement dosage is normally fixed in order to guarantee a reasonable amount of cement paste in the concrete, upper limits could also be fixed in order to limit effects of shrinkage, heat of hydration, ASR, etc. In this regard, it should be noted that increasing the cement content (at constant w/c ratio) may not be beneficial with respect to the properties of hardened concrete; in fact, it has been reported to be even detrimental (for instance, for the compressive strength [17]). Finally, the aggregate content can be evaluated by means of a volume balance.

The procedure of mix design should finish with a trial mix (or initial test) in order to check, before the production starts, if the concrete meets all the specified requirements in the fresh and hardened states. Subsequently, during the production the properties of concrete should be regularly tested in order to systematically examine the extent to which the product fulfils specified requirements. Concrete compositions shall be reviewed periodically to provide assurance that all concrete mixes are still in accordance with the actual requirements, taking account of the change in properties of the constituent materials and the results of conformity testing on the concrete compositions [3].

12.4 Concrete Production

Prevention of reinforcement corrosion is not limited to the design of a reinforced concrete structure or to the definition of the mix proportions. It should be further developed as the concrete is mixed, placed, compacted and cured. All necessary precautions must be taken during transport and placement of concrete so that the mixture does not undergo segregation or bleeding, and vibration leads to the maximum compaction possible [18, 19]. Spacers should be properly used to fix the reinforcement and thus ensure that the actual thickness of the concrete cover corresponds to the nominal value. Finally, curing should provide reasonable conditions of temperature and humidity for a sufficiently long period, in order to promote hydration of the cement paste. Mistakes or deficiencies in any of these stages will have adverse effects on durability and thus can compromise the effort made during the design stage.

Throughout all these phases all necessary quality control actions should be undertaken to obtain a product that fits the specifications. Since a number of individuals work on any phase of the project (e.g., design, mixing, construction, and maintenance), quality control is of primary importance in order to correctly attribute the responsibility for the quality of the product.

12.4.1 Mixing, Handling, Placement and Compaction

Mixing consists in blending the correct dosage of cement, water, aggregate, and admixtures. This is obtained by stirring the raw materials in a mixer for a suitable time, in order to obtain a uniform mass. Often, concrete is mixed in a specific plant and it is transported to the construction site (*ready-mixed concrete*). Segregation of concrete has to be prevented during transportation, discharging of the mixer and further handling of the concrete.

For ready-mixed concrete, it is essential that the class of consistence of concrete be specified. In fact, if the workability of the concrete that reaches the construction site is lower than required, there will be a tendency to add water to the mix in order to increase workability. Although it is deleterious, this is unfortunately a common practice on sites where quality control is lacking. While it increases the workability of the fresh concrete, it also increases the w/c ratio, leading to a

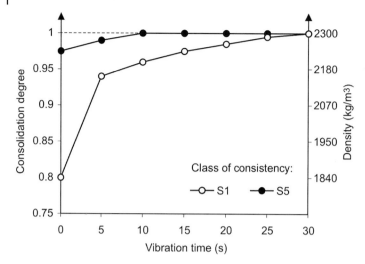

Figure 12.5 Influence of the duration of vibration on the degree of compaction of concrete of different consistence (degree of compaction was calculated as the ratio of the density of the concrete to the density of the fully vibrated concrete) [10].

lowering of strength and durability. On the other hand, a plasticizing admixture could be used in the mixing plant, to improve the workability without increasing the w/c ratio.[1]

Placing and compacting of concrete are aimed at filling the form and removing the entrapped air from the fresh concrete. Both operations, which are carried out almost simultaneously, may lead to segregation of concrete. To avoid segregation, concrete should not fall free or collide with the formwork, and should be placed in uniform layers. Each layer should be fully vibrated before placing the next one, in order to minimize the residual entrapped air (last row of Table 12.1 shows typical contents of entrapped air after vibration, as a function of maximum size of aggregate). The effort required for compacting concrete increases if the slump decreases (Figure 12.5); conversely, excessive vibration may lead to segregation of concretes with high slump [16].

Placing and compacting of concrete are extremely important with regard to durability. Excessive voids left after compaction, due to segregation or insufficient vibration, affect the permeability of the structure. Obviously, proper surrounding of the rebars is essential for their corrosion protection. It should be borne in mind that this regards the actual structure, but it cannot be detected by tests on cubes or cylinders made from samples of fresh concrete.

1) If the specification for concrete is correct, there should be no need for addition of water at the site. EN 206-1 forbids any addition of water and admixture at delivery, unless in special cases where, under the responsibility of the producer, it is used to bring the consistence to the specified value, providing the limiting values permitted by the specification are not exceeded and the addition of the admixture is included in the design of the concrete.

12.4.2
Curing

Curing of concrete consists in promoting hydration of the cement paste by controlling the moisture content and the temperature of concrete. Usually, it consists in maintaining the concrete saturated with water. In many cases preventing or retarding evaporation of water from concrete is sufficient to provide adequate curing. This can be achieved in several ways, such as delaying the removal of the formwork or covering the concrete with vapor-proof sheets or applying a curing compound. Once concrete has set, it can be sprayed or flooded with water or put in contact with wet sand, textiles or plastic sheeting. For concrete with low w/c ratio, self-desiccation can also take place, that is, consumption of water due to hydration leads to desiccation of pores in the cement paste, and thus internally consumed water should be resupplied as much as possible.

Curing affects both strength and durability, but the consequences of bad curing can be more serious for the latter. Figure 12.6 shows the role of wet curing on the compressive strength of concrete. If concrete is not kept moist at least in the early stage (3–7 days), hydration is interrupted and the 28-day compressive strength can be much lower than the strength that could be potentially reached. Actually, the concrete cover, that is, the concrete intended to protect the reinforcement, is more susceptible to drying out due to evaporation than the bulk concrete. Therefore, the effect of bad curing is even worse with regard to the durability of reinforced concrete structures.

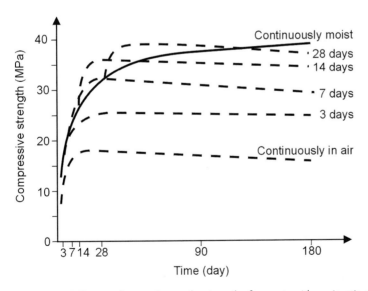

Figure 12.6 Influence of wet curing on the strength of concrete with a w/c ratio of 0.50 (after Price, from [1]).

Curing of concrete should proceed until it leads to segmentation of the capillary pores in the cement paste (Section 1.2.2), so that the cement paste becomes much less permeable to aggressive species. From a practical point of view, the duration of applied curing can be determined as a function of the development of the concrete properties in the surface zone. For instance, EN 13670 specifies that the duration of applied curing shall be a function of the development of the concrete properties in the surface zone and proposes four curing classes on the basis of the percentage of specified characteristic 28-day compressive strength to be reached. Indicative values of curing time, for class 3 (50% of f_{ck}), are shown in Table 12.4 on the basis of the temperature and the concrete strength development (measured by the ratio of the mean compressive strength after 2 days to the mean compressive strength after 28 days). When testing is not performed to determine the curing period, ACI 308.1 suggests a minimum period of 7 days provided that the concrete surface temperature is at least 10 °C.

Increasing the temperature accelerates the hydration of the cement paste. Therefore, a higher early strength can be achieved by increasing the curing temperature, although this may lead to a lower longer-term strength gain [2].

The influence of temperature on the rate of strength development of concrete can be used to increase the early strength of concrete by steam curing. This process is sometimes applied to precast elements that, after initial curing at room temperature, are slowly brought to high temperature (although lower than 65–70 °C to avoid *DEF*, Chapter 3) in wet conditions for several hours.

Table 12.4 Minimum curing period (days) for curing class 3 (corresponding to a surface concrete strength equal to 50% of the specified characteristic strength) according to EN 13670 [18].

Surface concrete temperature t (°C)	Concrete strength development ($r = f_{cm2}/f_{cm28}$)		
	Rapid $r \geq 0.50$	Medium $0.30 > r \geq 0.50$	Slow $0.15 > r \geq 0.30$
$t \geq 25$	1.5	2.5	3.5
$25 > t \geq 15$	2.0	4	7
$15 > t \geq 10$	2.5	7	12
$10 > t \geq 5$	3.5	9	18

Notes:
- any period of set exceeding 5 h should be added;
- for temperatures below 5 °C, the duration should be extended for a period equal to the time below 5 °C;
- the concrete strength development is the ratio of the mean compressive strength after 2 days to the mean compressive strength after 28 days determined from initial tests or based on the known performance of concrete of comparable composition;
- for very slow concrete strength development, special requirements should be given in the execution specification.

12.5
Design Details

Durability should be taken into consideration when details of a structure are designed [20]. Very often even simple details may prevent or promote significant deterioration of a reinforced concrete structure.

For instance, design may reveal itself inadequate if it favors the introduction of locally aggressive conditions, such as geometric details of the structure that lead to wetting of parts critical to the structure itself or favor the stagnation of water. Joints and every area of possible stagnation of water are to be considered as points of weakness as far as corrosion of steel is concerned. A frequent problem is related to a wrong design of the drainage system that allows water to flow on the surface of concrete; many piles of bridges suffer from early corrosion damage due to chloride-contaminated water drained onto them from the deck joints.

Unnecessary complex geometries, sharp corners, and layout of the rebars inappropriate for favoring the vibration of concrete should also be avoided. The structural design must be such as to minimize cracking of the concrete.

Figure 12.7 illustrates a few design details that experience has shown to frequently be the cause of failures, and proposes some improved alternatives. The examples provided demonstrate that the origin of corrosion can often be traced to ordinary errors that could have been avoided without any appreciable increase in cost.[2] In fact, the cost of adequate prevention carried out during the stages of design and building are minimal compared to the savings they make possible during the service life, and compared to the costs of rehabilitation that might be required at later dates. As far as possible, structures should be designed in order to be accessible to inspection and maintenance. Supplementary attention to durability should be given to structures or parts of structures that are not accessible. If the accessibility of the structure is very low or the structure is critical with regard to stability or serviceability, a monitoring system can be provided (Chapter 17).

12.6
Concrete with Special Properties

Some types of concrete have special properties that are useful with regard to the durability of reinforced concrete structures. Conventional concrete with mineral additions, high-performance concrete (*HPC*) and self-compacting concrete (*SCC*) are briefly discussed here.

---

2) It useful to observe that, although causes of corrosion can be trivial, consequences may be catastrophic. For instance, cases are reported where collapses of structures were caused by changes in the cross section of load-bearing elements, introduced for aesthetic reasons (analogous to that showed in the last example of Figure 12.7) that favor stagnation of chlorides and drastically reduce the concrete cover, leading to attack of the reinforcement. This is a typical example of failures that Peter Schießl calls SIC: *Stupidity-Induced Corrosion*.

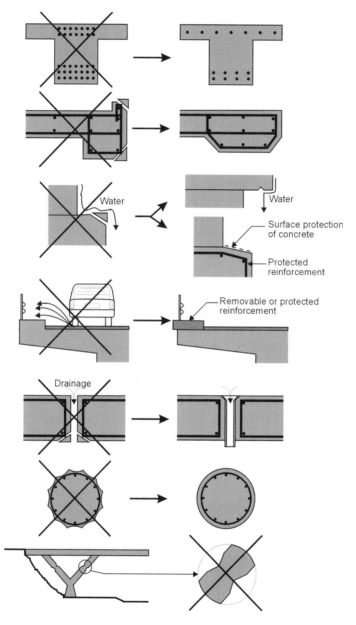

Figure 12.7 Examples of correct and incorrect design and some suggestions relative to vulnerable positions, in particular in bridges (most examples are taken from [20]).

12.6.1
Concrete with Mineral Additions

Studies in the early 1970s [21, 22] demonstrated that the addition of natural pozzolana to portland cement could reduce the chloride diffusion coefficient of concrete by three times. Additions of ground granulated blast furnace slag (GGBS) and fly ash (PFA) have an even more marked effect on the diffusion coefficient and raise the resistivity of concrete.

Today, it has been widely demonstrated that the initiation time for corrosion, and thus the service life of reinforced concrete structures operating in chloride-bearing environments, are much greater if a cement with pozzolanic or GGBS addition is used rather than pure portland cement. With regard to the use of cement with ground granulated blast furnace slag, the Dutch experience can be mentioned. In the Netherlands, blast furnace cements have been used in structures for coastal defense since the 1930s, and demonstrate that after more than 50 years of service, corrosion attack is practically nil, and chloride penetration very limited [23]. Concrete submerged for 16 years in the North Sea showed a diffusion coefficient that is up to 10 times lower and resistivity from 3 to 10 times greater [24] and a higher critical level of chlorides [25], in comparison to concrete made with portland cement and the same w/c ratio. Analogous improvements also have been seen in concrete operating in the splash zone [26, 27]. Indepth investigation in the early 2000s of six marine structures of up to 40 years of age showed that chloride penetration was consistently slow in slag structures [28]. As far as cement with fly ash is concerned, experience coming from Great Britain, where fly ash has been used since the 1950s, shows how its addition produces improvements analogous to those achieved with blast furnace slag cement [29].

It should be observed that nearly inert mineral additions, such as limestone, do not have the beneficial effect given by pozzolanic and hydraulic additions. For instance, Figure 12.8 shows the positive influence of fly ash (FA) and blast furnace slag (BF) on the apparent chloride diffusion coefficients obtained from rapid chloride migration tests and, conversely, shows the remarkable increase in the diffusion coefficient brought about by the use of limestone additions [17]. Therefore, the use of limestone portland cement should be discouraged in chloride environments.

Polder and Larbi [24] estimated depassivation times, and thus service lives, almost 7 times greater for cements with 70% GGBS than for portland cements (as was shown in Table 11.6).

Bamforth [29] suggested the recommendations given in Table 12.5 that show how the use of normal portland cement does not produce concrete adequate for a service life of 75 years in a chloride-containing environment, unless class C50/60 concrete and concrete covers of 100 mm are used. Cements with a high percentage of mineral additions allow the use of more reasonable concrete cover thicknesses and lower classes of strength.

In some countries, seawater is used for quenching molten blast furnace slag, resulting in a significant chloride content in the slag and thus in the blended cement. This practice is also indirectly permitted by EN 197-1 that allows

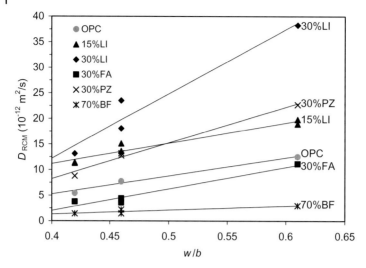

Figure 12.8 Chloride diffusion coefficients (D_{RCM}) at 28 days of curing obtained as an average value of two replicate specimens from the rapid chloride migration test as a function of w/b ratio, mineral addition, binder content (250 kg/m³ gray, 300 kg/m³ filled gray, 350 kg/m³ filled black, 400 kg/m³ black) and type of binder (OPC = portland cement CEM I 52.5R, 15%LI and 30% = 15% and 30% of ground limestone, 30%PZ = 30% natural pozzolana, 30%FA = 30% pulverized coal fly ash, 70%BF = 70% of ground granulated blast furnace slag) [17].

Table 12.5 Classes of strength[a] and composition recommended by Bamforth [29] for concrete operating in chloride-bearing environments (GGBS = granulated blast furnace slag; PFA = fly ash), based on 75 years time to initiation.

Type of cement	Recommended strength class for cover of:		
	50 mm	75 mm	100 mm
OPC	Not applicable	Not applicable	C50/60
OPC with:			
>50% GGBS	C40/50	C36/45	C24/30
>60% GGBS	C32/40	C24/30	C24/30
>70% GGBS	C24/30	C24/30	C24/30
>20% PFA	C50/60	C40/50	C32/40
>30% PFA	C40/50	C32/40	C24/30
>40% PFA	C32/40	C24/30	C24/30

a) Classes of strength are defined as in Table 12.3.

a maximum chloride content higher than 0.1% for blast furnace cements, although it requires to explicitly declare the actual chloride content. Indeed, the use of seawater for cooling the blast furnace slag should be discouraged. In fact, a chloride content significantly higher that 0.1% may be harmful for corrosion of steel after concrete undergoes carbonation (Section 5.3.1).

12.6.2
High-Performance Concrete (HPC)

High-performance concrete (*HPC*) may be defined as a concrete that, with particular care in the selection and proportioning of its constituents, shows a clear improvement in one or more of the properties with respect to ordinary concretes [1]. In practice, the term high-performance concrete is applied to concretes with a low water/binder ratio (0.3–0.35 or even lower), elevated cement content (400–550 kg/m^3) and the addition of 5–15% silica fume with respect to the total mass of binder. Sometimes, fly ash or GGBS are also added. In order to achieve a suitable workability with mixes typical of *HPC*, a high dosage of superplasticizer is required.

The low water/binder ratio, associated with the use of silica fume that, due to the very small dimension of its particles, fills the spaces between the cement grains, and leads to the development of a microstructure of extremely low porosity. The improvement of the porosity also occurs at the interface between the cement matrix and the coarse aggregate. The strength of these types of concrete is consequently very high; nowadays only concrete with 28-day compressive strength higher than 80 MPa is considered as *HPC* [1]. The term *high-strength concrete* that was used in the past has been substituted by *high-performance concrete* because this material is also expected to exhibit high durability in aggressive environments. There is not enough experience on *HPC* to quantify the advantage of its use in terms of durability of reinforced structures. Nevertheless, it should be observed that *HPC* is normally obtained by applying strictly all the procedures that favor the development of a dense microstructure with discontinuous pores, which are the basis for obtaining a durable concrete. In practice, *HPC* is able to ensure the most important requisite with regard to any type of degradation, that is, an extremely low permeability to aggressive species. It can then be expected that deterioration of concrete may be strongly reduced compared to ordinary concrete. If the experience confirms these expectations, in the future *HPC* will probably find applications in aggressive environments, in order to increase the service life, especially with regard to chloride-induced corrosion. There is still controversy in relation to the performance of *HPC* towards freeze–thaw attack. Curing of *HPC* is also critical because of autogenous shrinkage due to self-desiccation of capillary pores.

12.6.3
Self-Compacting Concrete (SCC)

Placing and compacting fresh concrete is a time-consuming, noisy, and heavy operation, especially in structures with complex geometry or congested reinforcement, which also negatively affects the health of the workers. Furthermore, lack of skilled workers dedicated to vibration of concrete is a worldwide problem, which often is the cause of poor durability. *Self-compacting concrete* (*SCC*) has been developed that can be placed and compacted into every corner of the formwork, purely by means of its own weight and fluidity, without any vibration [15, 30].

A self-compacting concrete, in the fresh state, is therefore characterized by:

- high deformability, defined as the ability to change its form under the sole action of its own weight, to match that of the mold where the mixture is introduced;
- high mobility in confined spaces, allowing the material to flow in the presence of narrowing of the cross section, where normally the flow of the concrete is stopped due to collisions between particles of coarse aggregate;
- high resistance to segregation, which allows maintenance of a uniform distribution of the constituents during the processing of fresh concrete.

Therefore, special mixes are required in order to achieve an extremely high flowability and self-leveling properties, to prevent blockage of coarse aggregate between the reinforcing bars and, in the meantime, to avoid segregation. A high dosage of superplasticizer is combined to a large quantity (usually more than $500 \, kg/m^3$) of powders with particles smaller than 100–150 μm. These include cement and addition of pozzolanic materials (fly ash or silica fume), GGBS, or ground limestone (filler). A large amount of paste phase can provide a combination of high fluidity and good cohesion to the fresh concrete. Specific admixtures (viscosity-modifying admixtures, *VMA*) have also been developed to improve the cohesion of the mix, similarly as the fine particles. The maximum size of aggregate should be limited to 16–20 mm.

The fluidity of *SCC* is so high that the usual tests for measuring workability cannot be used, and specific tests have been developed [31]. For instance, the slump-flow test measures the spread of the mixture introduced in the Abrams' cone. The slump flow should be greater than 600 mm and, typically, is between 650 and 750 mm. Also, the time needed to reach the diameter 500 mm is measured to evaluate the flow rate (it should be less than 12 s). Other tests, such as the J-ring test, V-funnel test, L-box test, and U-box test are used to investigate other properties such as the filling ability of the self-compacting concrete, the resistance to segregation, etc. [10].

The properties of hardened self-compacting concrete are influenced by its composition and raw materials. The high amount of fines may affect the compressive strength, steel–concrete bond, shrinkage and creep properties. The use of fine materials with pozzolanic or hydraulic properties (instead of inert fillers) may both increase the strength and reduce permeability, leading for instance to higher resistance to chloride penetration [32].

The use of self-compacting concrete in real structures is still modest. Nevertheless, the interest in this material is growing for several advantages in terms of: reduced time for casting, higher quality of concrete regardless of the skills of workers, reduction in health consequences especially due to vibration, better surface finishing of concrete. The cost of *SCC* is to some extent higher than conventional concrete, both due to the raw materials and the strict quality control required to achieve the expected properties. When *SCC* is used, special attention should also be dedicated to the formwork because of the high fluidity of the fresh concrete and the pressure exerted on it. It is interesting to note that the precast concrete industry in many European countries uses SCC on a wide scale, presumably for purely economic reasons.

References

1 Neville, A.M. (1995) *Properties of Concrete*, 4th edn, Longman Group Limited, Harlow.
2 Metha, P.K. and Monteiro, P.J.M. (2006) *Concrete: Microstructure, Properties, and Materials*, 3rd edn, McGrawHill.
3 CEN (2006) EN 206-1. Concrete – Part 1: Specification, Performance, Production and Conformity, European Committee for Standardization.
4 American Concrete Institute (2001) *ACI manual of concrete practice*.
5 CEN (2011) EN 197-1. Cement – Part 1: Composition, Specifications and Conformity Criteria for Common Cements, European Committee for Standardization.
6 CEN (2005) EN 450-1. Fly Ash for Concrete – Part 1: Definition, Specifications and Conformity Criteria, European Committee for Standardization.
7 CEN (2008) EN 12620. Aggregates for Concrete, European Committee for Standardization.
8 CEN (2002) EN 1008. Mixing Water for Concrete – Specification for Sampling, Testing and Assessing the Suitability of Water, Including Water Recovered from Processes in the Concrete Industry, As Mixing Water for Concrete, European Committee for Standardization.
9 Rixom, R. and Mailvaganam, N. (1999) *Chemical Admixtures for Concrete*, 3rd edn, E & FN Spon, London.
10 Collepardi, M. (2006) *The New Concrete*, Tintoretto, Villorba.
11 CEN (2008) EN 934-1. Admixtures for Concrete, Mortar and Grout – Part 1: Common Requirements, European Committee for Standardization, 2008.
12 CEN (2009) EN 934-2. Admixtures for Concrete, Mortar and Grout – Part 2: Concrete Admixtures – Definitions, Requirements, Conformity, Marking and Labelling, European Committee for Standardization.
13 CEN (2004) EN 1992-1-1. Eurocode 2: Design of Concrete Structures – Part 1-1: General Rules and Rules for Buildings, European Committee for Standardization.
14 Day, K.W. (1999) *Concrete Mix Design, Quality Control and Specification*, 2nd edn, E & FN Spon, London.
15 de Larrard, F. (1999) *Concrete Mixture Proportioning – A Scientific Approach*, E & FN Spon, London.
16 Collepardi, M. and Coppola, L. (1996) *Mix Design of Concrete*, ENCO, Spresiano (I) (in Italian).
17 Bertolini, L., Lollini, F., and Redaelli, E. (2011) Comparison of resistance to chloride penetration of different types of concrete through migration and ponding tests, in *Modelling of Corroding Concrete Structures*, RILEM Bookseries (eds C. Andrade and G. Mancini), Springer, pp. 123–135.
18 CEN (2009) EN 13670. Execution of Concrete Structures, European Committee for Standardization.
19 ACI (2011) ACI 222.3R-11. Guide to Design and Construction Practices to Mitigate Corrosion of Reinforcement in Concrete Structures, American Concrete Institute.
20 CEB (1992) *Durable concrete structures*, Bulletin d'information No.183.

21 Collepardi, M., Marcialis, A., and Turriziani, R. (1972) Penetration of chloride ions into cement pastes and concretes. *Journal of American Ceramic Society*, **55**, 534.
22 Collepardi, M., Marcialis, A., and Turriziani, R. (1970) La cinetica di penetrazione degli ioni cloruro nel calcestruzzo. *Il Cemento*, **67** (4), 157–164.
23 Wiebenga, J.G. (1980) Durability of concrete structures along the North Sea coast of the Netherlands, in Performance of concrete in marine environment, ASTM Special publication, SP-65, paper 24, 437, 1980.
24 Polder, R.B. and Larbi, J.A. (1995) Investigation of concrete exposed to North Sea water submersion for 16 years. *Heron*, **40**, 31–56.
25 Bakker, R.F., Wegen, G., and van der Bijen, G. (1994) Reinforced concrete: an assessment of the allowable chloride content, in Proc. of Canmet/ACI Int. Conf. on Durability of Concrete, Nice, 1994.
26 Bamforth, P.B. and Chapman-Andrews, J. (1994) Long term performance of RC elements under UK coastal conditions, in Proc. Int. Conf. on Corrosion and Corrosion Protection of Steel in Concrete, University of Sheffield, 24–29 July 1994, p. 139.
27 Polder, R.B., Bamforth, P.B., Basheer, M., Chapman Andrews, J., Cigna, R., Jafar, M.I., Mazzoni, A., Nolan, E., and Wojtas, H. (1994) Reinforcement corrosion and concrete resistivity–state of art, laboratory and field results, in Proc. Int. Conf. on Corrosion and Corrosion Protection of Steel in Concrete, University of Sheffield, 24–29 July 1994, p. 571.
28 Polder, R.B. and de Rooij, M.R. (2005) Durability of marine concrete structures–field investigations and modelling. *Heron*, **50** (3), 133–143.
29 Bamforth, P.B. (1993) Concrete classification for R.C. structures exposed to marine and other salt-laden environments, in Proc. of Structural Faults and Repair–93, Edinburgh, 29 June–1 July 1993.
30 Umoto, T. and Ozawa, K. (1999) *Recommendation for Self-Compacting Concrete*, Japan Society of Civil Engineers, Tokyo.
31 Okamura, H. and Ouchi, M. (1999) Self-compacting concrete. Development, present use and future, in Proc. Int. Conf. Self-Compacting Concrete, Stockholm, 13–14 September 1999, p. 3.
32 Bertolini, L., Carsana, M., and Redaelli, E. (2006) Influence of mineral additions on durability of self compacting concrete (SCC), in Proc. of the 2nd *fib* Congress, Naples, paper 15–29, 1–10, 5–8 June 2006.

13
Corrosion Inhibitors

In general, reinforced concrete has proved to be successful in terms of both structural performance and durability if the design rules (Chapter 11) and concrete technology (Chapter 12) were adequately considered. Additional protection measures will only be necessary in very aggressive environments or when a very long service life is required. Corrosion inhibitors, thus chemical compounds added as admixtures to the fresh concrete or applied on the surface of hardened concrete, are one possible way to improve the durability of concrete structures.

The use of corrosion inhibitors is of increasing interest as they are claimed to be useful in reinforced concrete not only as preventative measure for new structures (as addition to the mixing water) but also as surface applied inhibitors for preventative and curative purpose. In particular, the easy application from the concrete surface could be an economically interesting alternative to the traditional repair methods as it could increase the lifetime of structures that already show some corrosion attack.

The application of inhibitors on the concrete surface requires the transport of the substance to the rebar where it has to reach a sufficiently high concentration to protect the steel against corrosion or reduce the rate of the ongoing corrosion. In this context only corrosion inhibitors that prolong the service life due to chemical or electrochemical interaction with the reinforcement are considered. Any other substances that may prevent the onset of corrosion or reduce ongoing corrosion by other means, such as surface treatment (e.g., hydrophobation) or additions that reduce the porosity of the concrete (e.g., fly ash, silica fume, waterproofing admixtures, etc.), are not considered to be corrosion inhibitors and are treated in other chapters.

Ten years ago a first state-of-the-art report was published [1]. In this chapter the types of inhibitors presently available and their effectiveness in reinforced concrete structures are outlined, including recent research reports and papers that have been published on laboratory and an increasing number of field studies.

Corrosion of Steel in Concrete: Prevention, Diagnosis, Repair, Second Edition. Luca Bertolini,
Bernhard Elsener, Pietro Pedeferri, Elena Redaelli, and Rob Polder.
© 2013 Wiley-VCH Verlag GmbH & Co. KGaA. Published 2013 by Wiley-VCH Verlag GmbH & Co. KGaA.

13.1
Mechanism of Corrosion Inhibitors

Very often, the long experience with chemicals operating as corrosion inhibitors, for example, in the oil field, gas or petroleum industry, is taken as an example of the successful use of corrosion inhibitors for many decades. This undoubtedly is true and the overwhelming majority of literature on corrosion inhibitors deals with the effects of inhibitors on uniform corrosion, for example, of steel in acidic or neutral solutions, where they can be classified into [2]: (i) *adsorption inhibitors*, acting specifically on the anodic or on the cathodic partial reaction of the corrosion process or on both reactions (mixed inhibitor), (ii) *film-forming inhibitors*, blocking the surface more or less completely, and (iii) *passivators*, favoring the passivation reaction of the steel (e.g., hydroxyl ions).

Regarding inhibition of corrosion of steel in concrete, a completely different situation has to be considered (Chapter 4): steel in concrete usually is passive, and thus protected by a thin film of oxyhydroxides formed spontaneously in the alkaline pore solution (passive film). The mechanistic action of corrosion inhibitors is thus not against uniform corrosion (see above) but localized or pitting corrosion of a passive metal due to the presence of chloride ions or a drop in pH. It is thus obvious that the long and proven track record of inhibitors against general corrosion in acidic or neutral media cannot be the basis to assume that similar compounds should work also for steel in concrete. Indeed, inhibitors for pitting corrosion (the typical situation for steel in chloride-contaminated concrete) are by far less studied [3]. Inhibitors for pitting corrosion can act: (i) by a competitive surface adsorption process of inhibitor and chloride ions, (ii) by increasing or buffering of the pH in the local (pit) environment, and (iii) by competitive migration of inhibitor and chloride ions into the pit (so that low pH and high chloride contents necessary to sustain pit growth cannot develop).

Commercial inhibitors are frequently blends of several compounds, thus the mechanistic action can be multiple and difficult to identify. This will lead to major difficulties in the independent evaluation of corrosion inhibitors that are proposed in commercially available concrete-repair systems.

13.2
Mode of Action of Corrosion Inhibitors

According to the general service-life model of Tuutti (Figure 13.1), corrosion inhibitors *admixed* to the fresh concrete can act in two different ways: these inhibitors can extend the initiation period and/or reduce the corrosion rate after depassivation has occurred. From the point of view of service-life design and the desired extension of the initiation period, mixed-in inhibitors are more reliable, that is, it is easier and more secure to add the inhibitors to the mix then to apply them on the concrete surface and rely on their penetration.

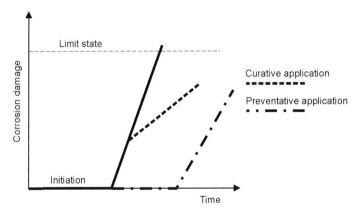

Figure 13.1 Mode of action of inhibitors: preventative application (mostly admixed inhibitors) prolongs initiation time and might reduce corrosion rate, curative application is done with surface-applied inhibitors.

When inhibitors are *applied on the surface* of hardened concrete during the initiation period (before depassivation occurs), the mode of action is in principle identical as for admixed inhibitors, provided the necessary concentration at the rebar is reached. If corrosion has already started, the only possible mode of action is to lower the corrosion rate.

From the point of view of the application, corrosion inhibitors can be used as a preventative or as a curative measure to increase the service life of a reinforced concrete structure. In *preventative* (also called *proactive*) applications inhibitors are used as admixtures to the fresh concrete (calcium nitrite, organic inhibitor blends) or applied on the surface of hardened concrete (MFP, organic inhibitor blends), in which case the inhibitor has to penetrate the concrete cover to reach the steel surface. In *curative* (also called *reactive*) applications inhibitors are applied on the surface of hardened concrete with the goal to reduce the corrosion rate of the rebars. In the following sections the two modes of action are presented.

13.3
Corrosion Inhibitors to Prevent or Delay Corrosion Initiation

The most frequently used technique is adding of the inhibitors as admixtures to the mixing water of concrete for new structures or in repair mortars in order to prevent or at least delay the onset of corrosion.

Calcium nitrite[1] is the most extensively tested admixed corrosion inhibitor [1, 4] and it has, when applied according to the specifications of the producer together

---
1) Commercial product name DCI, marketed by Grace.

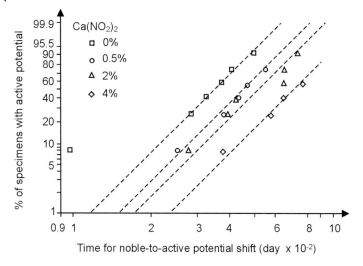

Figure 13.2 Statistical distribution of time to corrosion initiation of steel in mortar specimens with different admixed percentages of Ca(NO$_2$)$_2$ inhibitor with respect to mass of cement, exposed to seawater [5].

with high-quality concrete and sufficient cover, a long and proven track record in the USA, Japan and the Middle East. It is used in parking, marine and highway structures. Calcium nitrite acts as a moderate set accelerator of concrete and normally requires the additional use of water-reducing and retarding admixtures.

Calcium nitrite acts as a passivator due to its oxidizing properties and stabilizes the passive film due to its ability to oxidize ferrous ions (Fe^{2+}) to ferric ions (Fe^{3+}) that form poorly soluble iron oxides. All investigations reported in the literature [1] revealed a critical molar concentration ratio (threshold value) between inhibitor (nitrite) and chloride of about 0.6 (with some variation from 0.5 to 1) in order to prevent the onset of corrosion. This indicates that the inhibitor calcium nitrite has to be present in sufficiently high concentration in order to be effective. The delay in the onset of corrosion (time to corrosion initiation) is statistically distributed around a mean value and depends on the inhibitor dosage (Figure 13.2). This great variation in initiation time is shown also by a comparative study with different inhibitors (Figure 13.3), where only the commercial calcium nitrite with the highest dosage (CN 30) gave a significant improvement. For commercial applications a dosage of 30 l/m^3 of a 30% calcium nitrite solution (DCI) is recommended in order to act against chlorides penetrating from the concrete surface, for example, in bridge deck or parking garage situations. In carbonated concrete nitrite acts as a corrosion inhibitor as well.

Critical issues of the inhibitor calcium nitrite are the risk of increased local corrosion attack in the case of insufficient inhibitor dosage and the problem of leaching out over the long service life of a structure. Nitrite is not allowed as an inhibitor

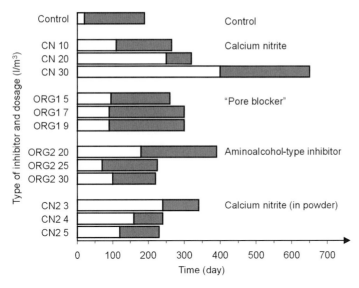

Figure 13.3 Time to corrosion initiation of four steel bars in mortar blocks exposed to cyclic ponding with chloride solutions for different inhibitors admixed to the mortar in three dosages [6]. Gray area indicates the interval of variation.

in all reinforced structures permanently immersed in water for environmental and health reasons.

Sodium mono-fluoro phosphate (Na_2PO_3F, MFP) can be used only as a surface-applied inhibitor due to adverse chemical reaction with fresh concrete [7]. MFP reacts with the calcium ions to form insoluble products such as calcium phosphate and calcium fluoride and the active substance, the PO_3F^- ion, disappears from the pore solution.

Laboratory studies of steel in mortar showed that by applying several intense flushings before the ingress of chlorides [8], the onset of corrosion during the test duration of 90 days could be prevented even at chloride concentrations as high as 2% by mass of cement. A critical molar concentration ratio MFP/chloride greater than 1 had to be achieved, otherwise the reduction in corrosion rate was not significant [8]. MFP acts as corrosion inhibitor in carbonated concrete as well.

The main problem using MFP as surface applied liquid is the penetration to the reinforcement in order to act as inhibitor. Contradictory results are reported [1]: in early field tests insufficient penetration of MFP has been found, partly due to a too dense concrete, a cover depth greater than 45 mm or an insufficient number of MFP applications on the surface. In more recent applications, for example, on the Peney Bridge near Geneva and on concrete buildings and balconies, MFP was applied onto cleaned, dry concrete surfaces in up to ten passes and the concrete was impregnated down to the reinforcement level in a few days or weeks [9]. More recently, the use of a MFP-containing gel on the concrete surface was proposed

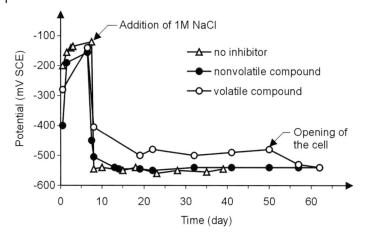

Figure 13.4 Corrosion potentials of rebar samples in saturated Ca(OH)$_2$ solution containing the two components (volatile dimethylethanolamine and nonvolatile benzoate) of a migrating inhibitor [11].

to improve the penetration of MFP. The MFP concentrations that have to be reached at the rebar level are still under discussion.

Since about twenty years ago a new group of organic molecules, especially *alkanolamines* and *amines* and their salts with organic and inorganic acids have been used as components in corrosion inhibitor blends of usually complex formulations that can also contain inorganic compounds [10]. Such inhibitors are produced and marketed by several producers under different trade names.[2] These blends often are not sufficiently well described and may even change in composition with time, so most of the published work has been undertaken with commercially available systems.

Several studies indicate that the inhibitor blends are effective in solutions, whereas pure solvents as dimethylethanolamine are not [1, 11]. A commercial "migrating inhibitor" blend could be fractionated into a volatile (dimethylethanolamine) and a nonvolatile (benzoate) component [11]. For complete prevention of corrosion initiation in saturated Ca(OH)$_2$ solution with 1 M NaCl added, the presence of both components at the steel surface in a concentration ratio of inhibitor/chloride of about one was necessary (Figure 13.4). Modern surface analytical techniques such as XPS have confirmed that for the formation of a significantly thick and protective organic film on iron in alkaline solutions, both components of the commercial inhibitor blend have to be present [12]. Experiments with inhibitor added to mortar showed similar results: the inhibitor blend admixed in the recommended dosage could delay the average time to corrosion initiation of passive steel in mortar subjected to chloride penetration by a factor

---

2) To this group belong, for instance, the commercial products Cortec MCI 2000 and MCI 2002 and Sika Ferrogard 900 and Ferrogard 903.

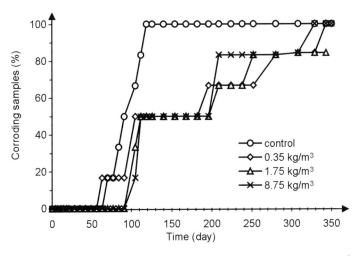

Figure 13.5 Percentage of corroding rebars in mortar with various dosages of admixed MCI 2000 inhibitor versus time of cyclic chloride treatment [11].

of 2–3 (Figure 13.5) but could not prevent completely the onset of corrosion. This result is confirmed by a comparative study where the alkanolamine-based inhibitor (ORG2, Figure 13.3) did not show a significant improvement [6]. The same alkanolamine based commercial inhibitor blend was tested as an admixture in mortar and concrete specimens exposed to chlorides [13]. After one year of testing, corrosion had started in specimens with w/c 0.6, the chloride threshold values for the inhibitor-containing specimens were in all cases higher (4–6% chloride by mass of cement) compared to the control samples (1–3% Cl$^-$). A prolongation of the initiation time of corrosion of about a factor 3 was reported for commercial inhibitor blends MCI 2022 and MCI 2021 applied on concrete (cover 25 mm) partially immersed in 3.5% NaCl solution [14]. A slight increase of the critical chloride content for corrosion initiation was reported for concrete exposed to three weeks ponding cycles in the presence of commercial inhibitor blends using the recommended dosage of producers [15]. On the contrary, results from a long-term study of admixed organic corrosion inhibitors (Ferrogard 903 and Rheocrete 222) in concrete with a w/c ratio of 0.41 exposed to saltwater showed that the admixed organic inhibitors did not or only slightly prolong the time to corrosion initiation [16]. Based on a ten-year exposure test with the same organic inhibitors in another study it was concluded that the inhibitors have not shown consistent indications of corrosion protection [17].

As summarized previously [18], mixed-in inhibitors can delay the initiation of pitting corrosion on passive steel in alkaline solutions or mortar and concrete. In order to be effective, all investigations regardless of the type of inhibitor seem to indicate that a critical molar ratio inhibitor/chloride of about 1 has to be exceeded. This implies that quite high inhibitor concentrations have to be present in the pore water in order to act against chlorides penetrating from the concrete surface. To have as little chloride ingress as possible and thus to avoid the use of excessively high inhibitor concentrations, the use of admixed inhibitors is recommended only combined with high-quality concrete [4]. In addition to the results presented above, after the initiation of corrosion the mixed-in organic inhibitors have in general been reported to be ineffective or only slightly effective in reducing the corrosion rate [19].

13.4
Corrosion Inhibitors to Reduce the Propagation Rate of Corrosion

A surface treatment with subsequent transport of the inhibitor to the corroding steel with the effect to stop or at least reduce ongoing corrosion would be an interesting and potentially cost-effective way to mitigate corrosion in existing structures. Several laboratory and field tests have been performed to investigate this particular situation. As surface applied inhibitors mainly organic "migrating inhibitors", MFP and recently also calcium nitrite are proposed [1].

In an independent study, 15% by mass solutions of *mono-fluoro phosphate* were applied repeatedly to reinforced concrete specimens (w/c 0.65, cover thickness 12 mm) with various levels of chloride contamination. The embedded bars, precorroded under cyclic wetting–drying conditions for about 6 months prior to the MFP treatment, showed a reduction in corrosion rate only at low chloride concentrations (Figure 13.6) [20]. Repeated drying and MFP-immersion cycles have been found to be a suitable method to allow the penetration of the inhibitor to the steel, but high concentrations and long treatments are needed to significantly reduce active corrosion.

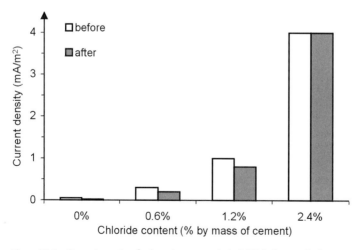

Figure 13.6 Corrosion rate of rebars in mortar (w/c 0.65) before and after treatment with MFP (sodium mono-fluoro phosphate) [20].

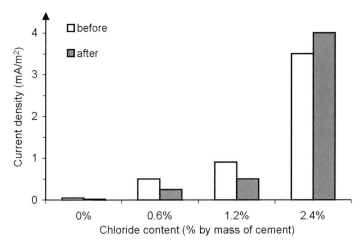

Figure 13.7 Corrosion rate of rebars in mortar (w/c 0.65) before and after treatment with proprietary alkanolamine inhibitor blend [20].

Alkanolamine-based inhibitors have been tested in similar conditions. For ongoing chloride induced corrosion in mortar specimens with a chloride level of about 1–2% by mass of cement, no reduction in corrosion rate was found (Figure 13.7) except at low chloride concentrations. This is confirmed by two other studies [1, 6, 21]: precorroded rebars in mortar ($w/c = 0.75$, cover thickness 25 mm) did not show any detectable effect on the corrosion rate of embedded steel once active corrosion had been initiated, despite the specimens having low cover and porous mortar [21]. In good-quality concrete with admixed chlorides surface applied inhibitors showed no significant reduction in corrosion rate. In a more recent study [22] three organic inhibitors on the basis of aminoalcohol, aminocarboxylate and siloxan, respectively, were applied on the surface of large concrete specimens with precorroded reinforcement (2 and 4% Cl^- by mass of cement) and different cover depth. None of the three inhibitor systems could reduce ongoing chloride-induced corrosion, not even at 10 mm cover depth. It seems that for "penetrating" or "migrating" inhibitors the favorable effects found in solution do not occur when applied to hardened mortar or concrete laboratory specimens with ongoing steel corrosion, as stated clearly in [23] "when applied to the concrete surface, neither inhibitor penetrated the concrete to reach the steel reinforcement". An extensive test and trial series with surface applied inhibitors in the framework of the EU SAMARIS (Sustainable and Advanced Materials for Road Infrastructure) project [24, 25] studied the inhibitors Ferrogard 903 and MFP and concluded "the circumstances under which corrosion inhibitors are most effective represent a combination of factors that include the ease with which the inhibitors may be surface-adsorbed, concrete permeability, chloride level and state of corroded reinforcement at time of repair. These factors can vary greatly from project to project. Thus inhibitors may be very effective in some cases but would represent an inappropriate strategy in other cases". It is thus necessary to look for information regarding the transport of inhibitor blends in mortar or concrete.

13.5
Transport of the Inhibitor into Mortar or Concrete

It is claimed for several inorganic and organic inhibitor blends that they will be carried by water or by vapor-phase migration into the proximity of the reinforcing steel [1], however great discrepancies in the measured transport rates were found. This might partially be due to the different experimental set-up (humidity) and measuring techniques used, but in general it is difficult to determine the transport rate of an inhibitor blend of unknown composition. A key issue in determining the transport of inhibitor blends in mortar and concrete is the analytical methodology, thus an accurate method for their detection and quantification. Ion chromatic analysis has been found to be well suited to determine a variety of amines, alkanolamines and associated anions in concrete/mortar pore solutions or digests [26]. In practice, the method of inhibitor determination is not standardized.

A detailed study on the transport of a proprietary aminoalcohol-based inhibitor into cement paste and mortar showed that both the amount and the rate of inhibitor ingress into alkaline cement paste is higher for the pure aminoalcohol compared to the inhibitor blend also containing phosphates (Figure 13.8) [27]. This discrepancy could be explained by a reaction of the inorganic phosphate component with calcium ions in the alkaline cement paste, blocking further ingress of

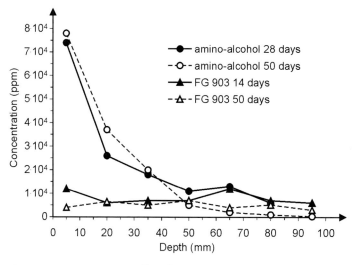

Figure 13.8 Concentration profiles of a proprietary aminoalcohol-based inhibitor (Sika Ferrogard 903) into alkaline cement paste [27].

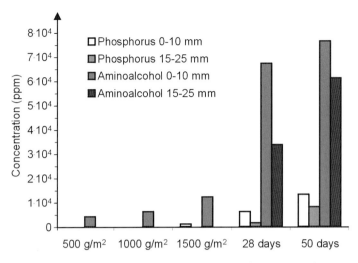

Figure 13.9 Concentration of inhibitor components (of Sika Ferrogard 903) at two depth intervals as a function of application amount (several brushes in order to obtain 500–1500 g/m²) or time of ponding in fully carbonated concrete cores [27].

the inhibitor. To avoid a reaction with calcium ions, the transport of the inhibitor was studied on cores taken from a 100-year-old fully carbonated concrete structure, varying the dosage and the way of inhibitor application (Figure 13.9). The recommended dosage (500 g/m²) and way of application (several brushings) led to only a moderate concentration of the aminoalcohol in the first 15 mm. An increase in the dosage to 1500 g/m² increased the aminoalcohol concentration, but the penetration depth remained low. Only ponding for 28 or 50 days resulted in the penetration of a significant inhibitor concentration (both amino alcohol and phosphate) at depths higher than 30 mm [27].

The penetration of the aminoalcohol based inhibitor Ferrogard 903 was studied in the framework of the SAMARIS project [24, 25]. The penetration was much easier in dry concrete; the concentrations were stable in the period from 7 to 117 days at about 12 000 ppm in a depth of 0–10 mm. In the depth range 13–23 mm the concentrations were always less than 10% of that in the depth 0–10 mm. It was not clear if a sufficiently high concentration remained at the level of reinforcement [24, 25].

Another problem arises because often only one, the most volatile, component of the inhibitor blend can be analyzed using an amine electrode, as in the case of proprietary migrating corrosion inhibitors [1, 11]. In this way, the diffusion of the volatile part of the inhibitor blend can be measured but no information on the diffusion of the nonvolatile fraction can be obtained. Both components of an inhibitor blend are needed at the steel surface to get an inhibiting effect (Figure 13.4). However, only the volatile component diffuses easily through the porous concrete. This may explain the discrepancy between solution experiments and mortar or field tests [1, 6, 11]. It has also been reported that the volatile component of organic inhibitor blends evaporates [1, 11].

Recently, it has been proposed to enhance the ingress of certain inhibitors into concrete through the application of an electric field, similarly as for electrochemical techniques (Chapter 20), with an anolyte consisting of an aqueous solution containing the inhibitor itself. This method is referred to as *electrochemical inhibitor injection (EII)* [28]. It has provided encouraging laboratory results when applied to carbonated concrete specimens. Presently, however, the effectiveness of this method requires further investigations.

13.6
Field Tests and Experience with Corrosion Inhibitors

Only calcium nitrite, introduced more than forty years ago, has a long and proven track record as a corrosion inhibitor for reinforced concrete [1, 4]. MFP and alkanolamine-based organic inhibitor blends are increasingly used but unfortunately most of the commercial applications lack rigorous control of the inhibitor effect. One of the very few comparative field tests on chloride-contaminated concrete studied MFP and a proprietary alkanolamine inhibitor added in the side walls of a tunnel [29]. Condition assessment prior to the inhibitor application showed chloride contents up to 2% at 15 mm depth at 1.5 m height above the traffic lane. High chloride contents were associated with negative half-cell potentials and corroded reinforcement. More than $500 g/m^2$ of Ferrogard 903 and about $2.4 l/m^2$ of MFP were applied by the providers. The measurements of macrocell currents and half-cell potential mapping (Figure 13.10) revealed that both inhibitors were virtually ineffective at the chloride concentrations of 1–2% by mass of cement present [30]; this occurred despite the fact that the inhibitor concentrations at the depth of reinforcement was higher than the threshold level.

Other field tests with proprietary vapor-phase inhibitors [31] in a parking garage with chloride-contaminated precast slabs did not show encouraging results. Corrosion-rate measurements showed a reduction of 60% in areas with initially intense corrosion but also an increase of corrosion rate in areas with low corrosion rates. On structures from 1960 with an admixed chloride content higher than 1% by mass of cement, featuring already patch repairs, a three-year corrosion rate survey showed lower corrosion rates in the treated areas compared to untreated ones, but cracking and spalling also increased in the treated areas [32].

13.7
Critical Evaluation of Corrosion Inhibitors

Assuming that the inhibitor action in laboratory experiments has been established, there remain two critical points for a successful and reliable application on reinforced concrete structures:

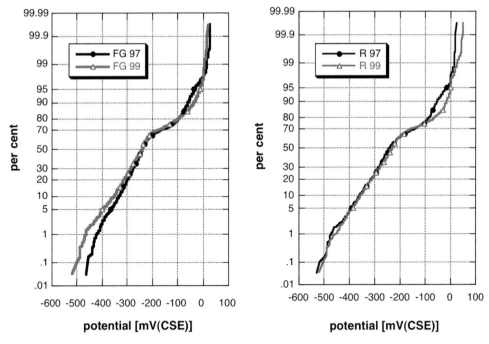

Figure 13.10 Cumulative frequency distribution of half-cell potentials measured on the chloride-contaminated test fields for surface-applied inhibitors before (97) and two years after (99) the application: Ferrogard 903 (a), reference field (b) [30].

- the inhibitor has to be present at the reinforcing steel in sufficiently high concentration with respect to the aggressive (chloride) ions over a long period of time;
- the inhibitor action on corrosion of steel in concrete should be measurable.

Concentration Dependence The available literature reports a concentration-dependent effect of inhibitors, suggesting that a critical ratio inhibitor/chloride has to be exceeded (as described above). For new structures, the inhibitor dosage thus has to be specified with respect to the expected chloride level for the design service life of the structure. Surface applied inhibitors on existing structures may present even more difficulties in achieving the necessary concentration at the rebar level: chloride contamination or carbonation may vary strongly along the surface, the cover thickness and permeability of the concrete may vary as well, and the inhibitor may react with pore solution components. It is crucial to specify the critical concentration to be achieved at the rebar level and not, as often occurs in the application notes of commercial surface applied inhibitors, simply an average volume (or weight) of inhibitor solution to be applied per square meter of concrete. This is usually omitted, in part due to the lack of analytical methods to measure the inhibitor concentration. Regarding long-term durability, it has also to be taken into account that inhibitors may be leached out from the concrete or evaporate.

Measurement and Control of Inhibitor Action   One of the main difficulties in evaluating the performance of inhibitors is to assess the inhibitor action on rebar corrosion "onsite". The interpretation of half-cell potential measurements (Chapter 16) may present difficulties due to changes in the concrete resistivity. Further, a reduction of the corrosion rate due to an inhibitor action may not be reflected straightforwardly in the half-cell potential: potentials may become more negative or more positive after inhibitor application, depending on the mechanism of the inhibitor action. Shifts in the half-cell potential may also occur due to wetting and drying of the concrete [33]. Linear polarization resistance measurements are considered suitable for onsite testing [32], but results of corrosion-rate measurements onsite depend on the type of device used for the measurements and can be interpreted so far only by specialists (Chapter 16). The main problems are the daily and seasonal changes of the corrosion rate with temperature and concrete humidity, making it difficult to evaluate the inhibitor action. Macrocell current measurements between isolated anodes (located and instrumented before inhibitor application) and the surrounding cathodes may give the most indicative results [29] but can be installed only at specific test sites.

13.8
Effectiveness of Corrosion Inhibitors

The available literature allows concluding that admixed inhibitors with the correct dosage in relation to the (anticipated) chloride content can strongly delay the onset of chloride-induced corrosion. Once corrosion has started, no significant reduction in corrosion rate has been found. The overall performance of surface applied organic and inorganic corrosion inhibitors intended to reduce ongoing chloride-induced corrosion cannot be considered positive. For the case of corrosion due to carbonation there remain at least some doubts.

The use of corrosion inhibitors could be a promising technique in restoring reinforced concrete structures, offering benefits such as reduced costs and inconvenience of repairs. It has, however, to be taken into account that the use of corrosion inhibitors in repair systems is far less well established than their application as admixtures in new structures. The performance of proprietary corrosion inhibitors in repair systems marketed under different trade names is not yet sufficiently documented by independent research work, especially when considering field tests.

References

1 Elsener, B. (2001) *Corrosion Inhibitors for Steel in Concrete – State of the Art Report*, European Federation of Corrosion Publication number 35, The Institute of Materials, Maney Publishing, London.

2 Trabanelli, G. (1986) Corrosion inhibitors, in *Corrosion Mechanism* (ed. F. Mansfeld), Marcel Dekker, New York, pp. 119–163, Chapter 3.
3 Sastri, V.S. (1999) *Corrosion Inhibitors, Principles and Applications*, John Wiley & Sons, pp. 567–703.
4 Berke, N.S. and Weil, T.G. (1994) World wide review of corrosion inhibitors in concrete, in *Advances in Concrete Technology* (ed. V.M. Malhotra), CANMET, Ottawa, pp. 899–1022.
5 Hartt, W.H. and Rosenberg, A.M. (1989) *Influence of $Ca(NO_2)_2$ on Sea Water Corrosion of Reinforcing Steel in Concrete*, American Concrete Institute, Detroit, SP 65-33, pp. 609–622.
6 Trépanier, S.M., Hope, B.B., and Hansson, C.M. (2001) Corrosion inhibitors in concrete. Part III: effect on time to chloride-induced corrosion initiation and subsequent corrosion rates of steel in mortar. *Cement and Concrete Research*, **31**, 713–718.
7 Hynes, M. and Malric, B. (1997) Use of migratory corrosion inhibitors. *Construction Repair*, **11** (4), 10–15.
8 Alonso, C., Andrade, C., Acha, M., and Malric, B. (1992) Preliminary testing of Na_2PO_3F as a curative corrosion inhibitor for steel reinforcements in concrete. *Cement and Concrete Research*, **22** (5), 869–881.
9 Annen, P. and Malric, B. (1996) Surface applied inhibitor in rehabilitation of Peney Bridge, Geneva (CH), in *Bridge Management 3* (eds E. Harding, G.A.R. Parke, and M.J. Ryall), E & FN Spon, London, p. 437.
10 Mäder, U. (1994) A new class of corrosion inhibitors, in *Corrosion and Corrosion Protection of Steel in Concrete*, vol. 2 (ed. N. Swamy), Sheffield Academic Press, p. 851.
11 Elsner, B., Büchler, M., Stalder, F., and Böhni, H. (1999) A migrating corrosion inhibitor blend for reinforced concrete – Part 1: prevention of corrosion. *Corrosion*, **55**, 1155–1163.
12 Rossi, A., Elsener, B., Textor, M., and Spencer, N.D. (1997) Combined XPS and ToF-SIMS analyses in the study of inhibitor function – organic films on iron. *Analysis*, **25** (5), M30.
13 Laamanen, P.H. and Byfors, K. (1996) Corrosion inhibitors in concrete – alkanolamine based inhibitors, *Nordic Concrete Research*, 19, 2/1996, Norsk Betongforengingk Oslo.
14 Bavarian, B. and Reiner, L. (2007) Corrosion protection of steel rebar in concrete using migrating corrosion inhibitors, in *Corrosion of Reinforcement in Concrete. Mechanisms, Monitoring, Inhibitors and Rehabilitation Techniques*, European Federation of Corrosion Publication number 38 (eds M. Raupach, B. Elsener, R. Polder, and J. Mietz), Woodhead Publishing Limited, Cambridge, pp. 239–249.
15 Bolzoni, F., Fumagalli, G., Lazzari, L., Ormellese, M., and Pedeferri, M.P. (2007) Mixed-in inhibitors for concrete structures, in *Corrosion of Reinforcement in Concrete. Mechanisms, Monitoring, Inhibitors and Rehabilitation Techniques*, European Federation of Corrosion Publication number 38 (eds M. Raupach, B. Elsener, R. Polder, and J. Mietz), Woodhead Publishing Limited, Cambridge, pp. 185–202.
16 Pereira, E.V., Figueira, R.B., Salta, M.M., and Fonseca, I.T.E. (2010) Long-term efficiency of two organic corrosion inhibitors for reinforced concrete. *Materials Science Forum*, **636–637**, 1059–1064.
17 Kessler, P.J., Powers, R.G., Paredes, M.A., Sagüés, A.A., and Virmani, Y. (2007) Corrosion inhibitors in concrete – results of a ten year study. CORROSION 2007, NACE conference, paper 07293.
18 Elsener, B. and Cigna, R. (2003) Mixed-in inhibitors, in *COST Action 521, Corrosion of Steel in Reinforced Concrete Structures, Final Report* (eds R. Cigna, C. Andrade, U. Nürnberger, R. Polder, R. Weydert, and E. Seitz), European Communities, Luxembourg, Publication EUR 20599, pp. 40–51.
19 Söylev, T.A. and Richardson, M.G. (2008) Corrosion inhibitors for steel in concrete: state of the art report. *Construction and Building Materials*, **22** (4), 609–622.
20 Page, C.L., Ngala, V.T., and Page, M.M. (2000) Corrosion inhibitors in concrete

repair systems. *Magazine of Concrete Research*, **52**, 25–37.
21 Elsener, B., Bürchler, M., Stalder, F., and Böhni, H. (2000) A migrating corrosion inhibitor blend for reinforced concrete – Part 2: inhibitor as repair strategy. *Corrosion*, **56**, 727.
22 Hunkeler, F., Mülan, B., and Ungricht, H. (2010) Korrosionsinhibitoren für die Instandsetzung chloridversuchter Stahlbetonbauten. *Beton- und Stahlbetonbau*, **106**, 187–196.
23 Sharp, S.R. (2004) Evaluation of two corrosion inhibitors using two surface application methods for reinforced concrete structures. Final Report Virginia Transportation Research Council VTRC 05-R16.
24 Richardson, M.G. and McNally, C. (2007) *Effective use of corrosion inhibitors in highway structures*, IABSE report vol. 93.
25 samaris.net (accessed 07 April 2011), deliverable D21, D25a.
26 Page, M.M., Page, C.L., Shaw, S.J., and Sawada, S. (2005) Ion chromatographic analysis of amines, alkanolamines and associated anions in concrete. *Journal of Separation Science*, **28** (5), 471–476.
27 Tritthart, J. (2003) Transport of a surface-applied corrosion inhibitor in cement paste and concrete. *Cement and Concrete Research*, **33** (6), 829–834.
28 Kubo, J., Sawada, S., Page, C.L., and Page, M.M. (2008) Electrochemical inhibitor injection for control of reinforcement corrosion in carbonated concrete. *Materials and Corrosion*, **59** (2), 107–114.
29 Schiegg, Y., Hunkeler, F., and Ungricht, H. (2000) The effectiveness of corrosion inhibitors – a field study, in Proc. IABSE Congress Structural Engineering for Meeting Urban Transportation Challenges, Lucerne 18–21 September 2000 (on CD).
30 Schiegg, Y., Hunkler, F., and Ungricht, H. (2007) Effectiveness of corrosion inhibitors – a field study, in *Corrosion of Reinforcement in Concrete. Mechanisms, Monitoring, Inhibitors and Rehabilitation Techniques*, European Federation of Corrosion Publication number 38 (eds M. Raupach, B. Elsener, R. Polder, and J. Mietz), Woodhead Publishing Limited, Cambridge, pp. 226–238.
31 Broomfield, J.P. (2000) Results of long term monitoring of corrosion inhibitors applied to corroding reinforced concrete structures, Corrosion/2000, paper 0791, NACE International Houston (TX) USA.
32 Broomfield, J.P. (1997) The pros and cons of corrosion inhibitors, Construction Repair, July/August 1997, 16.
33 Elsener, B. (2001) Half-cell potential mapping to assess repair work on RC structures. *Construction and Building Materials*, **15** (2–3), 133–139.

14
Surface Protection Systems

Surface protection systems are applied to new structures as a preventative measure, to existing structures where the need for future protection is anticipated, and to repaired structures in order to improve the service life of the repairs. Surface treatments can also be applied for aesthetical reasons, including to mask the visible effect of repairs. Excluding purely aesthetical applications and following the European standard EN 1504-2 [1], the term used here for these treatments is surface protection systems (SPS). Of the many types of surface protection systems of concrete, only those aimed at providing protection against corrosion of reinforcement will be mentioned here. This places coatings for protection against chemical attack and treatments for improving physical resistance outside the present scope.

14.1
General Remarks

Surface protection system is a term relating to a wide range of materials or treatments applied to the surface of a concrete structure, for which other terms are also used, such as concrete sealer [2–6]. Recently, these systems have been the subject of harmonized European standardization, in particular by EN 1504-2 [1]. It should be noted that EN 1504-2 describes principles and methods with regard to surface protection systems, and specifies properties, testing and performance requirements, based on testing by the manufacturer. EN 1504-2 itself does not contain test descriptions, but refers to a network of standards where those tests are defined, including various series of standards, such as EN 13579, 13580, and 13581 that describe test methods for hydrophobic treatment; and EN 1062-x and EN ISO 28yz that do so for paints and varnishes.

The composition of surface protection systems varies from polymeric to cementitious, their dimension from atomic layers to centimeters, their location from "on top" of the concrete surface to completely inside the pores. They share at least one of two objectives: to make the concrete cover zone less permeable to aggressive substances or to reduce its moisture content and thereby to increase the concrete resistivity. Both effects will prolong the service life of the structure. Within the

Corrosion of Steel in Concrete: Prevention, Diagnosis, Repair, Second Edition. Luca Bertolini,
Bernhard Elsener, Pietro Pedeferri, Elena Redaelli, and Rob Polder.
© 2013 Wiley-VCH Verlag GmbH & Co. KGaA. Published 2013 by Wiley-VCH Verlag GmbH & Co. KGaA.

scope of this chapter, EN 1504-2 distinguishes three principles: protection against ingress (abbreviated PI), moisture control (MC), and increasing resistivity by limiting moisture content (IR). The latter two are strongly related. These three principles can be achieved using either of three "surface protective methods", that is, types of surface treatment: principle PI by hydrophobic impregnation (denoted H), impregnation (I), and coating (C); and both principles MC and IR by hydrophobic impregnation (H) and coating (C).

These types of surface protection differ in almost any other aspect, which is why the different types are treated here in separate sections.

First, however, a central issue is raised: density versus openness. *Dense* surface treatments may be very effective in slowing down the penetration of aggressive substances (good barrier properties), but usually they also slow down the evaporation of water from the concrete. This may raise problems of vapor pressure, causing loss of bond between the concrete and the surface treatment layer (terminating or at least reducing the protection provided), or even frost damage to the concrete. On the other hand, surface protection systems that are *open* for water vapor, are also open to a certain extent for aggressive gases like carbon dioxide, so they are not very effective against carbonation. Surface protection systems open to water vapor may have a longer service life and have no or slower loss of bond, but their effectiveness may be less than for dense systems. It seems that there is a trade-off between density (good barrier properties) and openness (good durability) of the surface treatment material. Since the various surface protection systems behave so differently, it is possible that a particular surface treatment may be appropriate for a certain type of exposure but not suitable for a different set of aggressive conditions. The final choice will depend on various reasons related to the structure, the owner, the local market, etc. There also seems to be an element of fashion: in the 1980s, dense coatings (in particular epoxy based) were used to a large extent. In the 1990s, more open coating types like acrylates were used. The application of hydrophobic treatment (silanes and siloxanes) seems to have increased in the last twenty years.

As illustrated schematically in Figure 14.1, it is possible to distinguish four principal types of surface treatments for concrete. As mentioned above, EN 1504-2 identifies three so-called methods, with the fourth mentioned (cementitious overlays) here being outside its scope. These classes or methods are: (*a*) organic coatings that form a continuous film (method C according to EN 1504-2), (*b*) hydrophobic treatments or impregnations that line the surface of the pores (method H), (*c*) treatments that fill the capillary pores, also called impregnation (I) and (*d*) cementitious layers [7].

EN 1504-2 defines performance characteristics "for intended uses", that is, some characteristics are required for all types of use of the SPS, and some are only required for certain intended uses. Furthermore, it states that selection of SPS shall be based on assessment of the cause of deterioration, by consecutively considering the deterioration phenomena, selecting the principle for protection, selecting the method (type of SPS) and finally selecting the characteristics of the particular SPS to be used. Table 14.1 lists the performance characteristics as pro-

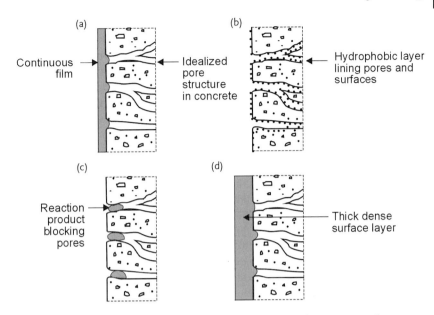

Figure 14.1 Schematic representation of the different types of surface treatment of concrete: (a) organic coating, (b) pore-lining treatments, (c) pore-blocking treatments, (d) thick cementitious coating, shotcrete or rendering [7].

vided by EN 1504-2. As this chapter is organized according to the principal types of SPS, these requirements will be discussed at relevant points in the text.

Table 14.2 lists the principal types of surface treatments and their subtypes [2, 3]. Combination of two types of surface treatment is possible, for example, by first applying a hydrophobic agent as a primer and subsequently an (organic) acrylic coating, providing what is called a dual system [6]. Detailed chemical information on polymeric surface protection systems is given by Cady [6].

14.2
Organic Coatings

Organic coatings are used to block the penetration of carbon dioxide or chloride ions. They form a continuous polymeric film on the surface of the concrete, of a thickness ranging from 100 to 300 μm. The binder can already be present in the liquid paint as a polymer, or the polymer can be formed due to chemical reactions between monomeric components that are mixed just before application. Modern coating systems are usually built up of several layers applied consecutively. They are compatible with the alkalinity of the concrete and are based on various types of polymers (e.g., acrylate, polyurethane, epoxy), pigments and additives, and are made suitable for application by the addition of solvents or diluents.

Table 14.1 Performance characteristics required by EN 1504-2 for surface protective methods H (hydrophobic impregnation), I (impregnation), and C (coating) [1].

	Principle	Ingress protection			Moisture control		Increasing resistivity	
Test methods defined in	Performance characteristics	H	I	C	H	C	H	C
EN 12617-1	Linear shrinkage			○		○		○
EN 1770	Coefficient of thermal expansion			○		○		○
EN ISO 2409	Adhesion by cross-cut test			○		○		○
EN 1062-6	Permeability to CO_2			●				●
EN ISO 7783-1 EN ISO 7783-2	Permeability to water vapor		○	●		●		●
EN 1062-3	Capillary absorption and permeability to water	●	●		●			●
EN 13687-1	Freeze–thaw cycling with deicing salt immersion		○	○		○		○
EN 13687-2	Thunder-shower cycling (thermal shock)		○	○		○		○
EN 13687-3	Thermal cycling without deicing salt impact		○	○		○		○
EN 1062-11	Aging: 7 days at 70 °C		○	○		○		○
EN 13687-5	Resistance to thermal shock			○				
EN ISO 2812-1	Chemical resistance		○	○				
EN 1062-7	Crack bridging ability			○		○		○
EN 1542	Adhesion strength by pull-off test		○	●		●		●
EN 13501-1	Fire classification of construction products and building elements – Part 1: Classification using test data from reaction to fire test		○	○		○		○
EN 13581	Resistance against freeze–thaw salt stress of impregnated hydrophobic concrete (Determination of loss of mass)	○			○		○	
EN 13036-4	Slip/skid resistance		○	○		○		○
various	Depth of penetration	●	●		●		●	
EN 1062-11	Behavior after artificial weathering			○		○		○
EN 1081	Antistatic behavior			○		○		○
EN 13578	Adhesion on wet concrete			○		○		
EN 13580	Water absorption and resistance to alkali test for hydrophobic impregnation	●			●		●	
EN 13579	Drying rate for hydrophobic impregnation	●			●		●	
subject to national standards and regulations	Diffusion of chloride ions	○	○	○				

● characteristic for all intended uses.
○ characteristic for certain intended uses within the scope of EN 1504-9.

Table 14.2 Classification of surface treatments [1, 2].

Classification material		Form of liquid paint	Curing
Film forming Coatings	Acrylic	Catalyzed, solvented (sealer)	Loss of solvent and chemical
		Solvented	Loss of solvent
		Aqueous dispersion	Drying
		Cementitious aqueous dispersion	Drying
	Butadiene copolymer	Aqueous dispersion	Drying
	Chlorinated rubber	Solvented	Loss of solvent
	Epoxy resin	Catalyzed, solvented (sealer)	Loss of solvent and chemical
		Catalyzed, aqueous dispersion	Drying and chemical
		Catalyzed, solventless	Chemical
	Oleoresinous	Solvented	Loss of solvent and oxidation
	Polyester resin	Catalyzed, solvented (sealer)	Loss of solvent and chemical
	Polyethylene copolymer	Solvented	Loss of solvent
		Aqueous dispersion	Drying
	Polyurethane	Solvented (sealer)	Loss of solvent and moisture
		Catalyzed, solvented (sealer)	Chemical
		Catalyzed	Chemical
	Vinyl	Solvented sealer	Loss of solvent
		Aqueous dispersion	Drying
Hydrophobic pore liner	Silicones	Solvented	Loss of solvent and reaction
	Siloxane	Solvented/solventless/aqueous	Loss of solvent and reaction
	Silane	Solvented/solventless	Loss of solvent and reaction
Pore blocker	Silicate	Aqueous dispersion or solvented	Reaction
	Silicofluoride	Aqueous solution, gas	Reaction
	Crystal growth materials	In cementitious slurry	Reaction
Rendering	Plain and polymer-modified cement-based mortars		Cement hydration

The effectiveness of a coating is related to the absence of pores or flaws and it increases in proportion to its thickness. The application requires thorough surface preparation, because a durable bond to the substrate is essential for coating effectiveness. The same applies to filling blowholes in the concrete surface.

Organic coatings may vary from very dense to rather open for water vapor. A dense coating (ideally without flaws), based on epoxy, polyurethane or chlorinated rubber polymers, may block ingress of aggressive species. Nevertheless, the presence of such a layer strongly hinders the evaporation of the moisture that is present in the concrete at the time of treatment. This situation can lead to loss of adhesion to the concrete and thus to a loss of effectiveness of the coating.

Nowadays, the most widely used coatings are relatively open, meaning that they strongly reduce the ingress of water from the environment, but allow the evaporation of moisture from the inside. Consequently the concrete (in the course of time) will reach lower values of humidity in pseudoequilibrium with the atmosphere in which it operates, in particular in wetting–drying situations, as illustrated in Figure 14.2. Acrylates are the most important materials in this group. Their service life under normal operating conditions may extend beyond 10 years. It seems that with this type of coatings, a favorable trade-off has been found between density and durability.

14.2.1
Properties and Testing

Organic coatings have been subjected to testing by a wide range of methods. Within EN 1504-2 there is a full list of performance requirements (Table 14.1), including separate standards describing the test methods. The following text

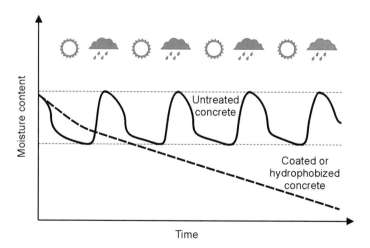

Figure 14.2 Schematic development of moisture content of concrete under wetting–drying cycles, for coated or hydrophobized and untreated concrete.

reports studies that provide background information, concentrating on some of the results reported for the effect on the penetration of carbon dioxide and chloride ions.

Bassi and Davies [8] described a method for testing carbon dioxide permeation through a ceramic tile coated with the material under study. The carbon dioxide resistance may be expressed in various units, of which the equivalent air thickness is most commonly used. An empirical criterion is that a coating should have the same resistance to CO_2 transport as 50 m of still air. A wide range of organic coatings (plus a silicate and a polymer-modified cementitious coating) was tested. Most coatings had an acceptable carbon dioxide resistance immediately after application. However, it dropped significantly during artificial or natural weathering. In fact, all coatings tested had a carbon dioxide resistance of less than 50 m equivalent air thickness after one year of natural weathering or after 1000 h of artificial weathering [8]. Polymeric emulsion coatings were less resistant to carbon dioxide permeation and suffered more strongly from weathering than chemically reacting materials. EN 1504-2 requires CO_2 permeability testing by the producer (according to EN 1062-2) for coatings used according to principle I (ingress protection) and "for all intended uses", thus as a general requirement.

Buenfeld and Zhang [9] tested a wide range of materials including two organic sealers, a silane, two organic coatings, and a polymer-modified cementitious coating applied to cement mortar discs, using steady-state chloride diffusion tests (Figure 2.4). It appeared that chloride diffusion was so slow with the polyurethane and acrylic coatings that it did not reach steady state within one year. Chloride diffusion was slowed down by one to three orders of magnitude by the silane, the polyurethane sealer, and the cementitious coating. The acrylic sealer did not have a positive effect.

Rodrigues and coworkers [10] tested polyacrylates and polymethacrylates with various pigments and textures borne in water or organic solvents. Methods used were water permeability (coating on concrete or brick) and steady-state chloride diffusion (coating on ceramic tile) in the lab and nonsteady state chloride diffusion (coating on concrete) up to two years at a marine exposure site. The coatings reduced the water permeability by a factor 4 to 50 and the steady-state chloride permeability from 2 to 10 000 times. In both tests, methacrylate coatings were less permeable than acrylate coatings. In the field, the chloride surface content was reduced 5 to 10 times by the methacrylate coating, but hardly by the acrylate coatings. The effect on the chloride diffusion coefficient was small.

EN 1504-2 requires capillary absorption and water permeability testing (according to EN 1062-3) and water-vapor permeability testing (according to EN ISO 7783-1 and -2) for coatings used according to principle I (ingress protection), MC (moisture control) and IR (increasing resistivity) and "for all intended uses", thus as general requirements. Water-vapor requirements are classified in three classes, going from permeable to not permeable. It requires chloride diffusion testing (no European standardized method) only for specific uses. This would be an obviously useful requirement where a coating has to protect concrete against chloride ingress, for example, from marine splash. Further requirements involve crack bridging ability, adhesion on wet concrete and resistance against artificial weathering.

An interesting method to demonstrate the density or openness to water vapor was published by Vogelsang and coworkers [11]. They prepared mortar specimens with embedded steel electrodes following Tritthart and Geymayer [12]. The specimens were either wet or dry, then coated with two types of coating and together with uncoated specimens, exposed in either a dry or wet environment. The resistance measured between the electrodes indicated a strong change of resistivity as a function of time, providing information on the rate of evaporation or water uptake from or by uncoated specimen or through the coatings. It was found that such resistivity and moisture changes occurred most rapidly in uncoated specimens, less rapidly in specimens coated with water-based acrylate, and most slowly in solvent-based epoxy-coated specimens.

Various researchers have tested the effect of coatings (and other surface treatments) on actively corroding reinforcement. Due to the extreme variation of materials (both coatings and concrete) and test methods and conditions, it is beyond the present scope to summarize that work. The overall impression is that coatings do not effectively reduce the rate of chloride-induced corrosion, see for example, [13]. Subject to many practical influences, it is possible that coatings reduce the rate of carbonation-induced corrosion.

14.2.2
Performance

It should be realized that surface treatments can have negative side effects, either in their effect on young concrete or later, due to loss of their efficacy.

It was demonstrated that coatings can slow down the progress of curing of (young) concrete by preventing water from penetrating, as expressed by high water sorptivity values for dry, sheltered, and coated specimens, as opposed to wet exposed uncoated specimens [5]. If hydration is not reasonably advanced before the coating is applied, the carbonation resistance may be permanently poor. Once deep carbonation has taken place, the system relies on the coating to keep the concrete dry. Subsequent degradation of the coating and increased water penetration may even promote corrosion propagation in extreme cases.

Negative effects of coating degradation (during later stages of service) are illustrated by a case study on a reinforced concrete bicycle bridge exposed to deicing salts [14]. The bicycle lane was coated with a dense epoxy wearing course with crushed stone strewn-in for optimal road surface properties. The parapets (side walks) were left uncoated. After about 30 years, chloride profiles were taken and analyzed using Fick's second law of diffusion (Section 6.2.2). Chloride penetration in the coated part was characterized by a chloride surface content of about 1.7% chloride by mass of cement and an apparent chloride diffusion coefficient of $(2-3) \times 10^{-12}$ m^2/s, which is typical for portland cement concrete in wet conditions. It appeared that the protective effect of the epoxy coating had practically been lost after about 15 years of service, probably due to penetration of the coating layer by the crushed stone particles in combination with UV and thermal loads. The degradation of the coating had allowed water to penetrate but despite its degradation, water evaporation remained restricted. This caused the concrete to remain relatively wet, resulting in relatively quick chloride transport and high corrosion rates after depassivation. The result was that after 30 years, heavy reinforcement corrosion and concrete delamination had developed and that the bridge had to be replaced. The noncoated parapets on the other hand, had not shown quick chloride penetration despite significant chloride loads. The chloride profiles in the parapets were characterized by chloride surface contents of about 1.6% chloride by mass of cement and an apparent diffusion coefficient of about $0.7 \times 10^{-12}\,m^2/s$. This relatively low diffusion coefficient is well below typical values for water-saturated portland cement concrete. This suggests that the absence of a coating on the parapets had allowed the concrete to remain dry most of the time, slowing down chloride transport. These results show that the protective effects of this type of epoxy coatings has a finite life. In this case, the coating should have been replaced after 10 to 15 years, or other preventative measures should have been taken. A simple but effective option might have been applying hydrophobic treatment of the concrete before applying the coating (as a dual system). Alternatively, the cover to the steel should have been increased and/or a blast furnace slag cement with higher resistance to chloride penetration should have been chosen (Chapters 11 and 12).

14.3 Hydrophobic Treatment

Application of hydrophobic treatment to a concrete surface aims at reducing the capillary absorption of water and dissolved aggressive substances. Hydrophobizing concrete leaves the pores open, so it does not affect the ingress of gaseous species.

Liquid water is transported rapidly into nonwater-saturated pores by capillary suction (Chapter 2), depending on the surface tension, the viscosity and the density of the liquid, on the angle of contact between the liquid and the pore walls and on the radius of the pores. The characteristics of the liquid (water) and the pore structure of the concrete are given constants. For normal concrete, the contact angle (θ) is low (<90°) due to molecular attraction between water and the cement paste (this is termed hydro*philic* behavior). A drop of water will spread on a flat surface, the level inside a capillary will rise above the surrounding liquid and the concrete will absorb the water. The molecular attraction between water and concrete can be weakened by impregnating the concrete with hydrophobic agents, such as silicones, making the concrete hydro*phobic*. The contact angle will be >90°, a drop takes the form of a sphere, while the capillary rise is negative, so the level of liquid in the capillary is lower than the surrounding liquid. The concrete is said to be water repellent. The two cases are shown in Figure 14.3.

From the silicone group of substances, silanes and siloxanes are most important for hydrophobizing concrete. Silanes are small molecules having one silicon atom;

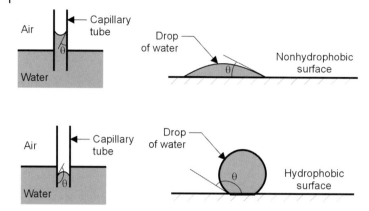

Figure 14.3 Interaction between water and a nonhydrophobic or hydrophobic material; illustrated for a capillary (left) and a concrete surface (right).

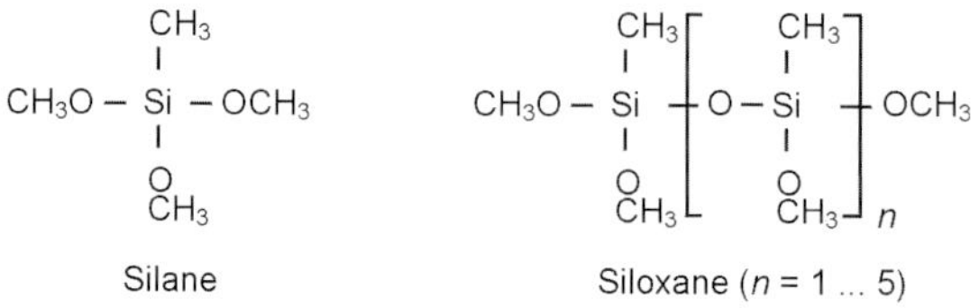

Figure 14.4 Silane and siloxane molecules.

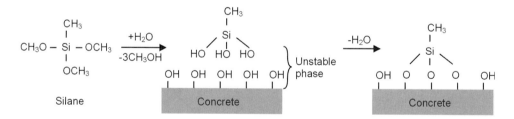

Figure 14.5 Reaction of (methyl-methoxy)silane with a concrete substrate [6, 7].

siloxanes are short chains of a few silicon atoms (Figure 14.4). Their molecules contain (organic) alkoxy groups linked to the silicon atoms, which can react with the silicates in the concrete pore-wall surface to form a stable bond as illustrated in Figure 14.5. The alkoxy groups can also react with other silane molecules, causing some degree of polymerization. Silanes and siloxanes also contain organic alkyl groups that have a fatty and water-repellent character. After reaction of the alkoxy groups with the substrate, the alkyl groups protrude from the pore surfaces [6]. As a result, water molecules will be repelled, the contact angle is greater than

90° and, ideally, water is no longer absorbed by capillary suction. In reality, capillary absorption is reduced to 10–20% of nontreated concrete.

Hydrophobic treatments form a layer of molecular thickness on the concrete pore walls. Because the pores are left open, such a treatment does not block transport of single water molecules, that is, water vapor. The effect, in particular in wetting–drying situations, is that hydrophobic treatment strongly reduces the ingress of liquid water from the environment, but allows the evaporation of moisture from the inside. Consequently, the concrete (under alternating wetting–drying exposure) will dry out compared to nontreated concrete, as was already illustrated in Figure 14.2. This process has been confirmed experimentally [13, 15]. At the same time, the pores of hydrophobized concrete are also open for penetrating gases such as carbon dioxide.

14.3.1
Properties and Testing

As for coatings, EN 1504-2 provides performance characteristics for hydrophobic impregnation (Table 14.1). Several properties are to be tested, such as penetration depth, water absorption and alkali resistance (according to EN 13580) and drying rate (EN 13579). Requirements for penetration depth and drying rate come in two classes, with class I having less-stringent requirements than class II.

Some of the tests and requirements are more or less based on dedicated studies carried out in the 1990s. Part of that work is summarized here to provide some background.

Since the 1990s, hydrophobic treatment has been used for concrete highway structures in the United Kingdom [16]. Because the advantages of such treatment were seen as promising, in particular as additional protection for bridge decks where open asphalt was going to be applied, research was carried out in The Netherlands into various aspects of performance and durability of the water-repellent effect [17]. Test methods and requirements were designed and nine commercially available products were tested. The primary objective of the treatment is to minimize water absorption; the requirement was set at less than 20% of control concrete (not hydrophobized). For a durable effect, the hydrophobic agent should penetrate sufficiently deep to prevent degradation by ultraviolet radiation; the requirement was a penetration depth of at least 2 mm. Water from within the concrete should evaporate freely to avoid damage, for instance due to freezing and thawing; the requirement was a level of evaporation through hydrophobic concrete that is at least 60% of the evaporation from control concrete. The hydrophobic material should withstand the alkalinity of the concrete pore solution; required was a water absorption lower than 20% of control when applied to strongly alkaline portland cement-sand mortar. Finally, when the hydrophobized concrete would be covered with an asphalt layer such as for bridge decks, the water-repellent effect should not be reduced significantly due to the temperature increase during asphalting. The requirement was a water absorption after heating at 160 °C lower than 30% of control concrete. All requirements apply for

Table 14.3 Test results of hydrophobic products on trowel-finished surfaces of portland cement concrete ($w/c = 0.50$) and compliance to The Netherlands Ministry of Public Works and Waterways (Rijkswaterstaat) requirements [17, 18].

Product	Product uptake (l/m²)	Relative water absorption (% of control)	Water absorption after heating at 160 °C (% of control)	Penetration depth (mm)	Change of evaporation (% of control)	Compliance
100% silane A	0.15	24	21	2–3	+40	ok
100% silane B	0.16	13	8	1–2	−16	ok
Silane/siloxane in water C	0.13	34	100	0	−2	Insufficient penetration
Silane/siloxane in water D	0.10	18	48	0	−12	Insufficient penetration
Silane/siloxane in water E	0.20	14	12	1–2	−10	ok
Siloxane in hydrocarbon F	0.16	15	81	0–1	−20	Insufficient penetration
Silane in ethanol G	0.12	29	59	0	−20	Insufficient penetration
Silane in hydrocarbon H	0.17	13	68	3	−32	ok except heating (asphalt)
Silane/siloxane in hydrocarbon I	0.17	23	96	0	−22	Insufficient penetration

hydrophobic products applied to two standard concrete mixes, one made with portland cement, CEM I, and one made with blast furnace slag cement with a high slag content, CEM III/B ($w/c = 0.50$, 340 kg/m³), after storage under standard conditions (at least four weeks in air of 20 °C and 65% R.H.). Several of the nine products tested complied with all requirements and some test results are shown in Table 14.3. Good performance was obtained with hydrophobic products that were either: 100% active substance (silanes), 20–40% active substance (silanes, siloxanes) in alkane solvents (white spirit) or ethanol, or at least 20% active substance (silane/siloxanes) in aqueous emulsion.

More recently, some well-performing products of the silane/siloxanes type have become available in the form of a water-based cream [15], which has practical advantages with respect to application in particular on soffit (overhead) surfaces.

Based on the results of additional tests, the following was advised with regard to the application. Two coats of the hydrophobic agent are given consecutively (wet-in-wet), for the investigated products resulting in an uptake of 200 to 500 g per square meter of concrete surface. The substrate should be sufficiently dry for optimal penetration, meaning that the hydrophobic agent should be applied at least 24 h after the concrete is wetted by rain. Application should not be carried out at concrete temperatures below 10 °C and not above 25 °C. Mold (formwork) surfaces require more attention than troweled surfaces. The substrate should be free of soiling, oil, and other contaminations and remains of curing compounds. From 24 h after application, asphalt can be applied.

A similar set of tests and requirements was designed in the United Kingdom [19]. A range of products was tested and pass/fail criteria were drawn up. Generally, a good agreement exists with the results mentioned above.

As mentioned above, EN 1504-2 requires performance testing for hydrophobic treatments with regard to depth of penetration, water absorption, alkali resistance, and drying rate as general requirements. In addition, it states as requirements for "certain intended uses" testing for resistance against freeze–thaw and salt action (according to EN 13581) for structures exposed to deicing salts, and diffusion of chloride in special cases (not further specified, supposedly for the same).

Several tests and requirements are similar to those reported above. However, as some tests and requirements differ to a certain extent from those previously used in The Netherlands [18], a study was carried out into the consequences of those differences. Without going into detail, some differences were considered important. For example, [18], does not require a freeze–thaw/salt action test, because it was felt that, apart from being costly, such a test would not be discriminating between various hydrophobic agents. This was based on the observation that concrete treated with all tested agents had much better resistance than untreated concrete. With regard to penetration depth, EN 1504-2 prescribes testing on concrete with w/c 0.70 and 260 kg/m^3 of CEM I 42.5R. [18] prescribes testing for penetration depth on two concretes with w/c 0.50, both made with CEM I and CEM III/B. It was felt that concrete with w/c 0.70 is not representative of concrete used for bridge decks. Consequently, in a document issued by the Dutch Ministry of Transport (RWS), in addition to EN 1504-2 hydrophobic products need to be tested for penetration depth on concrete made with w/c of 0.50, with the original requirement to the penetration depth (at least 2 mm). In [18] a test was required for determining the effect of high temperatures related to asphalt application. Such a test is absent in EN 1504-2. Considering the importance of such a property, RWS has added it as an additional test to EN 1504-2.

14.3.2
Performance

Onsite measurements using portable equipment for permeability testing (Autoclam, see Section 2.3) confirmed laboratory results in terms of low water absorption for hydrophobized concrete [20–22]. The absence of chloride penetration in hydrophobized concrete was found on a motorway pier after 7 years' service. Apparently field conditions, including thermal and load-related movement of concrete, do not harm the water-repellent effect of hydrophobized concrete on that time scale.

With a few of the products that complied with all requirements described above, further investigations were carried out [23, 24]. The conclusions apply to concrete of at least reasonable quality ($w/c = 0.50$) made with portland (CEM I) or blast furnace slag cement (CEM III B), which was thought to be representative of the upper parts of bridge decks. They can be summarized as follows:

- chloride ingress after intermittent (wetting–drying) contact up to one year with salt solution is reduced by a factor of 5 to 10 compared to nontreated concrete, in particular because the chloride surface content was reduced by a factor of two;
- absorption of water and salt solution during up to four weeks of permanent contact is reduced by a factor of ten;
- water absorption as a function of exposure time to outside conditions (horizontal, unsheltered) remains low for at least five years;
- carbonation depth after several years of relatively wet exposure (outside, unsheltered) is the same for hydrophobized as for nontreated concrete;
- effect on corrosion of bars in chloride-contaminated concrete is negligible, once corrosion has initiated; in laboratory tests, Vassie found a 37% reduction in corrosion rate of silane treatment [13], but stated that it is less likely that such results could be obtained in the field;
- corrosion rate in carbonated concrete may be reduced by hydrophobic treatment [25];
- scaling of concrete under freeze–thaw cycles with deicing salt application is reduced significantly.

From tests carried out on different concrete grades, it was found that it is increasingly difficult to obtain sufficient penetration of hydrophobic agents in concrete with increasing density and compressive strength. In "standard" concrete ($w/c = 0.50$ and $340 \, kg/m^3$ of cement, compressive strength = 43 MPa), penetration could be up to 6 mm. For concrete with $w/c = 0.40$ and $400 \, kg/m^3$ of cement (75% GGBS and 25% OPC), with an average cube strength of about 67 MPa, the average penetration depth was less than 2 mm. For concrete with $w/c = 0.30$ and $475 \, kg/m^3$ of cement (50% GGBS and 50% OPC) plus $25 \, kg/m^3$ silica fume, with an average cube strength of about 90 MPa, the penetration depth was negligible. It seems that hydrophobic treatment of high-strength/high-density concrete is less effective. On the other hand, where locally the concrete is more porous than intended, the hydrophobic agent may penetrate comparatively deep and so provide additional protection for otherwise "weak spots".

In general, it can be concluded that hydrophobic treatment of concrete is an effective and low-cost preventative measure against corrosion that may be caused by chloride ingress. The beneficial effect is mainly a strong reduction of chloride ingress, both in semipermanent contact and in wetting–drying situations. The effective life of the treatment in fully exposed atmospheric conditions is at least five years and more, probably much more, when protected from UV radiation by a layer of asphalt. Recently, some work on long-term behavior of hydrophobic treatment has been published. In a study on a repaired marine quay, several types of hydrophobic treatment (liquid, gel, cream) appeared to be effective in retarding chloride penetration up to 10 years, although the effect depended on the type of surface (sprayed concrete, vertical and horizontal cast surfaces) [26]. In another study, cores were tested after being obtained from motorway structures treated with silanes between 10 and 20 years before coring. Capillary suction showed that compared to untreated concrete, the silane-treated surfaces have a residual protection effect even after 20 years' service [27].

14.4
Treatments That Block Pores

Blocking of the pores in concrete surfaces may be achieved with materials like silicates or silico-fluorides [7]. Once these materials have penetrated the pores, they react chemically with the concrete, in particular with the calcium hydroxide present, and the reaction products block the pores, as seen in Figure 14.1c. Sodium silicate (water glass) -based pore blockers have been used in the past. Their penetration into concrete is superficial, except in the most porous concrete. A reaction with CO_2 is involved, forming alkali carbonates that cause efflorescence:

$$Na_2SiO_4 + Ca(OH)_2 + CO_2 \rightarrow C-S-H + Na_2CO_3 + H_2O \quad (14.1)$$

In addition, their effectiveness with regard to reducing the water absorption is relatively poor because essentially they form $C-S-H$ gel, which is hydrophilic. On the other hand, pores are blocked for water vapor present inside the concrete, so the risk of damage due to freezing may increase, in particular in porous, mechanically weak, concrete. It seems that the trade-off between "open" (durable) and "dense" (effective) for silicate-based pore blockers is unfavorable.

Aqueous solutions of metal-hexafluorosilicates can be applied to concrete surfaces, which form solid CaF_2, metal fluorides and silica(gel):

$$MgSiF_6 + Ca(OH)_2 \rightarrow CaF_2 + MgF_2 + SiO_2 + H_2O \quad (14.2)$$

all of which contribute to filling up the pores [7]. Treatment with gaseous silico-fluoride SiF_6 that transforms parts of the concrete surface layer into CaF_2 has been applied in the past to concrete sewer pipes to increase their resistance to acid attack.

No test results in terms of carbon dioxide or chloride-penetration resistance of the inorganic pore blockers mentioned above are available.

Some organic sealers may be considered to belong to the category of pore blockers [6]. They contain epoxy, polyurethane or acrylic resin, whose viscosity allows sufficient penetration into the pores (sometimes facilitated by vacuum saturation), where they set and thus exert their blocking action. Natural or synthetic gums and (raw or boiled) linseed oil can be included in this group as well [6]; they are mainly used in the US and less in Europe. As reported above, some organic sealers may be effective in reducing chloride penetration, while others are not [9].

As for coatings and hydrophobic agents, EN 1504-2 provides performance characteristics (Table 14.1). Several properties are to be tested, such as penetration depth, capillary absorption, and water permeability (by EN 1062-3) as a general test. Special requirements are permeability to water vapor, freeze–thaw cycling with deicing salt immersion (by EN 13687-1), thermal shock and cycling, and chloride-diffusion resistance, for use in special cases.

14.5
Cementitious Coatings and Layers

Cementitious coatings and layers are a wide category that ranges from true cement-based coatings of a few to less than 10 mm thick applied by brushing, to overlays of centimeters thickness applied by rendering or high-energy spraying like shotcrete, also called gunite (Figure 14.1d).

Shotcrete is in fact fine-grained concrete with a high cement content and a low w/c, sometimes with the addition of silica fume, which is very dense due to the high impact that goes with the application method. A layer of shotcrete is generally considered as additional concrete cover, because the high-density material has comparable or even better durability and mechanical properties than the original concrete cover. Shotcreting is in fact mainly a repair method (Chapter 19). Plastered renders are generally porous layers that are applied mainly for aesthetical reasons.

True cementitious coatings are layers of low permeability and a moderate thickness of a few millimeters. The mortar or paste used is generally fine grained and modified with polymers to decrease its permeability and to increase its bond to the concrete. The addition of silica fume also brings about a reduction in permeability. Plasticizers or superplasticizers are utilized as well, to lower the w/c ratio and maintain a good workability. Application may be executed by manual methods or spraying (with lower energy than used for shotcrete). Addition of expansive agents helps to avoid differential contraction of the coatings compared to the concrete base due to drying shrinkage, which would result in cracking or separation. Tests described above show that polymer-modified cementitious coatings may have good initial carbonation resistance, which may be lost due to weathering [8] and good chloride penetration resistance [9], but no effect on chloride-induced corrosion rate [13].

Pore fillers are a subcategory of cementitious coatings. They are applied to even out a rough surface, for instance when blowholes are present in the original surface. Usually they are applied before a final (aesthetical) finish with an organic coating is applied. Fillers have a high cement content and a high polymer content and contain very fine aggregate.

14.6
Concluding Remarks on Effectiveness and Durability of Surface Protection Systems

The beneficial action of surface protection systems generally lies in the fact that they prolong the period of initiation of corrosion. Once corrosion has begun, only those treatments that effectively obstruct the penetration of water, both liquid and vapor, will reduce the corrosion rate. This effect may be significant in particular if corrosion is due to carbonation [25]. Chloride-induced corrosion processes attract moisture so strongly that in general, surface treatments cannot stop ongoing corrosion [13, 24].

To block the penetration of carbon dioxide, organic coatings that form a continuous film are normally used. These coatings should obviously have a low permeability for carbon dioxide. The durability with respect to weathering of the barrier effect of many coatings may be questioned. In extreme cases, an insufficient barrier effect may make matters worse if drying out occurs at early ages, which prevents sufficient hydration of cement and may actually promote carbonation [5]. At the same time, to avoid disbonding from the concrete, coatings must allow the passage of water vapor. Published testing data is difficult to generalize because many different products have been tested using different methods, preventing direct comparison between references [28, 29]. However, comparative studies can be useful to get an impression of various materials behavior and performance, either in the laboratory or in the field [6, 30].

To reduce penetration by water and chloride ions, organic and cementitious coatings, and hydrophobic treatments are used. In particular, the latter seem to be most effective [29]. In order to evaluate the contribution to the durability of the structure, it is necessary to know how effective the surface treatment is in obstructing penetration of different aggressive species and how long it can maintain this effectiveness. It seems that water absorption or permeability is a good measure of the barrier effect against chloride penetration. The effectiveness of a coating may vary greatly even with small variations in the formulation of the product. Artificial aging of products may produce a decrease or increase of the barrier properties. Again, it is difficult to generalize the results. Degradation of a coating can contribute to increasing the corrosion risk.

Surface protection systems and in particular coatings have strongly different properties from concrete, for example, with regard to modulus of elasticity and expansion due to temperature changes and wetting–drying cycles. The resulting stresses in coatings may slowly degrade their barrier properties. (Micro)cracks appear and chloride or carbon dioxide can penetrate increasingly fast. Consequently, the European standard EN 1504-2 specifies artificial weathering for some types (Table 14.1). Concerning the ability of coatings to remain effective in time, there is relatively little data available at the moment to predict their durability beyond a period of 10 years under different conditions of exposure (with regard to: temperature, UV radiation, oxidation, humidity, freeze–thaw cycles, etc.). This is also because the materials used in this field have evolved rapidly in recent years.

In any case, a finite service life has to be attributed to all surface-treatment-based protection systems, after which they lose integrity and progressively allow chloride, water or carbon dioxide to penetrate. If the durable effectiveness of a coating is essential for the structure's service life, the coating should be reapplied after some time. This could be part of a proactive strategy of planned maintenance, chosen in the design stage (Chapter 11). Alternatively, the effectiveness should be monitored using embedded probes, for instance by regularly testing the electrical resistance of the concrete (Chapter 17). Resistance decrease after some time would indicate increased water penetration and reduced effectiveness of the coating.

References

1 CEN (2004) EN 1504-2. Products and Systems for the Protection and Repair of Concrete Structures. Definitions, Requirements, Quality Control and Evaluation of Conformity. Part 2: Surface Protection Systems for Concrete, European Committee for Standardization.

2 Keer, J.G. (1992) Surface treatments, in *Durability of Concrete Structures: Investigation, Repair, Protection* (ed. G. Mays), Chapman and Hall, London, pp. 146–165.

3 Mailvaganam, N. (1992) *Repair and Protection of Concrete Structures*, CRC Press Inc., Boca Raton, FL.

4 German Committee on Reinforced Concrete (1990) *Guidelines for the Protection and Repair of Concrete Component*, German Standard Institute, Berlin.

5 Ho, D.W.S. and Ritchie, D. (1992) Surface coating and their influence on the quality of the concrete substrate, in Proc. RILEM Conf. Rehabilitation of Concrete Structures, Melbourne, 1992, pp. 403–407.

6 Cady, P.D. (1994) *Sealers for Portland Cement Concrete Highway Facilities, Synthesis of Highway Practice 209*, National Cooperative Highway Research Program, National Academy Press, Washington, DC.

7 Bijen, J.M. (ed.) (1989) Maintenance and repair of concrete structures. *Heron*, **34** (2), 2–82.

8 Bassi, R. and Davies, H. (1996) *Testing anti-carbonation coatings for concrete*, BRE Information paper IP7/96.

9 Buenfeld, N.R. and Zhang, J.Z. (1998) Chloride diffusion through surface treated mortar specimens. *Cement and Concrete Research*, **28**, 665–674.

10 Rodrigues, M.P.M.C., Costa, M.R.N., Mendes, A.M., and Eusebio Marques, M.I. (2000) Effectiveness of surface coatings to protect reinforced concrete in marine environments. *Materials and Structures*, **33**, 618–626.

11 Vogelsang, J., Meyer, G., and Bepoix, M. (2007) Determination of coating permeability on concrete using EIS, in *Corrosion of Reinforcement in Concrete. Mechanisms, Monitoring, Inhibitors and Rehabilitation Techniques*, European Federation of Corrosion Publication number 38 (eds M. Raupach, B. Elsener, R. Polder, and J. Mietz), Woodhead Publishing Limited, Cambridge, pp. 250–262.

12 Tritthart, J. and Geymayer, H. (1985) Aenderungen des elektrischen Widerstandes in austrocknendem Beton. *Zement und Beton*, **1**, 74–79.

13 Vassie, P.R. (1990) Concrete coatings: do they reduce ongoing corrosion of reinforcing steel?, in *Corrosion of Reinforcement in Concrete* (eds C.L. Page, K. Treadaway, and P. Bamforth), Elsevier, pp. 456–470.

14 Polder, R.B. and Hug, A. (2000) Penetration of chloride from de-icing salt into concrete from a 30 year old bridge. *Heron*, **45** (2), 109–124.

15 Raupach, M. and Wolff, L. (2002) Investigation the long-term durability of hydrophobic treatment on concrete with respect to corrosion protection. 15th

International Corrosion Congress, Granada, 2002 (CD-ROM).

16 Department of Transport (1990) *Criteria and material for the impregnation of concrete highway structures*, (UK) Departmental Standard BD43/90.

17 de Vries, J. and Polder, R.B. (1995) Hydrophobic treatment of concrete. *Construction Repair*, **9** (5), 42–47.

18 Polder, R.B., Pijnenborgh, A.J., and de Vries, J. (1993) *Recommendation for testing hydrophobic agents for concrete according to the requirements of the Ministry of Transport*, BSW rapport 93-26 (in Dutch).

19 Calder, A.J.J. and Chowdury, Z.S. (1996) A performance specification for hydrophobic materials for use on concrete bridges, in *Proc. 4th Int. Symp. Corrosion of Reinforcement in Concrete Construction* (eds C.L. Page, P.B. Bamforth, and J.W. Figg), Society of Chemical Industry, Cambridge, UK, pp. 556–566.

20 Basheer, P.A.M. and Long, A.E. (1997) Protective qualities of surface treatments for concrete. *Proceedings of the Institution of Civil Engineers, Structures and Buildings*, **122**, 339–346.

21 Basheer, P.A.M. and Basheer, L. (1997) Surface treatments for concrete: assessment methods and reported performance. *Construction and Building Materials*, **11**, 413–429.

22 Basheer, L. and Cleland, D.J. (1998) Protection provided by surface treatment against chloride induced corrosion. *Materials and Structures*, **31**, 459–464.

23 Polder, R.B., Borsje, H., and de Vries, J. (2000) Corrosion protection of reinforcement by hydrophobic treatment of concrete, in *Corrosion of Reinforcement in Concrete. Corrosion Mechanisms and Corrosion Protection*, The European Federation of Corrosion Publication number 31 (eds J. Mietz, R.B. Polder, and B. Elsener), The Institute of Materials, London, pp. 73–84.

24 Polder, R.B., Borsje, H., and de Vries, J. (2001) Prevention of reinforcement corrosion by hydrophobic treatment of concrete. *Heron*, **46** (4), 227–238.

25 Sergi, G., Lattey, S.E., and Page, C.L. (1990) Influence of surface treatments on corrosion rates of steel in carbonated concrete, in *Corrosion of Reinforcement in Concrete* (eds C.L. Page, K. Treadaway, and P. Bamforth), Elsevier, pp. 409–419.

26 Rodum, E. and Lindland, J. (2012) Effect of different surface treatment products after 10 years of field exposure. International Congress of Durability of Concrete, Trondheim, June, 2012.

27 Christodoulou, C., Goodier, C., Austin, S., Glass, G.K., and Webb, J. (2012) Assessing the long term durability of silanes on reinforced concrete structures. International Congress of Durability of Concrete, Trondheim, June, 2012.

28 EN 1062 (2004) Paints and varnishes – Coating materials and coating systems for exterior masonry and concrete, part 2: Determination and classification of water vapour permeability; part 3: Determination and classification of liquid – water transmission rate (permeability) (2008); part 6: Determination of carbon dioxide permeability (2002); part 7: Determination of crack bridging properties (2004); part 11:2002: Methods of conditioning before testing (2002).

29 Rodrigues, P., Basheer, M., and Cleland, D. (2003) Surface treatments and coatings, in *COST Action 521, Corrosion of Steel in Reinforced Concrete Structures, Final Report* (eds R. Cigna, C. Andrade, U. Nürnberger, R. Polder, R. Weydert, and E. Seitz), European Communities, Luxembourg, Publication EUR 20599, pp. 159–182.

30 Thompson, D.M. and Leeming, M.B. (1992) *Surface treatments for concrete highway bridges*, TRRL Research Report 345, Transport Road Research Lab., Department of Transport.

15
Corrosion-Resistant Reinforcement

Reinforcing bars with a higher corrosion resistance than the common carbon steel rebars can be used as a preventative method under conditions of high environmental aggressiveness, or when a long service life is required. The corrosion resistance of rebars can be increased either by modifying the chemical composition of the steel or by applying a metallic or organic coating on their surface. Three families of corrosion-resistant bars are used in reinforced concrete structures, consisting respectively in stainless steels, galvanized steel, and epoxy-coated rebars.

Fiber-reinforced polymers (FRP) rebars, usually made of an epoxy matrix reinforced with carbon or aramide fibers, have also been proposed both as prestressing wires and reinforcement. Nevertheless, they are not discussed here, because these applications are still in the experimental phase and there is lack of experience on their durability. In fact, while they are not affected by electrochemical corrosion typical of metals, they are not immune to other types of degradation. FRP are also used in the form of laminates or sheets as externally bonded reinforcement in the rehabilitation of damaged structures; this application will be addressed in Chapter 19.

15.1
Steel for Reinforced and Prestressed Concrete

Before discussing corrosion-resistant reinforcement, mention will be made of the requirements for reinforcing bars and the characteristics of conventional carbon steel reinforcement.

15.1.1
Reinforcing Bars

Reinforcing bars should fulfill several requirements regarding strength, ductility, bond to concrete, weldability, etc. To achieve suitable anchorage when embedded in concrete, ribbed bars are normally used. Mechanical properties that are relevant for reinforcing bars are elastic modulus, strength, and ductility. For instance,

Corrosion of Steel in Concrete: Prevention, Diagnosis, Repair, Second Edition. Luca Bertolini,
Bernhard Elsener, Pietro Pedeferri, Elena Redaelli, and Rob Polder.
© 2013 Wiley-VCH Verlag GmbH & Co. KGaA. Published 2013 by Wiley-VCH Verlag GmbH & Co. KGaA.

European standards that deal with reinforcement for concrete structures [1, 2] prescribe requirements of strength based on the characteristic values of the tensile strength (f_{tk}) and of the yield strength (f_{yk}). The ductility of the steel is evaluated by means of the characteristic value of the ratio between the tensile strength (f_t) and the yield strength (f_y) or the characteristic value of the strain at the maximum load (ε_{uk}). According to Eurocode 2 [1], the allowed characteristic yield strength ranges from 400 to 600 MPa. Three classes of ductility are considered, from A (lower ductility) to C (higher ductility). Usually rebars of class B with characteristic yield strength of 500 MPa are used in Europe, although bars of lower strength and stricter ductility requirements (f_{yk} = 450 MPa, class C) are used in some countries such as Italy because of concern about the seismic behavior of structures.

As far as the chemical composition is concerned, restrictions have been placed on the content of certain elements (C, P, S, Mn, etc.) for weldable rebars. In the past, relatively high levels of alloying elements (i.e., C up to 0.5%, Mn, and Si) were used to reach the required strength with hot-rolled bars; this caused poor welding behavior. Today, to guarantee both greater strength and weldability onsite or during production of mesh, new manufacturing technologies are used. High-strength and ductile steels with low contents of carbon (C < 0.2%) and other alloying elements are produced with processes based on three principles: cold stretching of wires (only for lower ductility bars), addition of small amounts of alloying elements that cause precipitation hardening during hot rolling, or controlled cooling immediately following hot rolling of bars that produces a hardened outer layer and a ductile core.

15.1.2
Prestressing Steel

Prestressing steel, used for pretensioned or post-tensioned structures, can be in the form of wires, strands, or bars. The main requirement for these products is a high value of the yield strength (Figure 15.1). According to Eurocode 2, this type of steel is classified on the basis of the characteristic value of 0.1% yield strength ($f_{p0.1k}$), the ratio between ultimate tensile strength and the characteristic value of 0.1% yield strength ($f_{pk}/f_{p0.1k}$), and the relaxation behavior.[1] The strength of prestressing steel may vary according to the production technology, the chemical composition, and the geometry (small-diameter wires can reach higher strength values than bars). Compared to reinforcing steel, much higher strength levels must be obtained, and thus higher levels of carbon are used.

In order to obtain high yield strength values, tendons can be *cold worked, hot rolled*, or *quenched and tempered* (Table 15.1) [3, 4]. Cold-worked prestressing steel wires are obtained by drawing wires of steel with a ferritic–perlitic microstructure

---

1) Relaxation is reduction in stress in time under constant strain; it should be low for prestressing steel in order to reduce loss of prestressing load in time on the structure. It is conventionally measured by the percentage loss of load during a test in which a constant strain is applied for a predetermined period (e.g., strain corresponding to 70% of f_{pk} for 1000 h).

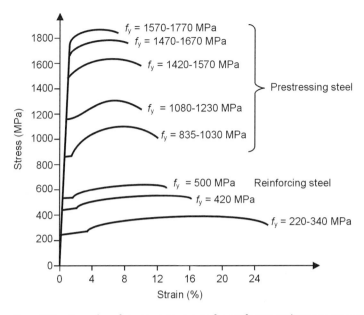

Figure 15.1 Examples of stress–strain curves for reinforcing and prestressing steels [3].

Table 15.1 Different types of prestressing steel [3].

Type	Shape, surface	Diameter (mm)	Anchorage system	Strength class European standard (MPa)	Production (ton/year)
Cold deformed Wire	Round-smooth	4–12.2	Wedge or button heads	1570–1860[a]	1 000 000
	Round-profiled	5–5.5			
Strand	Round-smooth (7wire)	9.3–15.5		1700–2060[a]	
Hot rolled Bar	Round-smooth	26–36	Thread (ends)	1030–1230	50 000 (Germany, UK)
	Round-ribbed	26.5–36	Thread (full length)		
Quenched and tempered wire	Round-smooth	6–14	Wedge	1570	5000 (Germany, Japan)
	Round-ribbed	5–14			
	Oval-ribbed	40–120			

a) In Germany max. 1770 MPa.

at room temperature, so that the reduction in the cross section leads to an increase of strength. Prestressing bars with diameter up to 36 mm are manufactured by controlled cooling after hot rolling. Their chemical composition leads to a perlitic microstructure with the presence of strengthening precipitates (due to the addition of vanadium); subsequently the bars are cold drawn. Finally, all cold-worked prestressing steels are annealed at moderate temperatures; wires and strands are usually heated under tensile stresses to improve the relaxation behavior. The quenched and tempered steels are heated to reach an austenitic microstructure and subsequently quenched (rapid cooling) and tempered to achieve a fine-grained microstructure with finely dispersed precipitates.

15.1.3
Corrosion Behavior

As far as corrosion behavior is concerned, prestressing steel needs to be distinguished from reinforcing steel with regard to hydrogen embrittlement, since it only affects the former; this has been illustrated in Chapter 10.

In noncarbonated and chloride-free concrete, the passivity of low-alloyed steels is not influenced appreciably by their composition, structure or surface conditions. Therefore, the usual thermal or mechanical treatments or the roughness of the surface of the rebars have negligible influence on their corrosion behavior.

Even the presence of magnetite scale that often covers the surface of the bars, which can cause dangerous localized attack on steel in contact with neutral solutions (such as fresh water or seawater), is not dangerous in concrete. In fact, noncarbonated and chloride-free concrete passivates all the surface of the steel. If adherent oxide films are present, they do not create problems. If the oxide layer contains chlorides, because for example, it is formed in a marine environment, it must be removed completely because it can hinder passivation.

Once the steel becomes active due to carbonation of concrete or chloride penetration, the influence of chemical composition, microstructure and surface finishing is still of secondary importance, because kinetic control of the corrosive process is of the ohmic type, or dependent on oxygen diffusion and thus on characteristics of the concrete (and in particular its moisture content) rather than on those of the metal. Nevertheless, it was shown that the susceptibility of steel to pitting corrosion in chloride-contaminated concrete may be slightly affected by the surface condition of the steel [5, 6].

15.2
Stainless Steel Rebars

Stainless steel is an extended family of steel types with a wide variety of characteristics with regard to physical and mechanical properties, cost, and corrosion resistance. They have a much higher corrosion resistance than carbon steel, which derives from a chromium-rich passive film present on their surface [7]. Stainless

15.2 Stainless Steel Rebars

Table 15.2 Approximate chemical composition and designation according to the European standard EN 10088-1 [8] of some grades of stainless steel commercially used for reinforcing bars.

Common name	Designation EN 10088-1	Microstructure	Approximate chemical composition (% by mass)			
			Cr	Ni	Mo	Other elements
304L	1.4307	Austenitic	17.5–19.5	8–10	–	–
316L	1.4404	Austenitic	16.5–18.5	10–13	2–2.5	–
22-05	1.4462	Duplex	21–23	4.5–6.5	2.5–3.5	0.1–0.22 N
23-04	1.4362	Duplex	22–24	3.5–5.5	0.1–0.6	0.05–0.2 N
21-01	1.4162	Duplex	21–22	1.4–1.7	0.1–0.8	4.0–6.0 Mn, 0.2–0.25 N

steel bars can be used as a preventative technique for structures exposed to aggressive environments, especially in the presence of chlorides. They can also be selectively used in those parts of structures where corrosion is most likely to occur and in the repair of corroding structures. Different available types of stainless steel allow engineers to select the most suitable in terms of strength, corrosion resistance and cost.

15.2.1
Properties of Stainless Steel Rebars

Stainless steels can be divided into four categories, based on their microstructure: ferritic, austenitic, martensitic and austenitic-ferritic (duplex). Only specific grades of austenic and duplex stainless steel are currently used in concrete (Table 15.2 [8]), although a ferritic type with 12% chromium has also been proposed [9–14]. In some countries clad bars, that is, bars with a carbon steel core and an external layer of stainless steel are also used; since these bars are susceptible to corrosion in the presence of defects in the external layer of stainless steel [15] and the consequences of defects on this type of bars are not clear, only bars with a bulk of stainless steel will be considered here. Similarly, alloys with chromium content far lower than 12% by mass, which thus do not belong to the family of stainless steel, will not be considered due to lack of knowledge on their corrosion behavior.

Chemical Composition and Microstructure The composition and microstructure of stainless steels is usually represented by specific designations, such as those provided by AISI in the United States and EN 10088-1 in Europe. The chemical composition of some commercially available stainless steel bars are reported in Table 15.2. Traditionally used stainless steel bars are of the austenitic grades 304L and 316L and duplex 22-05. The corrosion behavior of these steels has been studied by several authors since the 1970s [15–33]. In recent years the fluctuations and increase in the cost of nickel has led to an increment in the cost of traditional austenitic stainless steels. As a consequence, new austenitic or duplex stainless steels with low nickel and molybdenum contents have been proposed for reinforcing bars, possibly adding manganese to promote the formation of austenite [34–38]. Also, ferritic stainless steels are proposed on the market for mildly aggressive environments.

Mechanical Properties Stainless steel rebars are required to have mechanical properties at least equivalent to those of carbon steel rebars, in terms of characteristic yield strength (usually evaluated as 0.2% proof strength, as stainless steels do not exhibit a well-defined yield point), elastic modulus and ductility. The strength of annealed austenitic stainless steels is too low to comply with requirements for reinforcing bars (Section 15.1.1) and thus bars need to be strengthened. This is usually achieved by cold working for bars of lower diameter or by means of hot rolling for bars of higher diameter [10, 11, 14]. Sufficient yield strength can be achieved with duplex stainless steels even without any strengthening.

Weldability The weldability of stainless steel depends on its chemical composition: it is improved by decreasing the carbon content and increasing the nickel content. Although stainless steel bars are usually weldable, welding is not recommended under site conditions unless adequate control is maintained; in fact, welding may have some negative consequences with regard to mechanical properties and corrosion resistance. In general, the composition of stainless steel used for reinforcing bars is such that it is not susceptible to intergranular corrosion (i.e., precipitation of carbides at the grain boundaries, leading to corrosion of surrounding areas due to chromium depletion); for instance, bars of austenitic and duplex stainless steel are made of weldable types 304L (1.4307) and 316L (1.4404), and duplex 22-05 (1.4462) with low carbon content. However, some loss of strength can occur in the heat-affected zone around the weld of austenitic steels, since the hardening that gives the steel its strength can be lost. Furthermore, even for weldable steels, the corrosion resistance is adversely affected by the oxide scale that welding produces on the steel surface and that leads to interference colors. This oxide should be removed, for example, by pickling.

Other Properties The coefficient of thermal expansion of concrete is about $10^{-5}\,°C^{-1}$. That of ferritic steels is not very different (about 1.2×10^{-5} as in usual carbon steel reinforcement); that of austenitic steels is higher (about 1.8×10^{-5}); austenitic-ferritic steels are in an intermediate position. The higher thermal expansion of austenitic and duplex stainless steels is not believed to cause any problems in concrete, and no cases of damage due to differential expansion have been reported [10]. Furthermore, the thermal conductivity of austenitic stainless steel is much lower than that of carbon steel and thus increase in temperature throughout the steel is delayed.

Austenitic stainless steels are generally considered nonmagnetic, although the magnetic permeability can increase due to cold drawing of the steel. This property can be useful for specific applications [10, 14]; it should also be taken into consideration when conventional magnetic covermeters are used, since they fail to detect the rebars.

As far as practice at the construction site is concerned, it should be observed that compared with other types of corrosion resistant rebars (such as epoxy-coated or galvanized steel), corrosion resistance is a bulk property of stainless steel. Therefore, the integrity of stainless steel is unaffected if its surface is cut or damaged during handling. Obviously, this does not apply to clad bars, that is, usual carbon steel bars clad with a thin layer of stainless steel, that in some cases have been proposed as a cheaper alternative to solid stainless steel bars.

15.2.2
Corrosion Resistance

Although all types of stainless steel are passive in carbonated concrete, the use of stainless steel rebars is normally associated with chloride-bearing environments. In fact, for structures subjected only to carbonation, unless an extremely long service life is required (i.e., hundreds of years), prevention of steel corrosion can be achieved with a proper design of the concrete mix and the concrete cover or, if necessary, by using less-expensive additional protection measures (such as galvanized steel rebars).

Pitting is the only form of corrosion expected, in practice, on stainless steel in concrete. Intergranular corrosion induced by welding is normally avoided by using appropriate types of steel. Stress corrosion may take place only under conditions of high temperature, carbonated concrete, and heavy chloride contamination, which are very unlikely to occur concomitantly. Because of the alkalinity of the pore solution and the porosity of the cement paste, crevice corrosion is also unlikely on stainless steel embedded in concrete. The corrosion resistance of stainless steels is affected by the presence of a mill scale on their surface; this is, however, normally removed by pickling or sandblasting; pickling gives the best result.

Resistance to Pitting Corrosion In chloride-contaminated concrete, stainless steel can suffer pitting corrosion like carbon steel. As shown in Chapter 7 for carbon steel, also for stainless steels the susceptibility to pitting attack can be expressed in terms of both the pitting potential (i.e., for a given chloride content, the potential value above which pitting can initiate) or the critical chloride content (i.e., for a given potential, the threshold level of chloride content for the onset of attack). These two parameters are interrelated and depend on the chemical composition and microstructure of the steel, on the surface condition of the bars and on the properties of the concrete. Because of the higher stability of the passive film of stainless steel compared with carbon steel, their resistance to pitting corrosion is remarkably higher. This is due to the formation of chromium and nickel oxyhydroxides in contact with the alkaline pore solution [7].

As a result, stainless steel bars have a much higher chloride threshold level compared to conventional carbon steel bars. As a matter of fact, the diverse grades of stainless steel commercially available (see for instance Table 15.2) have significantly different corrosion performances; so they offer to designers a wide range of possibility, in terms of both corrosion resistance and costs. The selection of stainless steel bars should be made in the framework of a performance-based service-life design, where the actual critical chloride threshold could be considered in order to assess the expected service life of the structure.

Nevertheless, even small variations in the chemical composition, thermomechanical treatment or the surface condition may significantly affect the corrosion resistance of stainless steel bars in chloride-bearing concrete. Therefore, the chloride threshold should be measured for any specific type of stainless steel and its variability should also be evaluated.

Unfortunately, the evaluation of the chloride threshold for steel in concrete is rather difficult and there are no standardized or generally accepted methods for its evaluation (see Chapter 6). A large variety of techniques has been used by different researchers. Experimental details, such as the way chlorides are introduced in the concrete (e.g., added to the mixing water or penetrated by diffusion or migration), the means of measuring the chloride content (e.g., total acid-soluble chlorides expressed by mass of cement or free chlorides from pore solution expressed as concentration in solution), the surface preparation of the rebar sample or the technique used to detect the depassivation of steel, may have a large influence on the resulting value of the chloride threshold.

Often, electrochemical tests in solutions simulating the concrete pore liquid, such as potentiodynamic and potentiostatic polarization tests, are used to compare the corrosion resistance of steels of different composition. For instance, Figure 15.2 shows the pitting potential measured in alkaline solutions of saturated $Ca(OH)_2$ of different types of stainless steel and of carbon steel, as a function of chloride content. Carbon steel showed a reduction of about 750 mV/decade of chloride concentration, starting from potentials typical of oxygen evolution (about 600 mV SCE) to a value of about −400 mV SCE in 3% chloride ion by mass of solution. Pitting potentials higher than 300 mV SCE were found for stainless steels, even in the presence of 10% chloride solution.

Tests in solutions lead to a rapid evaluation of the resistance to pitting corrosion, allowing a preliminary assessment of the role of the chemical composition or the microstructure of the material. However, they do not permit the definition of a chloride threshold level in concrete, since chloride concentrations in solutions cannot be directly converted in chloride contents in concrete, and, moreover, they may not be representative of the real conditions of a reinforcing bar in concrete (oxygen content at the steel surface, steel/concrete interface, etc.). Moreover, solution tests are often carried out on polished surfaces of cross section of stainless steels for general use, while different results could be obtained when tests are performed on the lateral (worked and ribbed) surface of reinforcing bars. Although

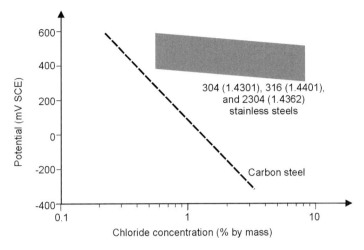

Figure 15.2 Schematic variation of pitting potential for austenitic 304 and 316 (with 2%Mo), and duplex 23-04 stainless steel, as a function of the chloride concentration in saturated Ca(OH)$_2$ solution, at 20°C [24, 26].

tests in concrete using real reinforcing bars are more time consuming, nevertheless they are indeed more representative of real conditions, in particular in considering the role of the steel/concrete interface [39]. Tests in concrete should be recommended especially when new types of stainless steel are used, since tests in solution may lead to misleading results [37].

Fields of Applicability In order to give an indicative picture of the order of magnitude of the chloride content at which traditional stainless steel bars can resist corrosion initiation, Figure 15.3 depicts safe fields of applicability in chloride-contaminated concrete exposed to temperatures of 20°C or 40°C. Fields have been plotted by analyzing the critical chloride values obtained by different authors from exposure tests in concrete or from electrochemical tests in solution and mortar and taking into consideration the worst conditions [11–32]. Nevertheless, it should be pointed out that values are indicative only, since the critical chloride content depends on the potential of the steel, and thus it can vary when oxygen access to the reinforcement is restricted as well as when stray current or macrocells are present. For instance, the domains of applicability are enlarged when the free corrosion potential is reduced, such as in saturated concrete. Furthermore, the values of the critical chloride limit for stainless steel with surface finishing other than obtained by pickling can be lower.

Since the pH for noncarbonated concrete is around 13, while in carbonated concrete it is near 9, the right-hand side of the graphs in Figure 15.3 is representative of alkaline concrete and the left-hand side of carbonated concrete. In alkaline concrete, austenitic steel 304L (1.4307) can be safely used in concrete up to 5% chloride by mass of cement, and 316L (1.4404) and duplex stainless steel (1.4462)

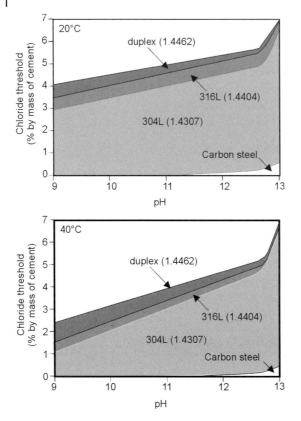

Figure 15.3 Schematic representation of fields of applicability of different stainless steel bars (pickled) in chloride-bearing environments for 20 and 40 °C. The threshold levels are indicative only. They can decrease if the oxides produced at high temperature, for example, welding or during manufacturing, are not completely removed, or the potential due to anodic polarization increases (e.g., due to stray current) or the concrete is heavily cracked. Conversely, they can increase when there is lack of oxygen or cathodic polarization.

even up to or higher than 5%, that is, for chloride contents that are rarely ever reached in the vicinity of the steel surface. In the presence of a welding scale on the surface of reinforcement, a lower critical chloride content of 3.5% has to be assumed [19]. Even this value will hardly ever be reached. The same reduction takes place if the surface is covered by the black scale formed at high temperature during thermomechanical treatments. Also in this case, a critical content of 3.5% has to be assumed for all types of steels. Pickling of rebars is more efficient in pushing up the critical chloride content than sand blasting, which does not completely free the surface from the oxide scale.

In carbonated concrete, or in the case where the concrete is extensively cracked, the critical chloride contents are remarkably lower. Situations where carbonated concrete and high chloride levels are simultaneously present are rare, but can be found, for instance, inside road tunnels [40]. The more highly alloyed stainless steels should be preferred in these more aggressive conditions. It is well known that for austenitic and duplex stainless steel, an increase in the content of chromium, molybdenum, and nitrogen improves the stability of the protective film. In neutral environments, the ability of a stainless steel to resist pitting attack is usually quantified by the so-called pitting-resistance equivalent index:

$$\text{PRE} = \%\text{Cr} + 3.3\%\text{Mo} + 16\%\text{N} \tag{15.1}$$

PRE is also a valid parameter when stainless steel parts are not completely embedded in concrete and are partially in direct contact with the aggressive environment. In these cases, it is more appropriate to specify austenitic stainless steels with molybdenum, since they provide additional corrosion resistance. A minimum molybdenum content of 2.5% is preferable to 2%, because of the resulting increase in corrosion resistance with only marginal increase in cost.

The critical chloride content decreases as temperature increases; for instance Figure 15.3 shows the expected variations between 20 °C and 40 °C. Thus, bars made of steels containing molybdenum should be preferred in hot climates.

15.2.3
Coupling with Carbon Steel

Often, the use of stainless steel reinforcement is limited to the outer part of the structure (skin reinforcement) or to its most critical parts for economical reasons. Furthermore, when stainless steel bars are used in the rehabilitation of corroding structures, they are usually connected to the original carbon steel rebars. Concern has been expressed with regard to the risk of galvanic corrosion of carbon steel induced by coupling with stainless steel bars. Actually, the galvanic corrosion that can arise when stainless steel is used in partial substitution of carbon steel has to be compared with that which takes place in the absence of stainless steels [41].

In Chapter 8 it was shown that macrocells may be formed in reinforced concrete, since carbonation or chlorides penetrated from the surface led to depassivation only of the outer layer of reinforcement and thus to a macrocell with the inner rebars of carbon steel that remain passive. Experimental studies clearly showed that the use of stainless steel in conjunction with carbon steel does not increase the risk of corrosion of the carbon steel. When both carbon steel and stainless steel rebars are passive and embedded in aerated concrete, macrocell action does not produce appreciable effects, since the two types of steel have almost the same free corrosion potential. Indeed, in this environment, carbon steel is even slightly nobler than stainless steel [42]. In any case, both carbon steel and stainless steel remain passive even after connection. Only when the carbon steel is already corroding does the macrocell current become significant. However, stainless steel is a poor cathode [41]. Figure 15.4 shows that the consequences of coupling corroding carbon steel with stainless steel are generally modest, and they are negligible with respect to those due to the coupling with passive carbon steel that always

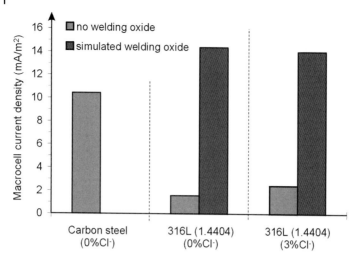

Figure 15.4 Macrocell current density exchanged between a corroding bar of carbon steel in 3% chloride-contaminated concrete and a (parallel) passive bar of: carbon steel in chloride-free concrete, 316L stainless steel in chloride-free concrete or in 3% chloride-contaminated concrete (20 °C, 95% R.H.). Also results on stainless steel bars with oxide scale produced at 700 °C (simulating welding scale) are reported [42].

surrounds the corroding area. Consequently, the increase in corrosion rate on carbon steel embedded in chloride-contaminated concrete due to galvanic coupling with stainless steel is significantly lower than the increase brought about by coupling with passive carbon steel. This behavior is explained by a higher overvoltage for the cathodic reaction of oxygen reduction on stainless steel with respect to carbon steel [19, 41], as shown for instance by the cathodic polarization curves of Figure 15.5. Only minor changes in the cathodic behavior occur due to variation of the chemical composition and strength of stainless steel bars [42], confirming that they can be considered a poor cathode indeed compared to carbon steel. For this reason, stainless steel has even been suggested as a better reinforcement material for use in repair projects when a part of the corroded reinforcement is to be replaced, compared to the usual carbon steel [10]. In fact, the stainless steel will minimize problems that could occur in neighboring corroding and passive areas after the repair (incipient anode effect, Section 18.3.1).

However, stainless steel with welding oxide or with the black scale formed at high temperature is a better cathode and increased galvanic corrosion can occur in the presence of these types of scale on stainless steel bars. A similar effect was observed in the presence of rust deposits from contact with carbon steel bars [43]. Obviously, in evaluating the effect of galvanic coupling, at least in the case of welds, it has to be considered that the area covered by the scale will normally be small compared to the total rebar area. The presence of these types of scale increases the macrocell current density generated by stainless steels, to the same

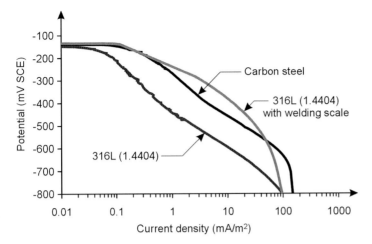

Figure 15.5 Cathodic polarization curves in saturated Ca(OH)$_2$ solution (pH 12.6) of 316L (1.4404) stainless steel compared with the curve of carbon steel. Results of stainless steel with simulated welding scale are also shown [42].

order of magnitude or even greater than that produced by coupling with carbon steel (Figure 15.4), as a consequence of a change in the cathodic behavior (Figure 15.5) [41]. This risk can be avoided by removing the scale. Also in this case, pickling is more effective than sand blasting.

15.2.4
Applications and Cost

In principle, stainless steel reinforcement can be a viable solution for preventing corrosion in a large number of applications. The chloride threshold is much higher than the chloride content that is normally found in the vicinity of the steel even in structures exposed to marine environment or deicing salts. There is no objection to using stainless steel only where its improved protection is necessary, combined with normal steel at other areas. Hence, stainless steel bars can be used in the more vulnerable parts of structures exposed to chloride environments, such as joints of bridges or the splash zone of marine structures. Similarly, they can be used when the thickness of the concrete cover has to be reduced, such as in slender elements.

Their use may have a significant impact on the cost of a structure. The cost of traditional stainless steels varies in time, especially due to fluctuations in the costs of nickel and molybdenum. The following rough indications can be provided: if 1 is the cost of carbon steel bars, 304L austenitic stainless steel bars cost 6–8, and 316L and 22-05 (duplex) cost 9–10 [9, 10].

The additional cost of using stainless steel can be drastically reduced by means of a selective use of stainless steel bars, that is, limited to the more vulnerable parts of the structure. In the past, structural designers were reluctant to use such an alternative because of the fears regarding galvanic coupling with carbon steel. Now, the combined use of stainless steel and carbon steel bars is encouraged in order to reduce cost, referring to an "intelligent" use of stainless steels. This additional cost must be compared to the cost of repair possibly needed in the future multiplied by the probability of its occurrence (Chapter 11). Several authors have shown that by applying life-cycle cost analysis to several types of structures exposed to a chloride environment, the choice of a suitable type of stainless steel in specific parts of the structure can allow savings on future maintenance expenses that can be much higher than the initial increase in cost [10–12, 29, 43–45].

The actual use of stainless steel bars in concrete is, however, still rather modest. Among the examples reported in the literature, some documented cases are: bridges subjected to the use of deicing salts [9], marine structures [46] or historical buildings. An interesting extreme example is the Guildhall Yard East in London [9, 47], which is a building hosting a Roman amphitheater. Stainless steel bars were used for new reinforced concrete walls in order to guarantee a design life of 750 years.

15.2.5
High-Strength Stainless Steels

The use of stainless steel has also been proposed for prestressing wires [48]. Cold-drawn high-strength wires made of austenitic stainless steels 1.4436 and 1.4401 showed a good resistance to various forms of corrosion, even in the presence of applied tensile stresses. Metallurgical or mechanical defects introduced on the surface by cold drawing may decrease the resistance to pitting corrosion. Although high-strength stainless steels could be subjected to chloride-induced stress corrosion cracking, the susceptibility to this form of attack is related to the resistance to pitting corrosion. A satisfactory resistance to this type of attack was found for austenitic steel 1.4401 in alkaline and/or carbonated concrete with high chloride content at ordinary service temperature, while higher-alloyed austenitic stainless steel are not recommended due to the higher cost [49]. Also, in the case of prestressing steel the use of duplex grades of stainless steel is being investigated [50].

15.3
Galvanized Steel Rebars

Galvanized steel rebars can be used as a preventative measure to control corrosion in reinforced concrete structures exposed to carbonation or mild contamination with chlorides, such as chimneys, bridge substructures, tunnels and coastal buildings.

Galvanized steel reinforcement may allow an increase of initiation time of corrosion and a greater tolerance for low cover thickness, for example, in slender (architectural) elements. Furthermore, corrosion protection is offered to the reinforcement prior to it being embedded in concrete or to parts exposed to the atmosphere.

15.3.1
Properties of Galvanized Steel Bars

Galvanized bars are produced by the hot-dip galvanizing process. Pickled steel bars or welded cages are dipped in a bath of molten zinc at a temperature of about 450°C [45]. This process produces a metallic coating composed of various layers of iron-zinc alloys, which has a metallurgical adhesion to the steel substrate. An external layer of pure zinc, left by the simple solidification of the liquid metal, is formed on top of a sequence of inner layers, increasingly rich in iron, which are the result of formation of brittle intermetallic compounds. The thickness of the iron-zinc layers depends on the composition of the steel, the temperature and composition of the zinc bath, and the immersion time. The silicon content in the steel has a great influence; usually it should be maintained between 0.16% and 0.20% to limit the thickness of these brittle layers. The total thickness of the coating should be at least 100 μm and it should not exceed 150 μm.

The proper execution of the galvanization process should guarantee that the temperature and the time of galvanization do not negatively affect the mechanical properties of the steel bars. The external layer of pure zinc is of primary importance with regard to the corrosion resistance of the bars. If galvanized steel is exposed to a neutral environment, such as the atmosphere, the duration of protection is primarily dependent on the thickness of the zinc coating, and its composition and microstructure has a negligible effect. For galvanized steel bars embedded in concrete, the protective properties of zinc coatings are due for the most part to the external layer of pure zinc, which can form a passive film if it has a sufficient thickness [51–54]. In fact, a loss of thickness of 5–10 μm is required prior to passivation, while, if the thickness is insufficient, the underlying layers of Zn–Fe alloy passivate with more difficulty.

The passivation of zinc depends on the pH of the pore solution. In contact with alkaline solutions, as long as the pH remains below 13.3, zinc can passivate due to formation of a layer of calcium hydroxyzincate. Figure 15.6 shows the typical corrosion rate of zinc as a function of pH; however, even at pH values higher than 12, in the presence of calcium ions, such as in concrete pore solution, zinc can be passive and has a very low corrosion rate. In saturated calcium hydroxide solutions it was found that for pH values up to about 12.8 a compact layer of zinc corrosion products forms, which will protect the steel even if the pH changes in a subsequent phase. For pH values between 12.8 and 13.3, larger crystals form that can still passivate the bar. Finally, for values above 13.3, coarse corrosion products form that cannot prevent corrosion.

Since the pH of the concrete pore solution may vary in the interval where remarkable changes in the behavior of zinc occur (Section 2.1.1), the behavior of galvanized steel may be influenced by the composition of the concrete and, especially by the cement type and its alkali content. In practice, however, the pH of the

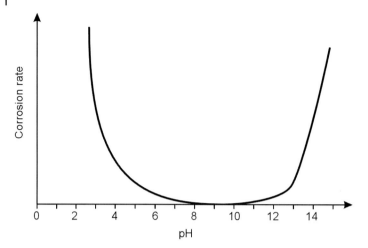

Figure 15.6 Effect of pH on the corrosion rate of pure zinc.

pore solution in concrete usually is below 13.3 during the first hours after mixing, due to the presence of sulfate ions from the gypsum added to the portland cement as a set regulator. A protective layer thus can be formed on galvanized bars.

The passive film that forms on zinc not only reduces the rate of the anodic process (zinc dissolution), but even hinders cathodic reactions of oxygen reduction and hydrogen development [55]. In conditions of passivity, the corrosion potential of galvanized steel is therefore much lower than that of carbon steel. Values typically measured are between −600 and −500 mV SCE compared to values above −200 mV SCE usually found for passive carbon steel reinforcement.

Bonding between reinforcement and concrete is essential for a safe and reliable performance of concrete structures. Several factors, such as concrete composition, placement, curing conditions, and age, may affect the bond between galvanized steel and concrete. At early age (the first days after casting) the bond strength may be lower than that of normal steel bars, due to the hydrogen evolution at the interface and the dissolution of the superficial layer of the zinc coating, which delays the hydration of the interfacial cement paste. However, after a few weeks the galvanized steel adheres well and its increased roughness improves adhesion to the concrete. A higher bond with respect to bare steel could be obtained, due to the formation of calcium hydroxyzincate crystals that fill the interfacial porosity of the cement paste and act as bridges between the zinc coating and concrete [56]. In practice, bond strength for ribbed black steel and galvanized steel bars is essentially the same because it is mainly provided by the mechanical interlocking between the ridges of the ribbed bars and the concrete [48, 57].

Often galvanized bars are chromate treated in order to inhibit zinc corrosion and to control hydrogen evolution (the presence of small amounts of Cr^{6+} in the cement also has a similar effect) [58]. New galvanizing baths are, however, being studied in order to improve corrosion resistance and reduce environmental impact [59, 60].

It should be observed that hydrogen evolution is possible on galvanized bars, first during pickling before galvanization, then in the first hours after casting and finally in hardened concrete in conditions of lack of oxygen. For this reason, galvanizing is usually not recommended as a protective measure of steel susceptible to hydrogen embrittlement (i.e., for prestressing steel).

Galvanized steel bars can be welded, but loss of the zinc coating may take place in the welded zone; the application of a zinc-rich paint should be recommended after cleaning of the welded area.

15.3.2
Corrosion Resistance

The passive film of galvanized rebars is stable even in mildly acidic environment, so that the zinc coating remains passive even when the concrete is carbonated. The corrosion rate of galvanized steel in carbonated concrete is significantly lower than that of carbon steel and it remains negligible even if a low content of chloride is present.

In chloride-contaminated concrete, galvanized steel may be affected by pitting corrosion, although the chloride threshold is somewhat higher than conventional carbon steel [61–64]. In general, from earlier studies, an approximate critical chloride level in the range of 1–1.5% by mass of cement has been suggested for galvanized steel (Figure 15.7), compared to the value of 0.4–1% normally considered for carbon steel reinforcement (Chapter 6). The slightly improved resistance to chloride attack is due, to a large part, to the lower value of the free corrosion potential of galvanized steel, leading to an increase in the chloride threshold (see Section 7.4.1).

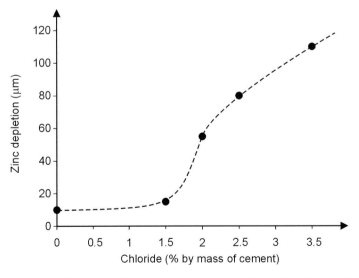

Figure 15.7 Depletion of the zinc coating of galvanized bars after 2.5 years of exposure in different chloride-contaminated concretes (the initial thickness was 160 μm) [58].

Even if pitting corrosion has initiated, the corrosion rate tends to be lower for galvanized steel, since the zinc coating that surrounds the pits is a poor cathode and thus it reduces the effectiveness of the autocatalytic mechanism that takes place inside pits on bare steel (Section 7.4.2). On the other hand, it can be observed that as long as the zinc coating is passive it is not able to provide active protection to steel (as happens for galvanized steel exposed to the atmosphere) and so it cannot reduce the corrosion rate in the areas of the steel that are not protected. Consequently, cracks in the zinc coating should be prevented and macroscopic defects have to be repaired prior to casting.

The price of galvanized bars is about 2 to 2.5 times the price of normal black steel bars.

15.3.3
Galvanized Steel Tendons

Tendons made of hot-dip galvanized wires, which are widely used for structural cables exposed to aggressive atmospheres, may also be used for prestressed concrete. Their advantage is the improvement of the corrosion resistance in the case of carbonation, mild chloride contamination, and unfilled ducts [48]. When they are used in contact with concrete or cement grout, the main concern is related to the risk of hydrogen evolution that takes place when the zinc coating comes in contact with fresh alkaline cementitious materials and the consequent risk of inducing hydrogen embrittlement. It has been suggested that the risk of hydrogen embrittlement is limited in real cases and it may be further reduced by using cement or grouts of controlled composition or by means of specific galvanizing baths.

15.4
Epoxy-Coated Rebars

Epoxy coating of reinforcing bars is a protective technique that was developed in the 1970s in North America. Earlier laboratory results confirmed the effectiveness of the epoxy-coated bars, in many cases, in preventing corrosion of reinforcement in carbonated or chloride-contaminated concrete [61–63]. Later, however, doubts arose about their long-term durability in very aggressive environments, doubts borne out above all by negative experience reported on structures in tropical environments [64–66].

15.4.1
Properties of the Coating

Protection of rebars by organic coatings is based on the principle of insulating the steel and protecting it from aggressive agents that penetrate the concrete cover. Nevertheless, the use of coated bars should not require any changes in the structural design or during the different phases of construction. The coating should be able to cover the reinforcement uniformly, be tough and well adherent, flexible enough to allow bending of the reinforcement, able to provide enough bond strength between concrete and the reinforcing bars. Of all organic coatings available, the only coating types able to satisfy all these conditions are those made with epoxy resins [65].

Beyond the need to insure good adhesion of the coating to the steel surface (essential in order to guarantee adequate resistance to corrosion and to allow safe bending of the rebar), it is of crucial importance to have a proper bond between the coated bar and the concrete. In fact, to avoid changes in the structural design procedures, it is necessary to obtain levels of bond strength with epoxy-coated bars comparable to those of bare reinforcement. Usually, coatings are less than 300 μm in thickness, and the reduction in bond strength to concrete of ribbed bars with epoxy coating with respect to uncoated bars of the same geometry is limited, at least for commonly used diameters.

Requirements for epoxy-coated reinforcing bars are reported in different international standards and recommendations; the first dates back to 1981 (ASTM A775-81) and more recent national standards in European countries are based on it, even though their requirements are much more rigorous.

The price of epoxy-coated bars is roughly twice the price of uncoated bars.

15.4.2
Corrosion Resistance

Even though it is not completely impermeable to oxygen, water and chlorides, epoxy coating of reinforcing bars can, in principle, guarantee protection against reinforcement corrosion in chloride-contaminated concrete. The protection provided by the coating improves as its thickness increases. There is, however, an upper limit fixed by the need to achieve adequate bond between the steel and the concrete. Thicknesses between 0.1 and 0.3 mm are recommended. Effective protection depends to a great extent on the integrity of the coating; in fact any damage will expose bare metal to the aggressive environment. In the case of chloride-contaminated concrete, the attack tends to penetrate below the coating and widen the area affected. In carbonated concrete, on the other hand, the attack tends to remain in the region of the defect.

Very high corrosion rates in the vicinity of defects in the coating can occur in the presence of macrocells. A typical situation is that of structures in which epoxy-coated reinforcement in contact with chloride-contaminated concrete is coupled to noncoated reinforcement embedded in concrete that is uncontaminated or contains a level of chlorides below the critical level (and electrical connection is provided, even unintentionally, between coated and uncoated bars). In this case, the passive noncoated reinforcement can act as an effective cathode of much greater size than the anodic areas corresponding to the defects in the coating, thereby determining a very unfavorable anode/cathode area ratio (Chapter 8). The corrosion rate will be particularly high if the resistivity of the concrete is low.

15.4.3
Practical Aspects

As indicated above, defects in epoxy coating on bars present the risk of strong local attack. Specifications and site practice must be aimed at obtaining coated bars without such defects, particularly with regard to production, handling, cutting, bending, storage, welding, and bonding. Furthermore, providing good chloride penetration resistance and high electrical resistivity to the concrete makes the overall protection more robust.

It is, furthermore, important to pay attention to those cases in which, for economical reasons, coated steel is used only in the most critical areas of the structure. In these cases it is important to ensure that epoxy-coated bars are electrically insulated from the uncoated reinforcement, in order to avoid macrocells.

15.4.4
Effectiveness

Very serious cases of corrosion damage have been reported in some structures in tropical areas in Florida where severe attack of epoxy-coated steel has been observed only a few years after construction [66–68]. These were explained by a combination of adverse factors [69, 70]. The epoxy-coated rebars contained a small number of initial coating imperfections, as permitted by the acceptance criteria at the time. Additional surface damage was introduced by shipping and handling, while fabrication introduced some disbonding. The bars were then exposed to the construction yard environment for a time that may have ranged from days to over a year. In summary, the presence of allowable (per specifications prevalent at the time of manufacturing) production imperfections that were then aggravated by shipping, handling, and a severe construction yard environment generated defects in the coating where corrosion could initiate when chloride ions could penetrate in the pore solution. Furthermore, structures were exposed to highly aggressive service environment which, in the absence of a thick cover of highly impermeable concrete, rapidly left the epoxy film as the only remaining corrosion protecting barrier on the steel bar. Given also the inherent vulnerability of the film to flaws and disbondment from the base metal, corrosion quickly ensued with electrochemical aggravating factors such as the formation of extended macrocells. Consequently, damage from corrosion of epoxy-coated bars continued to develop steadily.

This situation has shown that this protection system, at least in highly aggressive environments, is not robust enough with regard to the construction-site practice and led contractors and designers to reconsider the widespread use of this technique on structures exposed to chloride-bearing environment. Serious doubts have been expressed about whether epoxy coatings, even in the presence of low-permeability concrete, can insure long-lasting protection in heavily chloride-contaminated and hot environments, particularly when the concrete is frequently wetted [68].

It should also be observed that, because of the lack of electrical connection between the individual coated bars, if the coating would not be effective in protecting the bar, the application of electrochemical techniques (Chapter 20), such as cathodic protection, is not possible in practice. Even the inspection of structures is difficult, for example, potential mapping (Chapter 16) cannot be applied if bars are disconnected. To avoid this problem, for precast elements in Europe the coating has also been applied on welded mesh or complete reinforcement cages, so that electrical connection of bars was guaranteed [65].

References

1. EN 1992-1-1 (2004) Eurocode 2: design of concrete structures – Part 1-1: general rules and rules for buildings, European Committee for Standardization, 2004.
2. EN 10080 (2005) Steel for the reinforcement of concrete – Weldable reinforcing steel – General, European Committee for Standardization, 2005.
3. Nürnberger, U. (1995) *Korrosion und Korrosionschutz im Bauwesen*, vol. 1, Bauverlag, Wiesbaden.
4. Isecke, B. (2004) *Test Methods for Assessing the Susceptibility of Prestressing Steels against Hydrogen Induced Stress Corrosion Cracking*, European Federation of Corrosion, Special Publication No. 37.
5. Manera, M., Vennesland, Ø., and Bertolini, L. (2008) Chloride threshold for rebar corrosion in concrete with addition of silica fume. *Corrosion Science*, **50** (2), 554–560.
6. Angst, U., Elsener, B., Larsen, C.K., and Vennesland, Ø. (2009) Critical chloride content in reinforced concrete – A review. *Cement and Concrete Research*, **39** (12), 1122–1138.
7. Addari, D., Elsener, B., and Rossi, A. (2008) Electrochemistry and surface chemistry of stainless steels in alkaline media simulating concrete pore solutions. *Electrochimica Acta*, **53** (27), 8078–8086.
8. EN 10088-1 (2005) Stainless steels – Part 1: list of stainless steels, European Committee for Standardization, 2005.
9. Nürnberger, U. (ed.) (1996) *Stainless Steel in Concrete*, The European Federation of Corrosion Publication number 18, The Institute of Materials, London.
10. The Concrete Society (1998) Guidance on the Use of Stainless Steel Reinforcement, Technical Report No. 51, 1998.
11. Bauer, A.E. and Cochrane, D.J. (1999) The actual implication of stainless steel reinforcement in concrete structures, Euro Inox, 1999.
12. Hunkeler, F. (2000) Einsatz von nichtrostenden Bewehrungsstaehlen im Betonbau, Technical Report 80/00, Swiss Federal Department for Environment, Transports, Energy and Communications/Federal Office for Roads, Wildegg (CH), May 2000.
13. Bertolini, L. and Pedeferri, P. (2002) Laboratory and field experience on the use of stainless steel to improve durability of reinforced concrete. *Corrosion Reviews*, **20**, 129–152.
14. Kulessa, G. (1988) Stainless steel reinforcement for concrete. *Betonwerk und Fertigteiltechnik*, **54** (3), 58–63.
15. Cui, F. and Sagüés, A. (2003) Corrosion performance of stainless steel clad rebar in simulated pore water and concrete, Corrosion/2003, paper 03310, NACE International, Houston, 2003.
16. Brown, B.L., Harrop, D., and Treadaway, K.W.J. (1978) document 45/78. *Corrosion Testing of Steels for Reinforced Concrete*, Building Research Establishment, Garston.
17. Flint, G.N. and Cox, R.N. (1988) The resistance of stainless steel partly embedded in concrete to corrosion by seawater. *Magazine of Concrete Research*, **40**, 13–27.

18 Treadaway, K.W.J., Cox, R.N., and Brown, B.L. (1989) Durability of corrosion resisting steels in concrete. *Proceedings of the Institution of Civil Engineers, Part 1*, **86**, 305–331.

19 Sørensen, B., Jensen, P.B., and Maahn, E. (1990) The corrosion properties of stainless steel reinforcement, in *Corrosion of Reinforcement in Concrete* (eds C.L. Page, K.W.J. Treadaway, and P.B. Bamforth), Elsevier Applied Science, pp. 601–610.

20 Pastore, T., Pedeferri, P., Bertolini, L., Bolzoni, F., and Cigada, A. (1991) Electrochemical study on the use of duplex stainless steel in concrete, in *Proc. Int. Conf. Duplex Stainless Steel'91*, vol. 2, Beaune, pp. 905–913.

21 Rasheeduzzafar, Dakhil, F.H., Bader, M.A., and Khan, M.M. (1992) Performance of corrosion-resisting steels in chloride-bearing concrete. *ACI Materials Journal*, **89** (5), 439–448.

22 Callaghan, B.G. (1993) The performance of 12% chromium steel in concrete in severe marine environments. *Corrosion Science*, **35** (5–8), 1535–1541.

23 Nürnberger, U., Beul, W., and Onuseit, G. (1993) Corrosion behaviour of welded stainless reinforcing steel in concrete. *Otto Graf Journal*, **4**, 225–259.

24 Hewitt, J. and Tullmin, M. (1994) Corrosion and stress corrosion cracking performance of stainless steel and other reinforcing materials in concrete, in *Corrosion and Corrosion Protection of Steel in Concrete*, vol. I (ed. R.N. Swamy), Sheffield Academic Press, pp. 527–539.

25 Bertolini, L., Bolzoni, F., Pastore, T., and Pedeferri, P. (1996) Behaviour of stainless steel in simulated concrete pore solution. *British Corrosion Journal*, **31**, 218–222.

26 Cox, R.N. and Oldfield, J.W. (1996) The long term performance of austenitic stainless steel in chloride contaminated concrete, in *Corrosion of Reinforcement in Concrete Construction* (eds C.L. Page, P.B. Bamforth, and J.W. Figg), Society of Chemical Industry, pp. 662–669.

27 Pedeferri, P., Bertolini, L., Bolzoni, F., and Pastore, T. (1998) Behaviour of stainless steels in concrete, in *Repair and Rehabilitation of Reinforced Concrete Structures: The State of the Art* (eds W.F. Silva Araya, O.T. de Rincon, and L.P. O'Neill), American Society of Civil Engineering, pp. 192–206.

28 Bertolini, L., Pedeferri, P., and Pastore, T. (1998) Stainless steel in reinforced concrete structures, in *Concrete under Severe Conditions 2 – Environment and Loading*, vol. 1 (eds O.E. Gjørv, K. Sakai, and N. Bathia), E & FN Spon, pp. 94–103.

29 Knudsen, A., Jensen, F.M., Klinghoffer, O., and Skovsgaard, T. (1998) Cost-effective enhancement of durability of concrete structures by intelligent use of stainless steel reinforcement. Int. Conf. on Corrosion and Rehabilitation of Reinforced Concrete Structures, Federal Highway Administration, Orlando, 7–11 December 1998 (CD-ROM).

30 McDonald, D., Pfeifer, D., and Virmani, P. (1998) Corrosion resistant reinforcing bars – findings of a 5-year study. Int. Conf. on Corrosion and Rehabilitation of Reinforced Concrete Structures, Federal Highway Administration, Orlando, 7–11 December 1998 (CD-ROM).

31 Nurnberger, U. and Beul, W. (1999) Corrosion of stainless steel reinforcement in cracked concrete. *Otto Graf Journal*, **10**, 23–37.

32 Bertolini, L., Gastaldi, M., Pastore, T., and Pedeferri, M.P. (2000) Corrosion behaviour of stainless steels in chloride contaminated and carbonated concrete. *Internationale Zeitschrift für Bauinstandsetzen und Baudenkmalpflege*, **6**, 273–292.

33 Kouřil, M., Novák, P., and Bojko, M. (2010) Threshold chloride concentration for stainless steels activation in concrete pore solutions. *Cement and Concrete Research*, **40** (3), 431–436.

34 Alvarez, S.M., Bautista, A., and Velasco, F. (2011) Corrosion behaviour of corrugated lean duplex stainless steels in simulated concrete pore solutions. *Corrosion Science*, **53** (5), 1748–1755.

35 García-Alonso, M.C., González, J.A., Miranda, J., Escudero, M.L., Correia, M.J., Salta, M., and Bennani, A. (2007) Corrosion behaviour of innovative stainless steels in mortar. *Cement and Concrete Research*, **37** (11), 1562–1569.

36 Hartt, W.H., Powers, R.G., and Kessler, R.J. (2009) Performance of corrosion

37. Bertolini, L. and Gastaldi, M. (2011) Corrosion resistance of low-nickel duplex stainless steel rebars. *Materials and Corrosion*, **62** (2), 120–129.
38. Elsener, B., Addari, D., Coray, S., and Rossi, A. (2011) Nickel-free manganese bearing stainless steel in alkaline media – Electrochemistry and surface chemistry. *Electrochimica Acta*, **56** (12), 4489–4497.
39. Page, C.L. (2009) Initiation of chloride-induced corrosion of steel in concrete: role of the interfacial zone. *Materials and Corrosion*, **60** (8), 586–592.
40. Böhni, H., Haselmair, H., and Ubeleis, A.M. (1992) Corrosion-resistant fastening in road tunnels – field tests. *Structural Engineering International*, **2**, 253–258.
41. Bertolini, L., Gastaldi, M., Pastore, T., Pedeferri, M.P., and Pedeferri, P. (1998) Effects of galvanic coupling between carbon steel and stainless steel reinforcement in concrete. Int. Conf. on Corrosion and Rehabilitation of Reinforced Concrete Structures, Federal Highway Administration, Orlando, 7–11 December 1998 (CD-ROM).
42. Bertolini, L., Gastaldi, M., Pedeferri, M.P., and Pedeferri, P. (2000) Stainless steel in concrete. Annual Progress Report, European Community, COST 521 Workshop, Queens University, Belfast, 28–31 August 2000, pp. 27–32.
43. Qian, S. and Qu, D. (2010) Theoretical and experimental study of galvanic coupling effects between carbon steel and stainless steels. *Journal of Applied Electrochemistry*, **40** (2), 247–256.
44. Abbott, C.J. (1997) Corrosion-free concrete structures with stainless steel. *Concrete*, **31** (5), 28–32.
45. COST 521 (2003) Corrosion of steel in reinforced concrete structures, Final Report (eds R. Cigna, C. Andrade, U. Nürnberger, R. Polder, R. Weydert, and E. Seitz), European Commission, Directorate General for Research, EUR 20599, 2003.
46. Castro-Borges, P., de Rincon, O.T., Moreno, E.I., Torres-Acosta, A.A., Martinez-Madrid, M., and Knudsen, A. (2002) Performance of a 60-year-old concrete pier with stainless steel reinforcement. *Materials Performance*, **41**, 50–55.
47. Val, D.V. and Stewart, M.G. (2003) Life-cycle cost analysis of reinforced concrete structures in marine environments. *Structural Safety*, **25**, 343–362.
48. COST 534 (2009) New Materials, Systems, Methods and Concepts for Prestressed Concrete Structures (eds R. Polder, M.C. Alonso, D. Cleland, B. Elsener, E. Proverbio, Ø. Vennesland, and A. Raharinaivo), European Commission, Cost office, 2009.
49. Wu, Y. and Nürnberger, U. (2009) Corrosion-technical properties of high-strength stainless steels for the application in prestressed concrete structures. *Materials and Corrosion*, **60** (10), 771–780.
50. Mahmoud, H., Alonso, M.C., and Sanchez, M. (2012) Service life extension of concrete structures by increasing chloride threshold using stainless steel reinforcements. 3rd Int. Conf. on Concrete repair, Rehabilitation and Retrofitting, Cape Town, 3–5 September 2012.
51. Gonzales, J.A. and Andrade, C. (1982) Effect of carbonation, chlorides and relative ambient humidity on the corrosion of galvanized rebars embedded in concrete. *British Corrosion Journal*, **17**, 21–28.
52. Bianco, M.T., Andrade, C., and Macias, A. (1984) SEM study of the corrosion products of galvanized reinforcements immersed in solutions in the pH range 12.6 to 13.6. *British Corrosion Journal*, **19**, 41–48.
53. Sergi, G., Short, N.R., and Page, C.L. (1985) Corrosion of galvanized and galvannealed steel in solutions of pH 9.0 to 14.0. *Corrosion*, **41**, 618–624.
54. Maahn, E. and Sørensen, B. (1986) The influence of microstructure on the corrosion properties of hot-dip galvanized reinforcement in concrete. *Corrosion*, **42**, 187–196.
55. Tittarelli, F., and Bellezze, T. (2010) Investigation of the major reduction

reaction occurring during the passivation of galvanized steel rebars. *Corrosion Science*, **52** (3), 978–983.

56 Fratesi, R., Moriconi, G., and Coppola, L. (1996) The Influence of Steel Galvanization on Rebar Behaviour in Concrete. *Corrosion of Reinforcement Construction*, SCI, The Royal Society of Chemistry, Cambridge, pp. 630–641.

57 Yeomans, S.R. and Novak, M.P. (1990) Further studies of the comparative properties and behaviour of galvanised and epoxy coated steel reinforcement, International Lead Zinc Research Organization (ILZRO) project ZE-341, Progress report N° 4, Research Triangle Park, North Carolina, July, 1990.

58 Fratesi, R. (2003) Galvanized steel reinforcement. COST Action 521, Corrosion of Steel in Reinforced Concrete Structures, Final Report (eds R. Cigna, C. Andrade, U. Nürnberger, R. Polder, R. Weydert, and E. Seitz), European Communities, Luxembourg, Publication EUR 20599, pp. 28–40.

59 Ghosh, R. and Singh, D.D.N. (2007) Kinetics, mechanism and characterisation of passive film formed on hot dip galvanized coating exposed in simulated concrete pore solution. *Surface and Coatings Technology*, **201** (16–17), 7346–7359.

60 Bellezze, T., Malavolta, M., Quaranta, A., Ruffini, N., and Roventi, G. (2006) Corrosion behaviour in concrete of three differently galvanized steel bars. *Cement and Concrete Composites*, **28** (3), 246–255.

61 Macias, A. and Andrade, C. (1987) Corrosion of galvanized steel reinforcements in alkaline solutions. *British Corrosion Journal*, **22**, 119.

62 Nürnberger, U. and Buel, W. (1991) Einfluß einer Feuerverzinkung und PVC-Beschichtung von Bewehrungsstählen und von Inhibitoren auf die Korrosion von Stahl in gerissenem Beton. *Materials and Corrosion*, **42**, 537–546.

63 Maldonado, L. (2009) Chloride threshold for corrosion of galvanized reinforcement in concrete exposed in the Mexican Caribbean. *Materials and Corrosion*, **60** (7), 536–539.

64 Darwin, D., Browning, J., O'Reilly, M., Xing, L., and Ji, J. (2009) Critical chloride corrosion threshold of galvanized reinforcing bars. *ACI Materials Journal*, **106** (2), 176–183.

65 Schiessl, P. and Reuter, C. (1991) Coated reinforcing steels development and applications in Europe, Corrosion/91, NACE, Houston, Paper No. 556, 1991.

66 Clear, K.C. and Virmani, Y.P. (1983) Corrosion of non-specification epoxy coated rebars in salty concrete, Corrosion/83, NACE, Houston, Paper No. 114, 1983.

67 Darwin, A.B. and Scantlebury, J.D. (2002) Retarding of corrosion processes on reinforcement bar in concrete with an FBE coating. *Cement and Concrete Composites*, **24** (1), 73–78.

68 Clear, K.C. (1992) Effectiveness of epoxy-coated reinforcing steel. *Concrete International*, **5**, 58–62.

69 Sagues, A. and Powers, R.G. (1997) Corrosion and corrosion control of concrete structures in Florida. What can be learned? Proc. Int. Conf. Repair of Concrete Structures. From Theory to Practice in a Marine Environment, Svolvear (Norway), 28–30 May, 49.

70 Sagues, A.A., Lau, K., Powers, R.G., and Kessler, R.J. (2010) Corrosion of epoxy-coated rebar in marine bridges– part 1: a 30-year perspective. *Corrosion*, **66** (6), 0650011–06500113.

16
Inspection and Condition Assessment

There are many reasons why inspection or condition assessment of concrete structures is performed. Many large-scale owners of structures have a routine schedule according to which every structure is inspected every five to twelve years. Such activities are part of *asset management*. The assumption is that signs of deterioration will be observed by visual inspection and proper follow up can be taken before damage becomes too large. Another reason for the assessment is change of ownership, in which case the economical value has to be estimated, including any deterioration that may form a hidden liability. Furthermore, insurance companies may want the insured capital to be evaluated now and then. Structures with damage due to incidents such as fires and collisions need to be inspected for establishing proper corrective measures. Furthermore, inspection is needed any time when a structure shows signs of degradation, in order to investigate the causes of the phenomenon and its extent.

Concrete structures showing damage caused by reinforcement corrosion have to be repaired in order to reach their expected service life (see Part V). To find the optimum repair solution and to avoid corrosion problems in the future, a thorough condition assessment of the structure has to be performed, concentrating on the location of those areas already corroding, the identification of the cause of damage and the prediction of the expected progress of damage with time. A full evaluation normally is a two-step process: a *preliminary survey* (mainly based on a visual inspection) should identify the nature of the problem and give the basis for planning the *detailed survey*, where several investigation techniques described below are necessary (Figure 16.1). As corrosion is an electrochemical process, electrochemical techniques are especially well suited to assess the corrosion state of the reinforcement, to pinpoint corrosion sites and to measure the corrosion rate.

Recently, various aspects of full surface corrosion surveys, in particular with nondestructive and/or mapping techniques, were the subject of a Task Group of European Federation of Corrosion's Working Group 11 "Reinforcement corrosion", which were collectively published in a special issue of *Materials and Corrosion* [1].

Corrosion of Steel in Concrete: Prevention, Diagnosis, Repair, Second Edition. Luca Bertolini,
Bernhard Elsener, Pietro Pedeferri, Elena Redaelli, and Rob Polder.
© 2013 Wiley-VCH Verlag GmbH & Co. KGaA. Published 2013 by Wiley-VCH Verlag GmbH & Co. KGaA.

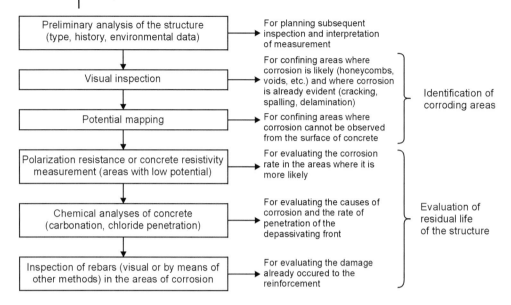

Figure 16.1 Example of integrated methodology of inspection for assessing corrosion of reinforcement.

16.1
Visual Inspection and Cover Depth

Visual inspection is the first step in any investigation. It should give a first indication of the extent of damage. Starting from a simple general impression of part of a structure it may end up with a careful registration of every defect visible on the concrete surface. Special attention should be given to the presence of cracks, concrete spalling, wet areas, signs of water run off or rust staining. A detailed visual survey has to be planned in advance, the results and its interpretation should always be confirmed by additional testing.

The presence of cracks can have many causes (Figure 16.2). Some types of cracks originate from early processes like plastic shrinkage of the fresh concrete, or drying out in the early life of the hardened product. They will have irregular patterns, called map cracking. Other cracks stem from differential shrinkage due to drying out or temperature gradients. Yet another type of cracks is due to plastic settlement or overload. All of these cracks may negatively influence the durability of the structure, in that they provide pathways for aggressive substances to reach the reinforcement. Cracks may also be the consequence of lack of durability: alkali silica reaction may cause cracks that have a map-like appearance. Freeze–thaw cycles may cause cracking and surface scaling. Finally, reinforcement corrosion may show up as rust staining, cracking parallel to reinforcement and spalling of parts of the cover. Many of these processes have been described in previous chap-

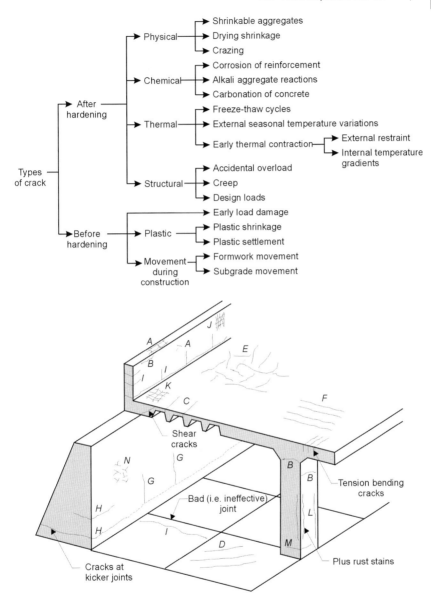

Figure 16.2 Types of cracks that can be encountered during visual inspection (from CEB [2]).

Type of cracking		Subdivision	Most common location	Primary cause (excluding restraint)	Secondary causes/factors	Time of appearance
Plastic settlement	A	Over reinforcement	Deep sections	Excess bleeding	Rapid early drying conditions	10 min to 3 h
	B	Arching	Top of columns			
	C	Change of depth	Waffle slabs			
Plastic shrinkage	D	Diagonal	Roads and slabs	Rapid early drying	Low rate of bleeding	30 min to 6 h
	E	Random	R.C. slabs			
	F	Over reinforcement	R.C. slabs	Rapid early drying, steel near surface		
Early thermal contraction	G	External restraint	Thick walls	Excess heat generation	Rapid cooling	1 day to 2-3 weeks
	H	Internal restraint	Thick slabs	Excess temperature gradients		
Long-term drying shrinkage	I		Thin slabs (and walls)	Inefficient joints	Excess shrinkage, inefficient curing	Several weeks or months
Crazing	J	Against formwork	'Fair-faced' concrete	Impermeable formwork	Rich mixes	1-7 days sometimes much later
	K	Floated concrete	Slabs	Over-trowelling	Poor curing	
Corrosion of reinforcement	L	Natural	Columns and beams	Lack of cover	Poor quality concrete	More than 2 years
	M	Calcium chloride	Precast concrete	Excess clacium chloride		
Alkali aggregate reaction	N		(Damp locations)	Reactive aggregate plus high alkali content		More than 5 years

Figure 16.2 (Continued)

ters. When a visual inspection is carried out, each type of cracking should be recognized and appropriate action should be taken. This chapter focuses on identifying reinforcement corrosion, and cracking can be an important sign that corrosion is going on—although chloride-induced localized corrosion might be detected by cracking only in a very advanced state.

A very important nondestructive test that is often combined with visual inspection is the measurement of the concrete cover. Low cover will favor the onset of corrosion because there, carbonation or chlorides reach the rebars more rapidly. The availability of oxygen and moisture usually is higher, resulting in a higher corrosion rate. A cover survey is essential to explain why some areas of a structure are corroding and to identify areas of future corrosion risk. An example of distribution of concrete cover measured onsite is shown in Figure 16.3, that highlights the great scatter that may be associated with this parameter [3]. Magnctic covermeters are available with programmed data logging capabilities, data storage, and printer or computer output. Nowadays, there is the possibility of actually scanning the surface and obtaining a map of cover depths and rebar layout. Nevertheless, care should be taken because congestion of rebars, different steel types and diam-

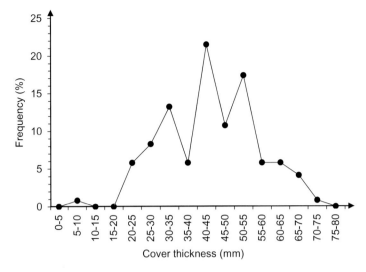

Figure 16.3 Frequency distribution of concrete cover thickness measured onsite on the walls of the reactor building in a nuclear power plant [3].

eters may give misleading information. Indepth analysis of covermeter performance including comparative testing on laboratory slabs is described in [4, 5].

16.2
Electrochemical Inspection Techniques

The electrochemistry of corrosion and the basics of electrochemical potentials and corrosion rate have been discussed in Chapter 7. Here, the principles and application of electrochemical inspection techniques for reinforced concrete structures are given. The different techniques will give different types of information (Figure 16.4).

16.2.1
Half-Cell Potential Mapping

Today potential mapping is the only widely recognized and standardized nondestructive method for assessing the corrosion state of rebars in concrete structures [6–8]. Half-cell potential mapping has proved a very useful, nondestructive means to locate areas of corrosion for monitoring and condition assessment as well as in determining the effectiveness of repair work [9]. As an early warning system, corrosion is detected long before it becomes visible at the concrete surface. Based on potential mapping, other destructive and laboratory analyses (e.g., cores to determine chloride content) and corrosion rate measurements can be performed more

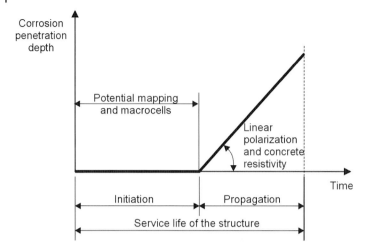

Figure 16.4 Types of information obtainable from different electrochemical techniques.

rationally [8]. In addition, the amount of concrete removal in repair works can be minimized because the corrosion sites can be located precisely.

Principle Corroding and passive rebars in concrete show a difference in corrosion potential of up to 0.5 V, thus a macrocell generates and current flows between these areas (Chapter 8). The electric field coupled with the corrosion current between corroding and passive areas of the rebars (Figure 16.5) can be measured with a suitable reference electrode (half-cell) placed on the concrete surface, resulting in equipotential lines (potential field) that allow the location of corroding rebars at the most negative values [10–13].

Procedure The procedure for measuring half-cell potentials is straightforward (Figure 16.6): a sound electrical connection is made to the reinforcement, an external reference electrode is placed on a wet sponge on the concrete surface and potential readings are taken with a high-impedance voltmeter (>10 MΩ) on a regular grid on the free concrete surface. Good electrolytic contact is essential to get stable readings, the point of measurement should be clean and wetted with a water-soaked sponge on the surface. The use of tap water with the addition of a small amount of a detergent is recommended.

The value of the potential depends on the type of reference electrode used for the measurement. A few types of reference electrodes are used for potential mapping, mainly copper/copper sulfate (CSE) and in chloride environments also silver/silver chloride (Ag/AgCl). They differ in their standard potential, which is the potential difference to the standard hydrogen electrode (SHE). Standard potentials of these reference electrodes are given in Table 16.1, together with some other types used as embedded probes in concrete (Chapter 17). CSE has a potential of

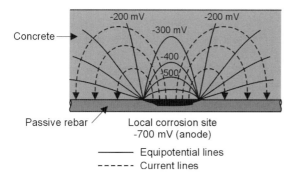

Figure 16.5 Schematic view of the electric field and current flow in an active/passive macrocell on steel in concrete.

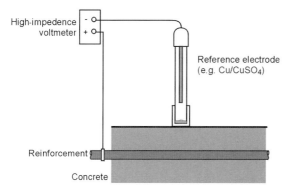

Figure 16.6 Schematic representation of the measurement of potential of steel reinforcement.

Table 16.1 Potentials vs. SHE for reference electrodes [14, 15].

Electrode	Potential (mV SHE)
Calomel (SCE)	+244
Silver/silver chloride (SSE)[a]	+199
Copper/copper sulfate (CSE)	+316
Manganese dioxide	+365
Graphite	+150/−20
Activated titanium	+150/−20
Stainless steel	+150/−20
Lead	−450

a) Value depends on internal KCl electrolyte concentration.

about +70 mV SCE, while Ag/AgCl has a potential that depends on the concentration of the KCl electrolyte and that is, −45 mV SCE for saturated KCl solution.

The grid size (interpoint distance) should be chosen in order not to miss typical potential dips, so should preferably be small, for example, 250 mm. On the other hand, the number of points to be measured should be economically small, so a grid size of 500 mm could be practical.

Data Collection and Representation The simplest configuration for potential measurement described above may be replaced by various microprocessor-controlled single or multiple electrode devices that have been developed and are commercially available, as are wheel electrodes that simultaneously record potential readings and position onsite.

To facilitate the potential survey of large bridge decks, walls, or parking decks, multiple-wheel measuring systems were developed [16] that allow surveys to be done of large areas with grid dimensions of for example, 150 mm using computer-assisted data acquisition and processing (Figure 16.7).

The way of data representation depends on the amount of potential readings: for small-sized elements (e.g., columns), a simple table might be appropriate. For large surfaces with thousand of readings the best way of representing the data has been found to be a map of the potential field, for example, a color map where every individual potential reading can be identified as a small square (Figure 16.8) or a map with isopotential curves. Alternatively, 3D plots can be generated.

Figure 16.7 Multiple-wheel electrode half-cell potential measuring instrument with computer assisted data acquisition. Note the slight wetting of the concrete surface at the wheels in order to achieve a good electrolytic contact between reference electrode and concrete.

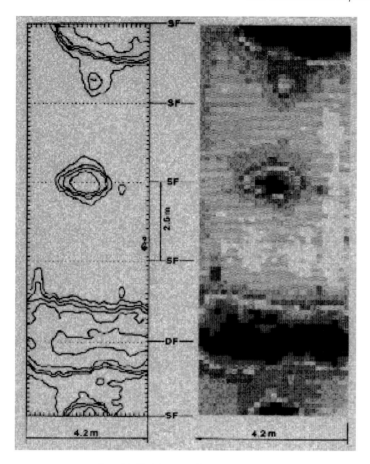

Figure 16.8 Examples of half-cell potential maps (riding deck in the San Bernardino tunnel). Data representation: isopotential plot (a) and color plot (b). DF: dilatation joint (every 25 m).

Interpretation Half-cell potential measurements allow areas of corroding reinforcement to be located, being the most negative zones in the potential field (Figure 16.8). However, the interpretation of the readings is not straightforward because the concrete cover and its resistivity in addition to the actual corrosion potential of the steel (Chapter 7) influence the readings at the concrete surface.

In atmospherically exposed reinforced concrete the potential of passive steel is between +50 and −200 mV CSE (Figure 16.9). If corrosion is ongoing the potential becomes more negative: chloride-induced pitting corrosion results in values from −400 to −700 mV CSE, corrosion due to carbonation usually results in values from +200 to −500 mV CSE, strongly depending on the presence of moisture (Figure 16.9). Problems in the interpretation of "negative" potentials may occur when dealing with structures that are completely immersed in the ground or in seawater, or are water saturated (or in any other condition of lack of oxygen); such

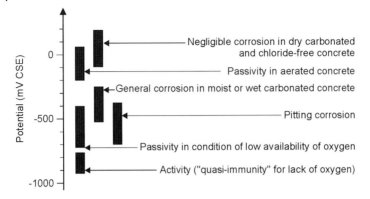

Figure 16.9 Correlation between potential and state of corrosion of carbon steel reinforcement.

Table 16.2 Interpretation of potential measurements according to the American Standard ASTM C876 [6].

Measured potential E (mV CSE)	Probability of corrosion
$E > -200$	<10%
$-200 > E > -350$	unknown
$E < -350$	>90%

conditions are characterized by very negative potentials (lower than −700 mV CSE). It is important to point out that despite these negative potential values the steel in such water-saturated concrete normally does not exhibit a significant corrosion rate [17].

For reinforced concrete structures exposed to the atmosphere (e.g., bridge decks), the American Standard ASTM C876 [6] provides criteria for interpretation, as summarized in Table 16.2. This way of interpretation was derived empirically from chloride-induced corrosion of cast-in-place bridge decks in the USA, thus the values reflect a specific condition, concrete type (based on portland cement) and cover and are not universally applicable. Indeed, as theoretical considerations and practical experience on a large number of structures [12, 18] have shown, there are no absolute potential values to indicate the probability of corrosion in a structure, in contrast to the interpretation given in the ASTM C876 that relies on a fixed potential value of −350 mV CSE. Depending on moisture content, chloride content, temperature, carbonation of the concrete and cover thickness, different potential ranges indicate corrosion of the rebars in different structures (Figure 16.10). In order to locate areas of corroding rebars the gradient between corroding and

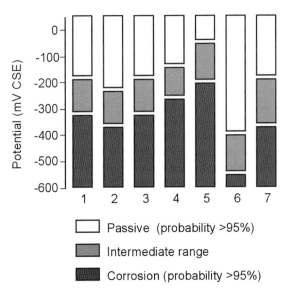

Figure 16.10 Experimentally determined potential range indicating active corrosion on different bridge decks compared to the ASTM standard: 1: Cugnertobel II, 2: Tunnel San Bernardino, 3: Rhinebridge Tamins, 4: Caslertobel bridge, 5: Morbio bridge, 6: column in seawater, 7: ASTM C876 standard [12, 19].

passive zones is more important than the absolute value of the potential. Following the pioneering work of the American Standard ASTM C876 [6], a RILEM recommendation was published [7] where the experience with potential mapping was incorporated. Several national guidelines (e.g., the Swiss SIA 2006 2012 [8]) describe the use and interpretation of half-cell potential measurements.

When large areas are to be surveyed, the huge number of potential readings can be examined statistically. Half-cell potential data represented in a cumulative probability plot (Figure 16.11) are especially well suited for this purpose. The different bridge decks represented in Figure 16.11 show similar curves but shifted along the potential axis. This shift is due to the influence of concrete temperature and humidity. The "San Bernardino curve" corresponds to the map in Figure 16.8 and it is interesting to note that in the potential range between −200 and −400 mV CSE very few readings are found (the horizontal part of the curve). This clear-cut distinction between corroding (negative) and passive (more positive) potentials is due to the huge difference in concrete humidity: the active zones are situated near (leaking) joints, contaminated with deicing salts, the passive zones in dry parts of the deck. Further examples of application of half-cell potential mapping and their interpretation are given in [20, 21].

Recently Gulikers and Elsener [22] have expanded on statistical analysis of potential mapping results, producing a method for obtaining the potential boundary between active and passive steel. Assuming normal distributions for active and

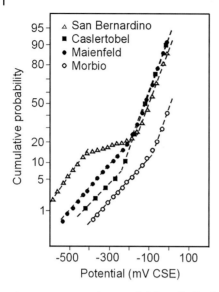

Figure 16.11 Cumulative probability of half-cell potential measurements on different bridge decks [19, 20].

passive steel, they provide equations for such analysis. The method results in a smaller range of potentials with "uncertain" corrosion as compared to ASTM C876. The authors state that a sufficiently large amount of readings is necessary, for example, a minimum of 1000 points, preferably with a significant portion of corroding steel. The procedure is illustrated using data from an example case.

Half-cell potentials cannot be correlated directly with the corrosion rate of the rebars. By excavating suitable inspection windows in the transition areas between very negative (active) and passive regions, the intensity of corrosion attack can be determined, and a correlation between potential and corrosion state can be established. This is valid only for the specific structure investigated.

All of the correlations seen above refer to situations of steel reinforcement in the free corrosion condition, that is, in the absence of factors that modify the potential of the system. They are in particular not applicable to: structures in concrete containing corrosion inhibitors; galvanized reinforcement (while on stainless steel it is essentially possible in the same way); structures subjected to electrical fields produced by stray current that induce current exchange between reinforcement and concrete (this case is dealt with in Section 9.4).

16.2.2
Resistivity Measurements

The resistivity of cement paste, mortar, and concrete depends on the pore volume and pore-size distribution of the cement paste, the pore-water composition (alkali

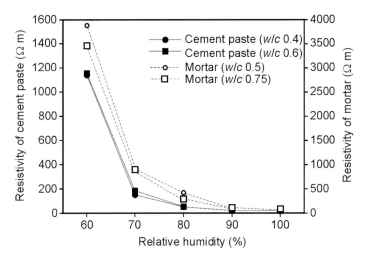

Figure 16.12 Dependence of resistivity of cement paste and mortar on water/cement ratio and relative humidity of the environment [26].

content, chloride content), and the moisture content of the concrete (Figure 16.12) [19, 23–26]. Depending on environmental conditions and concrete composition, resistivity may vary by several orders of magnitude, from 100 to 100 000 Ω m. High resistivities are found for dry concrete, concrete with low w/c ratio or with blended cements. Because the electrical current is transported only by the ions in the pore liquid in the cement paste (Chapter 2), concrete is not a homogeneous conductor. Aggregate particles are essentially isolating bodies. Coarse aggregates may have a similar size as the concrete cover and the spacing of the measuring electrodes has to be adjusted accordingly.

Temperature changes have important effects on concrete resistivity. A higher temperature causes the resistivity to decrease and vice versa (for a constant relative humidity). This is caused by changes in the ion mobility in the pore solution and by changes in the ion–solid interaction in the cement paste. As a first approach, an Arrhenius equation can be used to describe the effect of temperature on conductivity (inverse of resistivity):

$$\sigma(T_i) = \sigma(T_0) \cdot \exp\left(b \cdot \left(\frac{1}{T_0} - \frac{1}{T_i}\right)\right) \tag{16.1}$$

with:

σ = conductivity of concrete (1/Ω m),
T_0 = reference temperature (K),
T_i = actual temperature when conductivity is measured (K),
b = an empirical factor (K).

For steady-state conditions, b was found to be in the range of 1500 to 4500 [26–29]. The b value increases with decreasing relative humidity for a given cement paste, mortar or concrete and decreases with w/c ratio of the mix for a given relative humidity. The temperature dependence of the conductivity of bulk pore solution differs significantly from that of cement paste or mortar with the same ion concentration in the pores. This is due to strong ion–solid interactions. The humidity dependence of the temperature exponent in cement paste or concrete can be explained by the fact that at lower R.H. the pore solution becomes more concentrated and is present in more narrow pores, so the ratio of pore-wall surface area to liquid volume increases and consequently the degree of interaction between ions and solid increases. These interactions may be different for cements with different chemical compositions (e.g., they change with addition of GGBS or PFA). From the foregoing, it will be clear that accurate temperature correction of resistivity data is very complex. The concrete composition and the moisture content both influence the resistivity itself and its temperature dependence. For simplicity, it may be assumed that in the range of 0 °C to 40 °C, doubling of resistivity takes place for a decrease of 20 °C, or that a change of 3% to 5% per degree occurs [23].

According to the electrochemical nature of the corrosion process and the macrocell corrosion model (Chapter 8), a relationship may be expected between the concrete resistivity and the corrosion rate of the reinforcement after depassivation. Using a simplified approach, the corrosion rate of steel in concrete should be inversely proportional to the resistivity. This was confirmed in a general sense [30, 31], although the relationship is not universal but depends on the concrete composition [27].

Measurements at the Concrete Surface All methods for onsite measurement of concrete resistivity involve at least two electrodes (one of which may be a reinforcing bar). A voltage is superimposed between the electrodes, the resulting current is measured and the ratio gives a resistance (measured in Ω). The resistivity is obtained by multiplying the measured resistance by a geometrical conversion factor, the cell constant. This approach is valid only for a homogeneous material.

Resistivity can be measured using a four-point probe, called a *Wenner* probe after its inventor [32], placed directly on the concrete surface. The probe consists of four equally spaced point electrodes that are pressed on the surface (Figure 16.13). Usually a small amount of conducting liquid is applied locally to improve the contact between the electrode tips and the concrete surface. The two outer-point electrodes induce the measuring current, usually with AC current at a frequency of between about 100 and 1000 Hz (avoiding exact multiples of 50 Hz that may induce power-line interference), and the two inner electrodes measure the resulting potential drop of the electric field. The measured resistance R is given by $\Delta E/\Delta I$ and the resistivity of the concrete, ρ, is calculated as:

$$\rho = 2\pi \cdot a \cdot R \tag{16.2}$$

where a is the spacing between the electrodes. The applicability of this formula to concrete has been shown in [33, 34]. A good correlation was found between cali-

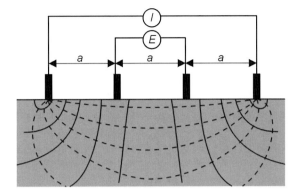

Figure 16.13 Scheme of the Wenner technique to determine the electrical resistivity of concrete from the surface. Spacing of the electrodes shall be bigger than the maximum aggregate diameter.

brated data measured with cast in electrodes and four-point resistivity obtained on the surface [35, 36].

Other types of resistance measurements, especially also involving the rebar network, have been applied. Some commercially available instruments combine half-cell potential mapping with resistance measurements between the electrode and the rebars, producing resistance maps. Nevertheless, conversion to true resistivity is much more difficult because the cell constant is also influenced by the cover depth to the steel bars and the size of the external electrode.

Procedure The detailed procedure for the measurement of resistivity of concrete is described in a RILEM recommendation [23]. The measuring system should be calibrated on concrete with known resistivity. As with half-cell potential measurements, concrete shall be clean and a good electrolytic contact between the electrodes and the concrete surface is important, but complete wetting of the surface should be avoided. When using the four-point method, measurements should be taken as far from the rebars as possible (e.g., diagonally inside the rebar mesh, Figure 16.14).

Interpretation The results of concrete resistivity measurements can be used for a quantitative or qualitative interpretation. Resistivity data measured on a structure and corrected for the temperature effect can be compared to reference data of similar concrete types (Table 2.4). Usually additional information is necessary. If for example, a wet structure made with OPC has a mean resistivity value of 50 Ω m, it means that the water/cement ratio and the porosity must be quite high. Consequently, the concrete will be susceptible to rapid chloride penetration and/or the corrosion rate after depassivation will be high.

If the concrete composition is relatively homogeneous, mapping the resistance may show wet (low resistance) and dry (high resistance) areas. The average

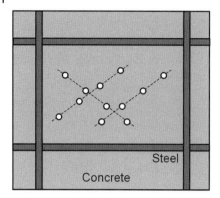

Figure 16.14 Positioning of Wenner probe electrodes on the concrete surface in order to stay as far as possible from the rebars after locating the reinforcement mesh.

Table 16.3 Interpretative criteria for measurement of electrical resistivity of concrete structures exposed to the atmosphere for OPC concrete [19].

Concrete resistivity (Ω m)	Corrosion rate
>1000	Negligible
>500	Low
200–500	Modest
100–200	High
<100	Very high (no ohmic control)

resistivities in the wet or dry areas can be interpreted quantitatively, as explained above. In addition a crude relationship between resistivity and corrosion rate can be obtained as suggested for OPC concrete (Table 16.3).

Important information can be obtained by combining results from half-cell potential mapping and resistivity measurements (Figure 16.15). Areas of positive potentials and high resistivity indicate dry, passive zones, negative potentials, and low resistivity indicates wet corroding zones. The points where low resistivity but still passive potentials are measured can be interpreted as the areas of risk that will corrode in the future.

16.2.3
Corrosion Rate

Quantitative information on the corrosion rate of steel in concrete is of great importance for the evaluation of repair methods in the laboratory, for service-life prediction and structural assessment of corroding structures as well as control of

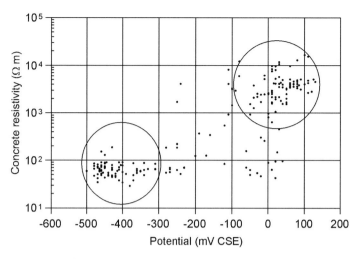

Figure 16.15 Relation between half-cell potential and concrete resistivity measured at the underside of a chloride-contaminated bridge deck [7].

repair work onsite. The only technique available today is the polarization resistance method; a RILEM recommendation covering this subject has been published [37].

When speaking about "corrosion rate" of steel in concrete, two different meanings, average corrosion rate and instantaneous corrosion rate, have to be distinguished:

- the *average corrosion rate* is the "engineering" value needed for implementation in models for service-life calculations or to predict the evolution of structural degradation (the slope of the propagation curve in Figure 16.4). It can be determined as the average value over a long period of time by measuring weight loss (limited to homogeneous corrosion, possible only in the laboratory) or loss of cross section of the steel on site. If the time of depassivation, that is, the start of corrosion, is not known (as is usually the case in practice), the calculated average corrosion rates will contain an unknown error. Furthermore, in real structures exposed to changing environmental conditions the average value is composed of periods with low and high corrosion rates [38–40];

- the *instantaneous corrosion rate* (i_{corr}) can be determined using electrochemical methods, in particular polarization resistance (R_p) measurements [41], determined by stationary or nonstationary electrochemical methods. The calculation of a corrosion rate from R_p measurements is straightforward and correct only for general corrosion attack; the implications for measurements on real structures and on locally corroding rebars are discussed below. It is obvious that "corrosion rate" values measured at different times during the life of a structure or on different structures have to be corrected for the influence of moisture, temperature, etc., in order to allow a reasonable comparison.

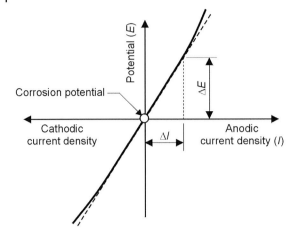

Figure 16.16 Polarization curve close to the corrosion potential.

Table 16.4 Principal relationships in linear polarization.

$R_p = (dE/i)$	R_p = polarization resistance ($\Omega\,m^2$)
	dE = polarization with respect to corrosion potential (mV)
	i = applied current density (mA/m^2)
$i = I/A$	I = applied current (A)
	A = polarized surface (m^2)
$i_{corr} = B/R_p$	i_{corr} = corrosion rate (expressed as corrosion current density, mA/m^2)
	B = Stern–Geary constant (mV)
$v_{corr} = K \cdot i_{corr}$	v_{corr} = corrosion rate (µm/year)
	K = 1.17 (µm/year)/(mA/m^2) for iron

Determination of the Polarization Resistance The polarization resistance method is based on the observation that the polarization curve close to the corrosion potential is linear (Figure 16.16), the slope $\Delta E/\Delta I$ (ΔE = step in potential, ΔI = resulting current) being defined as polarization resistance, R_p. The technique is called linear polarization resistance measurement (LPR). The R_p value is related to the corrosion current I_{corr} by the formula [42]:

$$I_{corr} = B / R_p \tag{16.3}$$

where: B is a constant containing the anodic and cathodic Tafel slopes, that is, the slopes of the polarization curves shown in Chapter 7 (Table 16.4). For actively corroding steel in concrete usually $B = 26\,mV$ is taken, for passive steel $B = 52\,mV$ [43]. R_p and thus I_{corr} is related to the area of the sample under test, resulting in the specific polarization resistance R_p^* ($\Omega\,m^2$), and the corrosion current density i_{corr} (mA/m^2), the instantaneous corrosion rate. Several precautions have to be

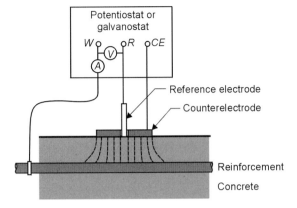

Figure 16.17 Schematic setup for linear polarization measurements.

taken in order to get reliable values, namely, compensation of the high electrical (ohmic) resistance of the concrete cover, achievement of a quasisteady state and linearity of the $\Delta E/\Delta I$ slope as discussed in more detail in [19, 44, 45]. When these precautions are taken into account, good correlations between the electrochemical weight loss, calculated by integrating R_p data from *LPR* measurements and gravimetric measurements [43] or measurement of corrosion pit volume [46] have been found.

Execution of the Measurements The device for measuring the polarization resistance onsite is illustrated in Figure 16.17. In its simplest version, it consists of a counterelectrode disk and a central reference electrode placed on the surface of the structure; the supply unit, usually a potentiostat, is connected to the reinforcement (the working electrode), to the counterelectrode, and to the reference electrode, as indicated in the figure. After having selected the measuring position (preferably based on half-cell potential mapping), proper connections to the rebar and between counterelectrode and concrete surface are made (wetting). The corrosion potential is measured and then a predetermined variation of potential (dE), in general equal to 10 mV, is applied and the current di that the system requires to achieve this variation is recorded. The measurement lasts about one minute.

From the polarization resistance, $R_p = dE/di$ ($\Omega\,m^2$), the corrosion current density i_{corr} (mA/m^2) is calculated using Eq. (16.3). Finally, the corrosion rate expressed as penetration rate (µm/year) is obtained, keeping in mind that for iron an anodic current density of 1 mA/m^2 corresponds to a corrosion rate of 1.17 µm/year.

Corrosion Rate Measurements Onsite The main difference in measuring the polarization resistance R_p onsite as opposed to in the laboratory is the geometrical arrangement of the electrodes. In the laboratory a uniform geometry (concrete, steel bars) and thus uniform current distribution can be achieved usually with small specimens, whereas onsite there is a nonuniform current distribution between the small counterelectrode (CE) on the concrete surface and the large rebar network (WE). The current fed by the external counterelectrode (Figure 16.18) does not only polarize the surface of the reinforcement below the counterelectrode itself but also spreads laterally or may reach deeper layers of the reinforcement. Due to the current spread-out the measured polarization resistance is related to an unknown rebar surface and cannot be converted directly to a corrosion rate with the Stern–Geary equation [47, 48]. The current spread-out, given by the critical length L_{crit}, where 90% of the current vanishes, depends both on the concrete resistivity and on the specific polarization resistance R_p^* of the rebar (thus on the corrosion state). Briefly summarizing the available literature [45, 49] it can be stated that in the case of homogeneous, *actively corroding* rebars (small specific polarization resistance R_p^*), the applied current is concentrated nearly completely beneath the counterelectrode (Figure 16.18). Conversely, if the steel is *passive* (high specific polarization resistance) a large current spread-out has to be expected and the measured apparent polarization resistance $R_{p,eff}$ is related to an area of rebars that may be up to 100 times larger than the counterelectrode surface. The concept of a sensorized guard ring [50] usually implemented in commercial instruments to avoid the current spread-out on passive rebars does not always seem to work correctly [51].

Anyway, the calculation of a *local corrosion rate* (penetration rate) is intrinsically difficult because the area of the localized attack is not known [45].

Interpretation of the Results The interpretation has to be done by corrosion specialists with experience in reinforced concrete structures, in the context of other

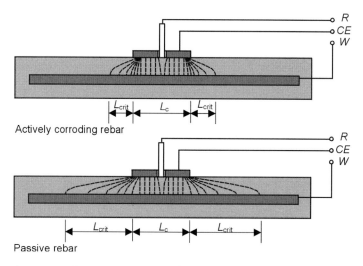

Figure 16.18 Onsite measurements of corrosion rate.

information from condition assessment. The results can be used to pinpoint sites with high corrosion activity (in addition to half-cell potential measurements), to predict future deterioration of the structure or to assess residual service life. Daily and seasonal variations in the corrosion current due to changes in temperature and relative humidity will induce variations in I_{corr} values by more than a factor two. To predict future deterioration usually several measurements in time are needed. When only one single value is available, it is recommended to assume a variation in I_{corr} between 50% and 200% of the measured value. The interpretation of single values can be improved by taking cores, measuring their resistivity "as received", wetting them and measuring again. If the resistivity at the time of measurement was high (compared to the wet value), the corrosion rate is on the low side of the temporal (seasonal) variation. If the resistivity "as received" is low and comparable to the wet value, the corrosion rate measured at that point in time is more representative for wet and high corrosion rate parts of the seasonal cycle. This approach was proposed by Andrade [52].

Several examples of application are given in the literature [46], for example, for predicting the residual service life of a corroding bridge and testing the efficiency of protection with a corrosion inhibitor [53].

16.3
Analysis of Concrete

Carbonation (Chapter 5) or ingress of chloride ions (Chapter 6) are the two principal causes of rebar corrosion. Inspection, condition assessment, and prediction of the future service life of a structure requires information on the carbonation depth or chloride content of concrete. The areas on a structure where these chemical analyzes are performed should be based on a problem-oriented approach, for example, on visual inspection or better on potential mapping. Other types of chemical or microstructural analyses that may help in the characterization of the concrete and its resistance to penetration of aggressive agents are outside the scope of this chapter.

16.3.1
Carbonation Depth

Carbonation of concrete is associated with loss of alkalinity of the pore solution. This change in pH can be revealed by a suitable indicator that changes color near pH 10. A phenolphthalein solution will remain colorless where concrete is carbonated and will turn pink where concrete is still alkaline. The best indicator solution for maximum contrast of the pink coloration is a solution of phenolphthalein indicator in alcohol and water, usually 1 g of indicator in 100 ml of 50:50 or more alcohol to water mix.

To determine the carbonation depth it is essential to spray freshly broken concrete surfaces with the indicator solution. Fresh concrete surfaces can be obtained by breaking off pieces of concrete or by coring and splitting the core in the laboratory. Care has to be taken that dust from drilling, coring or cutting does not get on the treated surface, otherwise already carbonated zones can show up as alkaline.

Carbonation depth sampling should allow the average and standard deviation of the carbonation depth on a structural element to be calculated [54, 55]. Comparing this value to the average reinforcement cover, the amount of depassivated steel can be estimated. A method for comparing the statistical distribution of carbonation depth and concrete cover thickness was proposed by Mattila [54]. The progress of the carbonation front cannot be calculated on a single carbonation depth measurement because the time law of carbonation progress usually is not known (Chapter 5). Using a square-root law, the future carbonation depth on atmospherically exposed concrete will generally be overestimated. In addition, considerable scatter may be present in carbonation depths [54, 55].

16.3.2
Chloride Determination

Chlorides in concrete structures can be present as cast-in chlorides (in cases where chloride-containing admixtures or seawater as mixing water were used) or can be transported into concrete from the environment (seawater, deicing salts). To distinguish between these two cases and to be able to get information on the future service life of a structure, chloride depth profiles should be determined. This is essential because the actual chloride level at the reinforcement determines the present probability of corrosion, but the profile determines the future development of corrosion.

Above a "critical chloride content" pitting corrosion of the steel will start (Chapter 6). As mentioned earlier, it is thus essential to determine the location for chloride profile analysis on the basis of a half-cell potential map, for example, cores should be taken at corroding areas, passive areas, and in the transition region. On several structures an empirical correlation between half-cell potential and chloride content at different depths in the cover has been found even for passive reinforcement. This can be explained by the influence of diffusion potentials affecting half-cell potential readings.

Chloride contents can be measured by different methods, described in several recommendations or standards. Field methods are rapid but usually of poor accuracy, whereas laboratory methods are more accurate, but more time consuming and costly. All methods (except the "dry" method of X-ray fluorescence) require four working steps: (i) sampling, (ii) crushing to powder, (iii) dissolution in acid, and (iv) chloride analysis. As concrete is an inhomogeneous material and chlorides are present only in the cement paste, a minimum diameter of 50 mm for cores or three to five points for 20 mm diameter dust drillings is recommended in order to get representative chloride depth profiles. Recently, profile grinding, that is, grinding powder for slices of one or two millimeter thickness has become widely used for measuring accurate penetration profiles. In order to obtain sufficient powder, specimens usually are 100 mm cores or $100 \times 100 \, mm^2$ faces of specimens.

Relatively simple methods based on spraying the surface of split cores with silver nitrate have been proposed; a color change indicates the chloride penetration front [56]. The result may depend on the original color of the concrete and its particular chemistry. More complex tests are based on cutting slices from cores and pressing out the pore solution for determination of the free chloride concentration; this technique was developed for hardened cement paste [57] but is also used for concrete [58].

Chloride Profile Based on Cores or Powder Drilling Cores are cut in slices of maximum 10 mm thickness that are then crushed and milled. The resulting powder should have a controlled fineness typically of 0.25 mm maximum (usually obtained with constant time of crushing and milling). The equipment has to be carefully cleaned after every slice, in order to avoid crosscontamination of powders from different depths or samples.

The location of *powder drilling* should be determined as for cores. Three to five 20-mm diameter holes should be drilled to the required depth (e.g., in steps of 10 mm) and the powder sampled. It is essential to mix the powder of the individual depths and to store it dry. After every depth sample is drilled, the equipment and the borehole have to be cleaned carefully in order to avoid crosscontamination between different depth samples.

Dissolution of the Powder After homogenizing the powder, obtained either directly by drilling or by crushing core slices, a part is weighted precisely and dissolved in a constant amount of liquid, most frequently in concentrated nitric acid solution. In this way the total or acid soluble chloride content is determined. The dissolution process is the most important step determining the reproducibility of the chloride analysis. Alternatively, the Soxhlet extraction technique (reflux of boiling water on the concrete powder for 24 h) is used to dissolve chloride ions. Both methods result in a similar total chloride content.

Considerable work has been done to differentiate between bound and free chlorides, using cold-water dissolution, alcohol dissolution, etc. The results of such chloride extraction processes strongly depend on the level of powder fineness, time of dissolution, etc. and no sufficient reproducibility has been obtained. For routine analysis an acid-soluble procedure is recommended [59]. Recently, a method was proposed for determining water-soluble chloride [60].

Chemical Analysis Chloride ions in the extraction solution described above can be determined by various analytical techniques such as color-based titration, potentiometric titration, chloride-ion selective electrodes, etc. All of the analytical techniques have been shown to give comparable results with good accuracy, provided that frequent calibration with standard solutions is carried out. For very small amounts of liquid (as e.g., obtained by pore-water expression) the ICP technique (inductively coupled plasma spectroscopy) has been used successfully. However, this instrumentation is available only in specialized analytical laboratories.

Interpretation The presence of an above-critical amount of chloride ions at the rebars leads to depassivation and in the presence of oxygen and water to corrosion attack. From chloride profiles information on the transport of chlorides into the concrete (Chapter 6) can be obtained. In combination with results from potential mapping, the critical chloride content for the specific structure can be obtained. On chloride-contaminated structures an empirical correlation between chloride content and half-cell potential could be established, thus the chloride distribution can be roughly obtained from the potential map.

References

1 Raupach, M., Gulikers, J., Reichling, K. et al. (2013) Special issue on "corrosion surveys". *Materials and Corrosion*, in press.
2 CEB (1989) Durable concrete structures, CEB Design Guide, Bulletin d'information 182, 1989.
3 Bertolini, L., Manera, M., and Anselmi, F. (2004) Investigation on reinforced concrete structures of Caorso nuclear power plant in Italy. CNSI/RILEM Workshop on Use and performance of concrete in NPP fuel cycle facilities, Madrid.
4 Andrade, C., Alexander, M., Basheer, M., Beushausen, H., Fernández Luco, L., Gonçalves, A.F., Jacobs, F., Neves, R., Podvoiskis, J., Polder, R., and Romer, M. (2007) Comparative test–part II: covermeters, in *Non-Destructive Evaluation of the Penetrability and Thickness of the Concrete Cover – State-of-the-Art Report of RILEM Technical Committee 189-NEC* (eds R. Torrent and L. Fernández Luco), REP 40, RILEM, Bagneux, pp. 187–199.
5 Fernández Luco, L. (2005) Comparative test–part II–comparative test of "covermeters". *Materials and Structures*, **38** (10), 907–911.
6 ASTM C876-09 (2009) *Standard Test Method for Half-Cell Potential of Reinforcing Steel in Concrete*, American Society for Testing and Materials.
7 Elsener, B., Andrade, C., Gulikers, J., Polder, R., and Raupach, M. (2003) Recommendations on half-cell potential measurements–potential mapping on reinforced concrete structures. *Materials and Structures*, **36** (7), 461–471.
8 SIA 2006 (2012) *Richtlinie zur Potentialfeldmessung*, Schweiz Ingenieur und Architektenverein, Zürich.
9 Schiegg, Y. (1995) *Potentialfeldmessungen nach der Instandsetzung*, SIA Dokumentation D0126. Schweiz Ingenieur und Architektenverein, Zürich.
10 Stratfull, R.F. (1957) The corrosion of steel in a reinforced concrete bridge. *Corrosion*, **13** (3), 43–48.
11 Elsener, B. and Böhni, H. (1987) Location of corroding rebars in RC structures. *Schweiz Ingenieur und Architekt*, **105**, 528. (in German).
12 Elsener, B. and Böhni, H. (1990) Potential mapping and corrosion of steel in concrete, in *Corrosion Rates of Steel in Concrete*, ASTM STP 1065 (eds N.S. Berke, V. Chaker, and D. Whiting), American Society for Testing and Materials, Philadelphia, pp. 143–156.
13 Hunkeler, F. (1991) Bauwerksinspektion mittels Potentialmessung. *Schweizer Ingenieur und Architekt*, **109**, 272.
14 Vennesland, Ø., Raupach, M., and Andrade, C. (2007) Recommendation of Rilem TC 154-EMC: electrochemical techniques for measuring corrosion in concrete–measurements with embedded probes. *Materials and Structures*, **40** (8), 745–758.
15 Myrdal, R. (2007) *The Electrochemistry and Characteristics of Embeddable Reference Electrodes for Concrete*, European Federation of Corrosion Publication number 43, Woodhead Publishing Limited, Cambridge.
16 Elsener, B., Müller, S., Suter, M., and Böhni, H. (1990) Corrosion monitoring

of steel in concrete – theory and practice, in *Corrosion of Reinforcement in Concrete* (eds C.L. Page, K.W.J. Treadaway, and P.B. Bamforth), Elsevier Applied Science, London, p. 348.
17 Arup, H. (1983) The mechanisms of the protection of steel by concrete, in *Corrosion of Reinforcement in Concrete Construction* (ed. A.P. Crane), Ellis Horwood Ltd, Chichester, pp. 151–157.
18 Elsener, B., Flückiger, D., Wojtas, H., and Böhni, H. (1996) Methoden zur Erfassung der Korrosion von Stahl in Beton, VSS Forschungsbericht Nr. 520, Verein Schweiz, Strassenfachleute Zürich.
19 COST 509 (1997) Corrosion and Protection of Metals in Contact with Concrete, Final Report (eds R.N. Cox, R. Cigna, Ø. Vennesland, and T. Valente), European Commission, Directorate General Science, Research and Development, Brussels, EUR 17608 EN, 1997.
20 Elsener, B. (2001) Half-cell potential mapping to assess repair work on RC structures. *Construction and Building Materials*, **15** (2–3), 133–139.
21 Hunkeler, F. (1994) Grundlagen der Korrosion und der Potentialmessung bei Stahlbetonbauten. ASB Brückenunterhaltsforschung, Verein Schweizer Strassenfachleute (VSS) Zürich, Report No. 510.
22 Gulikers, J. and Elsener, B. (2009) Development of a calculation procedure for the statistical interpretation of the results of potential mapping performed on reinforced concrete structures. *Materials and Corrosion*, **60** (2), 87–92.
23 Polder, R., Andrade, C., Elsener, B., Vennesland, Ø., Gulikers, J., Weydert, R., and Raupach, M. (2000) Test methods for on site measurement of resistivity of concrete. *Materials and Structures*, **33** (10), 603–611.
24 Frederiksen, J.M. (ed.) (1996) HETEK, Chloride Penetration into Concrete. State of the art. Transport Processes, Corrosion Initiation, Test Methods and Prediction Models, The Road Directorate, Report No. 53, Copenhagen, 1996.

25 Tuutti, K. (1982) *Corrosion of Steel in Concrete*, Swedish Cement and Concrete Research Institute, Stockholm.
26 Bürchler, D., Elsener, B., and Böhni, H. (1996) Electrical resistivity and dielectrical properties of hardened cement paste and mortar, in *MRS Proceedings "Electrically Based Microstructural Characterization"*, vol. 411 (eds R.A. Gerhardt, S.R. Taylor, and E.J. Garboczi), p. 407.
27 Bertolini, L. and Polder, R.B. (1997) Concrete Resistivity and Reinforcement Corrosion Rate as a Function of Temperature and Humidity of the Environment, TNO Report 97-BT-R0574, 1997.
28 Elsener, B. (2000) Corrosion of steel in concrete, in *Corrosion and Environmental Degradation*, vol. 2 (ed. M. Scutze), Materials Science and Technology Series, Wiley-VCH Verlag GmbH, pp. 389–436.
29 Elkey, W. and Sellevold, E.J. (1995) *Electrical Resistivity of Concrete*, Norwegian Road Research Laboratory, Publication No. 80.
30 Gonzalez, J.A., Lopez, W., and Rodriguez, P. (1993) Effects of moisture availability on corrosion kinetics of steel embedded in concrete. *Corrosion*, **49**, 1004–1010.
31 Polder, R.B., Bamforth, P.B., Basheer, M., Chapman-Andrews, J., Cigna, R., Jafar, M.I., Mazzoni, A., Nolan, E., and Wojtas, H. (1994) Reinforcement corrosion and concrete resistivity – state of the art, laboratory and field results, in *Proc. Int. Conf. on Corrosion and Corrosion Protection of Steel in Concrete* (ed. R.N. Swamy), Sheffield Academic Press, 24–29 July, pp. 571–580.
32 Wenner, F. (1916) Bulletin of US Bureau of Standards, Science Paper 12, 1916
33 Millard, S.G., Harrison, J.A., and Edwards, A.J. (1989) Measurements of the electrical resistivity of reinforced concrete structures for the assessment of corrosion risk. *Journal of Non-destructive Testing*, **31**, 11.
34 Elsener, B. (1988) Electrochemical methods for monitoring of structures, SIA Documentation D020, in *Non-Destructive Testing of Reinforced Structures*, Schweiz Ingenieur- und Architekten-Verein, Zürich, pp. 27–37. (in German).

35 Polder, R.B., Valente, M., Cigna, R., and Valente, T. (1992) Laboratory investigation of concrete resistivity and corrosion rate of reinforcement in atmospheric conditions, in Proc. RILEM/CSIRO Int. Conf. on Rehabilitation of Concrete Structures, Melbourne (eds D.W.S. Ho and F. Collins), pp. 475–486.

36 Weydert, R. and Gehlen, C. (1998) Electrolytic resistivity of cover concrete: relevance, measurement and interpretation. 8th Conf. on Durability of Materials and Components, 1998.

37 Andrade, C., Alonso, C., Gulikers, J., Polder, R., Cigna, R., Vennesland, Ø., Salta, M., Raharinaivo, A., and Elsner, B. (2004) Recommendations on test methods for on-site corrosion rate measurement of steel reinforcement in concrete by means of the polarization resistance method. *Materials and Structures*, **37** (9), 623–643.

38 Andrade, C., Sarria, J., and Alonso, C. (1996) Statistical study on simultaneous monitoring of rebar corrosion rate and internal RH in concrete structures exposed to the atmosphere, in *Corrosion of Reinforcement in Concrete Construction* (eds C.L. Page, P. Bamforth, and J.W. Figg), Society of Chemical Industry, London, pp. 233–242.

39 Zimmermann, L., Schiegg, Y., Elsner, B., and Böhni, H. (1997) Electrochemical techniques for monitoring the conditions of concrete bridge structures, in *Repair of Concrete Structures* (ed. A. Blankvoll), Norwegian Road Research Laboratory, Svolvaer, Norway, pp. 213–222.

40 Schiegg, Y., Audergon, L., Elsner, B., and Böhni, H. (2007) On-line monitoring of corrosion in reinforced concrete structures, in *Corrosion of Reinforcement in Concrete. Mechanisms, Monitoring, Inhibitors and Rehabilitation Techniques*, European Federation of Corrosion Publication number 38 (eds M. Raupach, B. Elsner, R. Polder, and J. Mietz), Woodhead Publishing Limited, Cambridge, pp. 133–145.

41 Gonzales, J.A., Algaba, S., Andrade, C., and Feliu, V. (1980) Corrosion of reinforcing bars in carbonated concrete. *British Corrosion Journal*, **15**, 135.

42 Stern, M. and Geary, A.L. (1957) Electrochemical polarization I. A theoretical analysis of the shape of polarization curves. *Journal of the Electrochemical Society*, **104**, 56.

43 Andrade, C. and Gonzales, J.A. (1978) Quantitative measurements of corrosion rate of reinforcing steels embedded in concrete using polarization resistance measurement. *Materials and Corrosion*, **29**, 515–519.

44 Andrade, C., Castelo, V., Alonso, C., and Gonzales, J.A. (1986) The determination of the corrosion rate of steel embedded in concrete by the polarization and AC impedance methods, in *Corrosion Effect of Stray Currents and the Techniques for Evaluating Corrosion of Rebars in Concrete*, ASTM STP 906 (ed. V. Chaker), ASTM, Philadelphia, pp. 43–63.

45 Elsner, B. (1998) Corrosion rate of steel in concrete – from laboratory to reinforced concrete structures, in *Corrosion of Reinforcement in Concrete. Monitoring, Prevention and Rehabilitation* (eds J. Mietz, B. Elsner, and R. Polder), The European Federation of Corrosion Publication number 25, The Institute of Materials, London, pp. 92–103.

46 Polder, R.B., Peelen, W.H.A., Bertolini, L., and Guerrieri, M. (2002) Corrosion rate of rebars from linear polarization resistance and destructive analysis in blended cement concrete after chloride loading. ICC 15th International Corrosion Congress, Granada, 22–27 September 2002 (CD-ROM).

47 Feliù, S., Gonzales, J.A., Andrade, C., and Feliu, V. (1987) On site determination of the polarization resistance in a reinforced concrete beam, Corrosion/87, Nace, Paper No. 145, San Francisco, March 1987.

48 Feliù, S., Gonzales, J.A., Andrade, C., and Feliu, V. (1988) On site determination of the polarization resistance in a reinforced concrete beam, Corrosion/88, Nace, Paper No. 44, 1988.

49 Feliù, S., Gonzales, J.A., and Andrade C. (1996) Electrochemical methods for on-site determinations of corrosion rates of rebars, in *Techniques to Assess the Corrosion Activity of Steel Reinforced Concrete Structures*, ASTM STP 1276 (eds

N.S. Berke, E. Escalante, C.K. Nmai, and D. Whiting), American Society for Testing and Materials, p. 107.

50 Broomfield, J.P., Rodriguez, J., Ortega, L.M., and Garcia, A.M. (1993) Corrosion rate measurements in reinforced concrete structures by a linear polarization device. Int. Symp. on Condition Assessment, Protection Repair and Rehabilitation of Concrete Bridges Exposed to Aggressive Environments, ACI Fall Convention, Minneapolis, 1993.

51 Nygaard, P.V., Geiker, M.R., and Elsener, B. (2009) Corrosion rate of steel in concrete: evaluation of confinement techniques for on-site corrosion rate measurements. *Materials and Structures*, **42** (8), 1059–1076.

52 Andrade, C. (2003) Measurement of polarization resistance on-site, in COST Action 521, Corrosion of Steel in Reinforced Concrete Structures, Final Report (eds R. Cigna, C. Andrade, U. Nürnberger, R. Polder, R. Weydert, and E. Seitz), European Communities, Luxembourg, Publication EUR 20599, pp. 82–98.

53 Broomfield, J.P. (1997) The pros and cons of corrosion inhibitors. Construction Repair, July/16–18 August.

54 Mattila, J. (2003) Durability of patch repairs. in COST Action 521, Corrosion of steel in reinforced concrete structures, Final Report (eds R. Cigna, C. Andrade, U. Nürnberger, R. Polder, R. Weydert, and E. Seitz), European Communities, Luxembourg, Publication EUR 20599, pp. 183–196.

55 Pentti, M. (1999) *The Accuracy of the Extent-of-Corrosion Estimate Based on the Sampling of Carbonation and Cover Depths of Reinforced Concrete Facade Panels*, Tampere University of Technology, Publication 274.

56 Collepardi, M. (1995) Quick method to determine free and bound chlorides in concrete, in *Proc. RILEM. Int. Workshop on Chloride Penetration into Concrete* (eds L.-O. Nilsson and P. Ollivier), St-Remy-les-Chevreuses, 15–18 October.

57 Page, C.L. and Vennesland, Ø. (1983) Pore solution composition and chloride binding capacity of silica fume cement pastes. *Materials and Structures*, **16**, 91.

58 Polder, R.B., Walker, R., and Page, C.L. (1995) Electrochemical desalination of cores from a reinforced concrete coastal structure. *Magazine of Concrete Research*, **47**, 321–327.

59 Rilem TC 178-TMC Recommendation (2002) Analysis of total chloride content in concrete. *Materials and Structures*, **35** (9), 583–585.

60 Rilem TC 178-TMC Recommendation (2002) Analysis of water soluble chloride content in concrete. *Materials and Structures*, **35** (9), 586–588.

17
Monitoring

17.1
Introduction

A number of techniques can be used to carry out a condition survey on reinforced concrete structures to detect ongoing corrosion of the reinforcement or to locate areas of future corrosion risk (Chapter 16). These techniques, alone or in combination, provide a picture of the structure at a particular point in time. However, if one is interested in the progress of the corrosion risk (before corrosion actually starts) or the rate of deterioration of the structure, it is necessary to monitor the condition change with time. According to the codes for durability design [1–3], monitoring becomes a part of the design and construction process, which is thought to include maintenance.

Monitoring of concrete structures can be done at different levels [4], starting from a simple *low-level* form of periodic visual inspections. This type of "monitoring", although it is still the basis of most bridge-management programs, can detect deterioration only in a rather late stage, when rust staining, spalling or cracking have occurred. An *intermediate level* of monitoring combines periodic visual inspection with destructive testing. Visual inspection plus nondestructive testing is rarely applied, because nondestructive techniques such as half-cell potential mapping require direct access to the concrete surface – a condition very often only given with expensive and time-consuming installation of scaffolds. The measurements themselves are time consuming and labor intensive. These limitations might be overcome with an automated climbing robot for corrosion monitoring, shown in Figure 17.1 [5]. The use of this robot can implement full surface half-cell potential mapping into bridge management systems when data elaboration and interpretation are automated, too. A *high-level* form of monitoring, treated in this chapter, is the use of embedded sensors that are built in the structure at the time of construction or inserted later. Optical fiber sensors, strain gauges, etc. can monitor the structure deformation or local temperature changes [6]. Other sensors that will be described further on, can detect the depassivation of the reinforcement, changes in resistivity of the concrete cover or the ingress of chlorides and carbonation.

The benefits of such a long-term monitoring strategy are obvious. First, if corrosion-risk conditions for the reinforcement are detected sufficiently early,

Corrosion of Steel in Concrete: Prevention, Diagnosis, Repair, Second Edition. Luca Bertolini,
Bernhard Elsener, Pietro Pedeferri, Elena Redaelli, and Rob Polder.
© 2013 Wiley-VCH Verlag GmbH & Co. KGaA. Published 2013 by Wiley-VCH Verlag GmbH & Co. KGaA.

Figure 17.1 Automated climbing robot for corrosion monitoring [5].

relatively simple maintenance measures or protection systems can be programmed in a cost-effective way and damage can be avoided. Secondly, monitoring the progress of condition changes allows to predict the future development of the structure and to determine the optimum time of an intervention. The installation of monitoring systems on new structures is to be recommended, especially where access is difficult, when the durability of the structure is a major issue and for structures with very long service lives; in all these cases adequate maintenance should be defined in the design stage. It should, however, be defined at the time of installation of the monitoring system what action has to be taken if one or more sensors indicate critical values. The benefit is that monitoring allows anticipating the development of deterioration, such that adequate actions can be taken, which allows for cheap preventive measures as opposed to the need for expensive corrective interventions in the case of full development of corrosion and concrete damage. Installing monitoring on existing structures works similar, but poses additional challenges of embedding probes in hardened concrete. Monitoring of existing structures may be triggered by signs of beginning deterioration, change of ownership or increased awareness of the risks involved in structure management. Economic benefits of installing corrosion monitoring has been discussed for case study in ref. [7].

17.2
Monitoring with Nonelectrochemical Sensors

In recent years great progress has been made in the development of embedded sensors and remote data acquisition. Some of the sensors are already well proven

 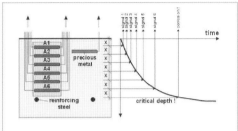

Figure 17.2 Monitoring time-to-corrosion system (anode-ladder) developed by P. Schiessl [8, 9].

in the field, others are highly sophisticated research instruments and not (yet) used on a large scale. In this section sensor types are described that do not need any reference electrode to be embedded (essentially macrocell and resistivity sensors).

Sensors Based on Macrocell Measurements Since about 1990, macrocell sensor systems, for example, the so-called *anode-ladder system*, have been used to monitor the corrosion risk of new concrete structures [8]. This sensor system (Figure 17.2) consists of carbon steel anodes mounted at different depths from the surface. A cathode (stainless steel or activated titanium) is placed in the concrete near the anodes. By connecting the individual anodes with the cathode a galvanic current can be measured: as long as a carbon steel bar is still in passive condition, the electrical current that it exchanges with the cathode is very small. When either ingress of chlorides or carbonation has reached the depth of the first carbon steel and leads to depassivation (and sufficient water and oxygen are available for significant corrosion), the macrocell current increases markedly (Chapter 8). In the course of time the deeper carbon steel anodes will be depassivated one by one, allowing to monitor the time to depassivation and the critical depth of the chloride or carbonation front. In this way, the time until the depassivation front reaches the reinforcement can be estimated using simplified equations, for example, the square-root formulas for penetration of chloride or carbonation (Chapters 5 and 6).

Although the macrocell currents measured with the anode-ladder system can detect the time to depassivation, no information regarding the corrosion rate of the steel can be obtained due to the small cathode/anode ratio. This macrocell monitoring system has been installed since 1990 into tunnels, bridges, foundations and other structures exposed to aggressive environments [9]. Similar systems have been developed and are on the market [10]. It should be noted that the interpretation of signals from macrocell sensors installed in very wet (i.e., submerged or underground) structures may be difficult [11, 12].

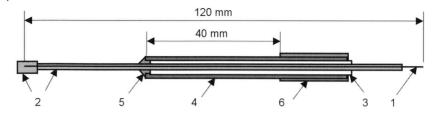

Figure 17.3 Combined chloride and resistivity sensor element. (1) silver wire, (2) AgCl coating, (3) teflon tube (isolation), (4) stainless steel tube (1.4301), (5) epoxy sealing, (6) insulation. Dimensions are in mm [15].

Sensors Based on Indepth Resistivity Measurements As has been shown in Chapter 2, changes in the resistivity of the concrete cover are related to changes in the humidity and/or ion content in the pore solution. Sensors that allow monitoring of changes in the resistivity allow (indirectly) getting information on the corrosion risk. Since several years a *combined chloride/resistivity sensor* is used to monitor the ingress of water and chloride ions into mortar and concrete [13, 14]. The basic sensor element is shown in Figure 17.3. It consists of a central chloride-sensitive element mounted electrically isolated in a stainless steel tube. When arranging several sensor elements at multiple depths (Figure 17.4), changes in resistivity can be followed for any depth interval by measuring the AC resistance between two adjacent stainless steel tubes.

For monitoring existing structures, cores are taken from the position to be monitored and the sensor elements are inserted in the laboratory by drilling small holes. Sound electrical connections are made and the core is remounted at its original position in the structure. More than one hundred of such resistivity sensors were installed at different exposure conditions, mainly in bridges, bridge abutments and side walls (Figure 17.5). The results obtained [16] underline the marked temperature influence on concrete resistivity that has to be compensated correctly, otherwise erroneous conclusions on the wetting–drying process will be drawn.

A different sensor type based on indepth resistivity measurements is the so-called *multiring-electrode sensor* [17]. Many of these devices (Figure 17.6) have been installed in new structures. As an extension of the original design, multiring sensors have been embedded in existing structures in bore holes that are then injected [18]. The space between sensors and concrete is kept as small as possible. A further development, the expansion-ring system, allows a hole to be drilled in the concrete and the indepth distribution of the concrete resistance on existing structures to be measured as well. Combining the multiring device with rings made of carbon steel with a nearby cathode, macrocell currents between the different rings and the cathode can be measured (as in the anode-ladder described above) and time to depassivation can be determined [8]. The different possibilities of onsite monitoring of the concrete resistivity are discussed in ref. [19].

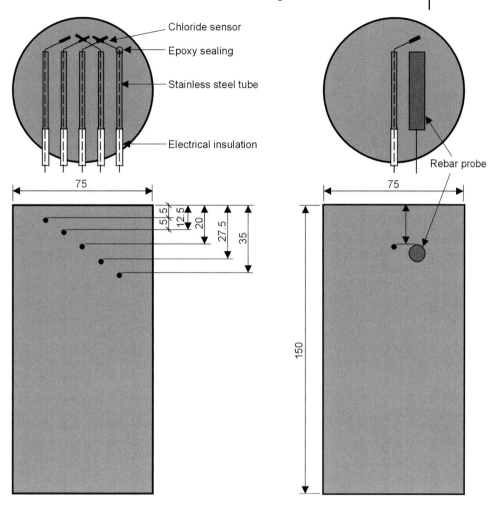

Figure 17.4 Sensor arrays mounted in a concrete core taken from a structure [13].

Macrocell Corrosion Monitoring Monitoring the macrocell current by connecting a microamperemeter between a small piece of corroding reinforcing steel (as an inserted probe or as a piece saw cut apart from the rebar network) with the surrounding passive reinforcement allows information to be obtained on the rate of the ongoing (macrocell) corrosion, the influence of temperature and humidity, etc. In addition, the potential difference and the resistance between corroding and passive rebars can be measured. This type of measurement (Figure 17.7) has been used frequently to monitor the efficiency of repair methods onsite, for example, the effect of shotcrete on a 17-year-old jetty [21], the application of inhibitors on a side wall in a tunnel [22] or to follow the drying out of a bridge substructure after repair [23].

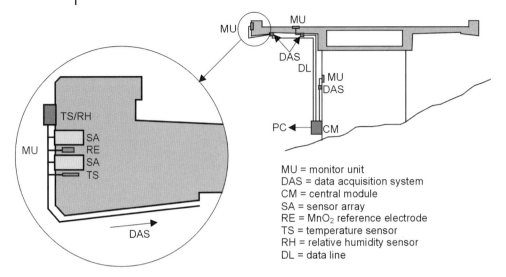

Figure 17.5 Installation of monitoring units on different parts of a bridge to study different exposure conditions [13].

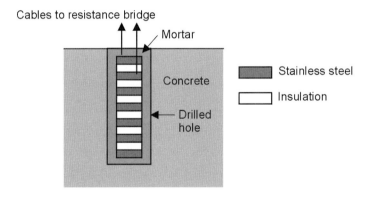

Figure 17.6 Multiring electrode to measure in-depth resistance profiles in concrete, as mounted in an existing structure.

Instead of cutting a piece of rebar to get an isolated probe, an external preconditioned corroding piece of steel can be mounted into the structure, usually contained in a concrete core (Figure 17.4) that is put into a hole drilled in the structure. This measurement arrangement allows getting the same information as from the isolated piece of rebar. Experience from several years (Figure 17.8) has shown that in this way correct corrosion rates (when compared to mass-loss measurements) can be obtained and the prediction of future corrosion rates for a given exposure condition becomes possible [16, 20].

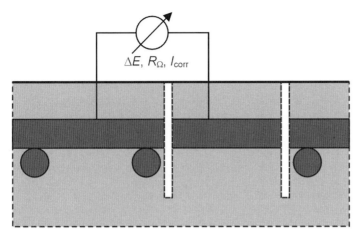

Figure 17.7 Schematic view of macrocell current measurements between an isolated piece of rebar (anode) and the surrounding rebar network (cathode). In the open-circuit condition the potential difference and the resistance between the two electrodes can be monitored [20].

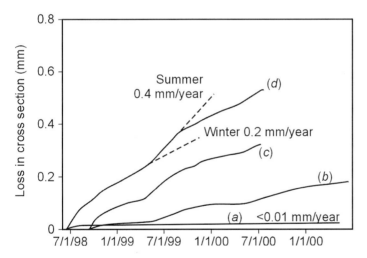

Figure 17.8 Loss of cross section of rebars in concrete contaminated by about 1.5% chloride by mass of cement for different exposure classes. (a) bridge Nanin Kon-solkopf sheltered, (b) underpass Malabarba bridge abutment, (c) Tunnel Cianca Presella, column exposed, (d) bridge Nanin Kon-solkopf, exposed [20].

Relative Humidity Sensors Instead of measuring the concrete resistivity it is interesting to determine the relative humidity (R.H.) of concrete because it gives direct information on moisture content and allows the degree of capillary saturation to be determined. R.H. monitoring over time can indicate critical changes in the degree of saturation of concrete as a function of temperature, for example, daily, monthly, and seasonal variations. Such moisture sensors are available and their reliability is discussed in ref. [24]. Onsite it is usually necessary to drill holes into the concrete, clean them carefully and mount the R.H. sensor carefully sealed. Procedure and results on R.H. monitoring as well as a critical evaluation of the R.H. influence on corrosion rate is given in ref. [25].

17.3
Monitoring with Electrochemical Sensors

If electrochemical data (steel potential, linear polarization resistance for corrosion-rate measurements), the chloride content or the pH of the concrete pore solution shall be monitored, an embedded *reference electrode* is an essential element of the sensor system. Of the commonly used "true" reference electrodes, only the MnO_2 (manganese dioxide, Figure 17.9) or Ag/AgCl (silver/silver chloride) reference electrodes are suitable for embedding into concrete (Section 16.2.1). Sometimes, pseudoreference electrodes such as activated titanium, carbon (graphite) or lead electrodes are used [26]; however, their long-term stability is a critical issue. Information on the long-term behavior of commercial reference electrodes embedded in concrete is reported in ref. [27]. MnO_2, activated titanium, graphite, and lead electrodes are sensitive to pH changes and/or oxygen content [28].

Corrosion Potential Any embedded reference electrode allows measurement of the electrochemical potential of the adjacent rebars. This allows detection of depassivation of the rebars or of any other steel sensor element by a drop in half-cell potential. The corrosion potential will be influenced by concrete humidity and oxygen content (Chapter 7). The depassivation of the steel probe located in the outermost cover concrete will present an early warning and suitable indepth distribution of a set of steel probes allows evaluation of the corrosion risk or to calculate the time of depassivation of the rebars. Monitoring the corrosion potential requires stable ("true") reference electrodes, like MnO_2 and Ag/AgCl.

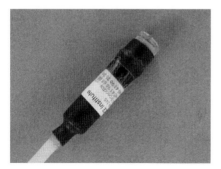

Figure 17.9 Manganese dioxide reference electrode.

Linear Polarization Resistance (LPR) These measurements allow monitoring the actual corrosion rate of embedded probes or of the reinforcing bars over time. The measurement principle is described in Section 16.2.3. In addition to the reference electrode, a counterelectrode of a corrosion-resistant material (e.g., stainless steel or activated titanium) has to be embedded. Several compact *LPR* (linear polarization resistance) sensor systems were developed and installed in structures such as precast deck elements in a road tunnel or cast in situ bridge decks [10, 29]. When existing structures have to be monitored for corrosion rate, a corroding piece of rebar can be isolated (by cutting) to get the working electrode and only a reference and a counterelectrode have to be embedded; alternatively, cores are taken and provided with reference and counterelectrodes and connecting cables in the laboratory [16, 20, 30]. Due to the short duration of the measurement, long term stability is nonessential and reference electrodes of all types can be used, including activated titanium.

This type of corrosion-rate monitoring can be used to control the effect and durability of repair work, to follow the onset of corrosion, the influence of climatic variations and to predict the future corrosion rate (loss in cross section) of the reinforcement.

Chloride Content By embedding the combined chloride/resistivity sensor elements mentioned above (Figure 17.3), the activity of the free chloride ions in the pore solution of concrete can be monitored over time at different depths. The potential of the embedded chloride sensors is measured versus a MnO_2 reference electrode and converted by Nernst's law to chloride concentration. In several field applications, hundreds of chloride sensors worked well over several years [20]. A more detailed description of the chloride sensor, its calibration and long-term stability is given in references [15, 29]. As a reference electrode is necessary to measure the sensor potential, ohmic potential drop caused by macrocell currents and diffusion potentials caused by differences in pH and chloride content of the concrete surrounding the chloride sensor and the reference electrode can lead to erroneous readings [31, 32]. Thus, a combined reference electrode/chloride sensor would be necessary.

pH Monitoring The pH of the pore solution of concrete is an important information; it can indicate the risk of corrosion due to carbonation and it is important for the evaluation of the correct $[Cl^-]/[OH^-]$ ratio leading to the depassivation of the rebars due to chloride ingress. Commercial sensors for pH monitoring are not available so far, but several materials have been tested in the laboratory. There is some evidence on the performance of activated titanium (coated with metal oxides, see also Chapter 20) as a pH electrode in concrete. It appears that the alkaline and carbonated state of the surrounding concrete could clearly be distinguished by a potential difference of about 200 mV (Figure 17.10) [26, 33, 34]

Oxygen-Transport Monitoring In some circumstances it might be important to know the oxygen concentration in concrete, especially when dealing with

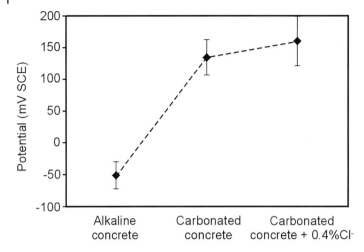

Figure 17.10 Average value in time and standard deviation of potential of activated titanium electrodes embedded in alkaline and carbonated concrete (portland cement, w/c 0.65, outside exposure, unsheltered) [33, 34].

submerged structures or with protection or repair systems based on organic coatings on the concrete. Oxygen-concentration measurement is based on the cathodic oxygen reduction current at constant potential (Section 7.2.2). The higher the cathodic current measured, the higher the oxygen transport. Laboratory tests and field trials have been performed in Norway [35]. Oxygen-transport measurements allowed evaluating various repair measures on areas where the rebars were still passive and noncorroding.

17.4
Critical Factors

Monitoring of reinforced concrete structures exposed to aggressive environmental loads with regard to the time evolution of their durability can provide important information for the management of structures. Ideally, such information can be used to optimize maintenance and safeguard reliability and safety at desired levels. However, the monitoring strategy has to be clearly defined in terms of technical aspects (types, numbers, and locations of probes) and of management issues (structural performance, availability), otherwise heavily instrumented structures produce large amounts of data with little or no practical value. Presently, integrating technical possibilities and managerial policies is still a great challenge. The points made below are discussed in more detail in refs. [26, 36].

Objective of Monitoring A monitoring system, eventually with computerized data acquisition, should meet specifically defined objectives, such as: (i) to monitor the durability of the structure and its condition in order to make timely decisions for preventive and/or repair actions, (ii) to monitor the effect of preventive or repair actions, (iii) to monitor the condition of structures based on new materials and/or new technology (including service-life prediction models), and (iv) to follow the time development in areas where access is difficult.

Monitoring Design In order to meet one or several of the objectives the *number and location* of permanently installed sensors should be defined based on the geometry of the structure, structural and environmental loads, previous experience with similar types of structures and the planned level of maintenance of the structure. The representativeness of the individual sensor location should be evaluated. Early-warning probes should be located in critical areas for the initiation of corrosion, for example, below expansion joints that are prone to develop leakage.

Procedures and responsibilities regarding operation and service of the monitoring system and quality assurance and control of the recorded data have to be established. Data collected should be presented in a clear way, understandable for the end-user.

Choice of Sensors and Probes The choice of the sensors should take into account their stability as a function of time, so that the necessary calibrations and maintenance can be planned. Initial sensor calibration has to be documented. For remote or inaccessible locations only robust and durable sensors should be installed.

17.5
On the Way to "Smart Structures"

The installation of sensors in a reinforced concrete structure is a feasible way to obtain information on durability-related changes, for example, indication of the corrosion risk. There are, as discussed above, several problems and doubts related with the long-term stability of some sensor types. Depending on the measurement method, sensors can be more or less sensitive to loss of stability. In general, "passive" methods (corrosion potentials, pH, chloride sensing, which all use a reference electrode) tend to be more sensitive; "active" methods (*LPR*, resistivity) tend to be less sensitive to "drift" or poor stability.

An interesting example of an "active" method has been applied to post-tensioning structures where the high-strength steel strands are placed in (grouted) polymer ducts with complete electrical isolation (Figure 17.11). These new post-tensioning systems have the advantage of reduced friction, increased fretting fatigue resistance and a tight and noncorroding duct that avoids ingress of aggressive substances from the environment. In addition, because of the electrically isolated anchor heads, it offers the possibility of monitoring the electrical impedance between the steel strands in the duct and the rebar network [38]. Monitoring of the electrically isolated tendons requires only sound electrical connections to the rebar network and to one of the end anchorages. The AC impedance is measured

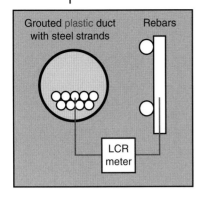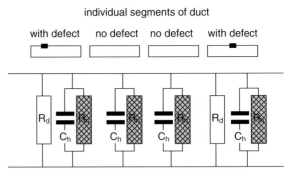

Figure 17.11 Principle of measuring the electrical impedance of a tendon with the LCR meter (a) and electrical equivalent circuit for an electrically isolated tendon with small defects: R_d defect resistance, R_h, C_h resistance and capacitance of the intact duct (b) [37].

Table 17.1 Experimental and derived specific values of the ohmic resistance R and the capacitance C from the flyover "P.S. du Milieu" (length 100 m) [38].

Tendon No.	Experimental		Specific values	
	R (kΩ)	C (nF)	R (Ω m)	C (nF/m)
1	7.234	234.0	723	2.34
2	13.70	233.0	1370	2.33
3	20.87	235.0	2087	2.35
4	17.81	237.2	1781	2.37
5	28.25	234.7	2825	2.35
6	0.006	–	Short circuit	–

at a fixed frequency (120 Hz or 1 kHz) with a portable instrument and the ohmic and capacitive impedance is recorded. The experimentally measured impedance data of six individual tendons (length 100 m, PT Plus duct, diameter 59 mm) are presented in Table 17.1. As can be observed, the ohmic resistance of five out of six of the tendons is very high (Nos. 1–5) indicating electrical isolation. Tendon No. 6 shows a short circuit. The capacitance values of the polymer duct (with 100 m length) are very reproducible at 235 nF. This capacitance value agrees with the laboratory tests of the same duct where 2.34 ± 0.03 nF/m were determined [39].

Different AC resistance values indicate different amounts of very small sized defects along the duct. More interesting than the actual value of the resistance is its evolution over time (Figure 17.12): due to hydration and drying out of the cement grout and the concrete, the electrical resistance increases approximately with the square root of time. A steadily decreasing resistance of one of the tendons

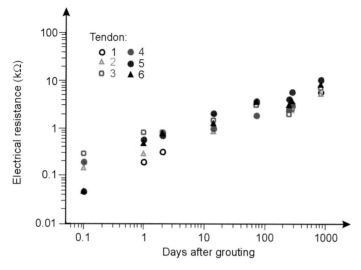

Figure 17.12 Evolution of the electrical resistance with time of six electrically isolated post-tensioned strands of the flyover "Pré du Mariage" [37].

at a certain time would indicate the infiltration of water (and potentially also chlorides) at one of these defects long before corrosion of the high-strength steel starts.

In the case of post-tensioned structures with electrically isolated tendons in polymer ducts a simple, robust and highly efficient monitoring system is built, producing a "smart structure". It allows, together with simple data treatment, quality control of the electrical isolation during the construction process, monitoring of the corrosion protection of the high-strength steel over time during service and detecting of a corrosion risk situation of individual tendons well before actual corrosion of tendons may occur. Because no special sensors are needed, the monitoring system can be used during the whole service period of the structure without any issue of electrode stability. Results of the application in a number of viaducts and railway bridges in Switzerland or within the new high-speed train system in Italy are reported [37, 40, 41]. Electrically isolated tendons are the highest level of tendon protection strategy (PL3) according to the *fib* guidelines [42, 43].

17.6 Structural Health Monitoring

Although not strictly related to this book, the topic shall be mentioned briefly. Structural health monitoring is an important methodology in evaluating the "health" of a structure by assessing the level of deterioration and remaining service life of civil infrastructure systems [43]. One approach is to use "global" inspection methods, thus ideally the assessment of the structural performance characteristics of a whole bridge or building. The idea of vibration-based damage detection is to measure dynamic characteristics such as *eigenfrequencies*, damping ratios and mode shapes on a regular basis. The state, and eventually degradation, of the structure is – in principle – reflected in the evolution of these characteristics [44].

Another approach in structural health monitoring is the use of different sensor technologies and methods of sensing to monitor changes in structural performance characteristics [43]. Deformation or mechanical stress are recorded by piezo-electric, magnetoelastic or fiber-optic sensors that have to be connected by cable or wireless sensor network to a central unit. Signal processing and evaluation with statistical pattern recognition are critical issues for damage detection.

Sonic and ultrasonic NDT methods are used for health monitoring and assessment of concrete structures [45]. They include sonic/ultrasonic transmission method, tomography, sonic/ultrasonic reflection methods, impact echo system or impulse resonance tests.

Acoustic emission frequently is used to monitor reinforced and prestressed concrete structures, especially to detect and locate wire breaks in grouted post-tensioning bridges [46, 47]. An overview of methods for the assessment and structural health monitoring for prestressed and post-tensioned structures is presented in [48].

References

1 EN 1992-1-1 (2004) Eurocode 2: Design of Concrete Structures – Part 1-1: General Rules and Rules for Buildings, European Committee for Standardization.
2 *fib*, International Federation for Structural Concrete (2010) *Model code 2010 – first complete draft – vol. 1*, Bulletin 55, Lausanne, 2010.
3 *fib*, International Federation for Structural Concrete (2010) *Model code 2010 – first complete draft – vol. 2*, Bulletin 56, Lausanne, 2010.
4 Jensen, J.S. and Andersen, E.Y. (1994) Monitoring of structural performance – a must for the future, in *Proc. Int. Conf. Concrete Across Borders*, Odense, Denmark, pp. 585–594.
5 Leibbrandt, A., Caprari, G., Angst, U., Flatt, R., Siegwart, R.J., and Elsener, B. (2012) Climbing robot for corrosion monitoring of reinforced concrete structures. 2nd Int. Conf. on Applied Robotics for the Power Industry CARPI 2012, 11–13 September 2012, Zurich.
6 Inaudi, D., Casanova, N., Vurpillot, S., Glisic, B., Kronenberg, P., and Loret, S.L. (2000) Lessons learned in the use of fibre optic sensors for civil engineering monitoring, in *Present and Future of Health Monitoring* (eds P. Schwesinger and F.H. Wittmann), Aedificatio Publishers, Freiburg, pp. 79–92.
7 Polder, R.B., Peelen, W.H.A., Klinghoffer, O., Eri, J., and Leggedoor, J. (2007) Use of advanced corrosion monitoring for risk based management of concrete structures. *Heron*, **52** (4), 239–250.
8 Raupach, M. and Schiessl, P. (2001) Macrocell sensor systems for monitoring the corrosion risk of the reinforcement in concrete structures. *NDT&E International*, **34**, 435–442.
9 Raupach, M. and Schiessl, P. (1997) Monitoring system for the penetration of chlorides, carbonation and the corrosion risk for the reinforcement. *Construction and Building Materials*, **11** (4), 207–214.
10 Broomfield, J.P., Davies, K., and Hladky, K. (2002) The use of permanent corrosion monitoring in new and existing reinforced concrete structures. *Cement and Concrete Composites*, **24** (1), 27–34.

11 Raupach, M., Polder, R., Frolund, T., and Nygaard, P. (2007) Corrosion monitoring at submerged concrete structures–macrocell corrosion due to contact with aerated areas? EUROCORR 2007, Freiburg im Breisgau, Germany, 9–13 September 2007 (CD-ROM).

12 Polder, R.B., Peelen, W.H.A., and Leegwater, G. (2008) Corrosion monitoring for underground and submerged concrete structures–examples and interpretation issues, in *Tailor Made Concrete Structures* (eds J.C. Walraven and D. Stoelhorst), fib Symposium, May 19–21, Amsterdam, Taylor & Francis Group, London, pp. 187–192.

13 Zimmermann, L., Schiegg, Y., Elsener, B., and Böhni, H. (1997) Electrochemical techniques for monitoring the conditions of concrete bridge structures, in *Repair of Concrete Structures* (ed. A.B. Svolvaer), Norwegian Public Roads Administration, pp. 213–222.

14 Elsener, B., Zimmermann, L., Flückiger, D., Bürchler, D., and Böhni, H. (1997) Chloride penetration–nondestructive determination of the free chloride content in mortar and concrete, in *Proc. RILEM Int. Workshop Chloride Penetration into Concrete* (eds L.O. Nilsson and J.P. Ollivier), pp. 17–26.

15 Elsener, B., Zimmermann, L., and Böhni, H. (2003) Nondestructive determination of the free chloride content in cement based materials. *Materials and Corrosion*, **54** (6), 440–446.

16 Schiegg, Y., Audergon, L., Elsener, B., and Böhni, H. (2007) On-line monitoring of corrosion in reinforced concrete structures, in *Corrosion of Reinforcement in Concrete, Mechanisms, Monitoring, Inhibitors and Rehabilitation Techniques*, European Federation of Corrosion Publication number 38 (eds M. Raupach, B. Elsener, R. Polder, and J. Mietz), Woodhead Publishing Limited, Cambridge, pp. 133–145.

17 Schiessl, P. (1996) "Einsatz von Sonden für die Bauwerksüberwachung", 4. Internationales Kolloquium "Werkstoffwissenschaften und Bauinstandsetzen", Techn. Akademie Esslingen, 17–19 December 1996, pp. 3–20.

18 Borsje, H., Peelen, W.H.A., Postema, F.J., and Bakker, J.D. (2002) Monitoring alkali-silica reaction in structures. *Heron*, **47** (2), 95–110.

19 Hunkeler, F. (1997) Monitoring of repaired reinforced concrete structures by means of resistivity measurements, in *Repair of Concrete Structures* (ed. A.B. Svolvaer), Norwegian Public Roads Administration, pp. 223–232.

20 Schiegg, Y. (2002) *Online-monitoring to detect corrosion of rebars in RC structures*, PhD Thesis Nr. 14583 ETH Zurich (in German).

21 Broomfield, P., Davies, K., and Hladky, K. (2000) Permanent corrosion monitoring in new and existing reinforced concrete structures. *Materials Performance*, **39** (7), 66–70.

22 Schiegg, Y., Hunkeler, F., and Ungricht, H. (2007) The effectiveness of corrosion inhibitors–a field study, in *Corrosion of Reinforcement in Concrete, Mechanisms, Monitoring, Inhibitors and Rehabilitation Techniques*, European Federation of Corrosion Publication number 38 (eds M. Raupach, B. Elsener, R. Polder, and J. Mietz), Woodhead Publishing Limited, Cambridge, pp. 226–238.

23 Bindschedler, D. and Hunkeler, F. (1997) Korrosionsuntersuchungen an der Europabrücke. *SI+A*, **19**, 28–32.

24 Nilsson, L.O. (1997) Assessing moisture conditions in marine concrete structures, in *Repair of Concrete Structures* (ed. A.B. Svolvaer), Norwegian Public Roads Administration, pp. 273–282.

25 Andrade, C., Alonso, C., and Sarria, J. (2002) Corrosion rate evolution in concrete structures exposed to the atmosphere. *Cement and Concrete Composites*, **24** (1), 55–64.

26 COST 521 (2003) Corrosion of Steel in Reinforced Concrete Structures, Final Report (eds R. Cigna, C. Andrade, U. Nürnberger, R. Polder, R. Weydert, and E. Seitz), European Commission, Directorate General for Research, EUR 20599, 2003.

27 Videm, K. (1997) Instrumentation and condition assessment performed on Gimsoystraumen bridge, in *Repair of Concrete Structures* (ed. A.B. Svolvaer),

Norwegian Public Roads Administration, pp. 375–390.

28 Myrdal, R. (2007) *The Electrochemistry and Characteristics of Embeddable Reference Electrodes for Concrete*, European Federation of Corrosion Publication number 43, Woodhead Publishing Limited, Cambridge.

29 COST 509 (1997) Corrosion and Protection of Metals in Contact with Concrete, Final Report (eds R.N. Cox, R. Cigna, O. Vennesland, and T. Valente), European Commission, Directorate General Science, Research and Development, Brussels, EUR 17608 EN, 1997.

30 Gulikers, J. (1996) Development of a galvanic corrosion probe to assess the corrosion rate of steel reinforcement, in *Proc. 4th Int. Symp. on Corrosion of Reinforcement in Concrete Construction* (eds C.L. Page, P.B. Bamforth, and J.W. Figg), Society of Chemical Industry, Cambridge, UK, 1–4 July, pp. 327–336.

31 Angst, U., Elsener, B., Larsen, C.K., and Vennesland, Ø. (2010) Potentiometric determination of the chloride ion activity in cement based materials. *Journal of Applied Electrochemistry*, **40** (3), 561–573.

32 Angst, U., Vennesland, Ø., and Myrdal, R. (2009) Diffusion potentials as source of error in electrochemical measurements in concrete. *Materials and Structures*, **42** (3), 365–375.

33 Bertolini, L., Bolzoni, F., Pedeferri, P., and Pastore, T. (1998) Cathodic protection of reinforcement in carbonated concrete, NACE International, Corrosion/98 Conference, paper 98639.

34 Bertolini, L., Pedeferri, P., Redaelli, E., and Pastore, T. (2003) Repassivation of steel in carbonated concrete induced by cathodic protection. *Materials and Corrosion*, **54** (3), 163–175.

35 Vennesland, Ø. (1997) Electrochemical parameters of repaired and non-repaired concrete at Gimsoystraumen bridge, in *Repair of Concrete Structures* (ed. A.B. Svolvaer), Norwegian Public Roads Administration, pp. 253–262.

36 Espelid, B., Markey, I., and Videm, K. (1997) Strategies for monitoring durability of a concrete structure, in *Repair of Concrete Structures* (ed. A.B. Svolvaer), Norwegian Public Roads Administration, pp. 233–242.

37 Elsner, B. and Büchler, M. (2011) Quality Control and Monitoring of Electrically Isolated Post-Tensioning Tendons in Bridges, VSS Report 647. download at http://www.ifb.ethz.ch/corrosion/research/reports.

38 Elsener, B. (2005) Long-term monitoring of electrically isolated post-tensioning tendons. *fib Journal Structural Concrete*, **6** (3), 101–106.

39 Toller, L. (2000) Elektrische Widerstandsmessung von Spanngliedern in Kunststoffhüllrohren. Master Thesis, Department of Civil Engineering, ETH Zürich.

40 Della Vedova, M. and Elsener, B. (2006) Enhanced durability, quality control and monitoring of electrically isolated tendons. Proc. of the 2nd fib Congress, Naples, 5–8 June 2006.

41 Elsener, B., Della Vedova, M., Büchler, M., and Ormellese, M. (2009) New Systems for post-tensioned structures, in COST 534, New Materials, Systems, Methods and Concepts for Prestressed Concrete Structures, Final Report (eds R. Polder, M.C. Alonso, D. Cleland, B. Elsener, E. Proverbio, Ø. Vennesland, and A. Raharinaivo), European Commission, Cost Office, TNO Delft, pp. 89–116.

42 *fib*, International Federation for Structural Concrete (2006) *Durability of post-tensioning tendons* (eds J.P. Fuzier, H.R. Ganz, and P. Matt), Bulletin 33, Lausanne.

43 Kharbari, V.M. (ed.) (2009) *Structural Health Monitoring in Civil Infrastructure Systems*, Woodhead Publishing.

44 Peeters, B., Maeck, J., and De Roeck, G. (2001) Vibration-based damage detection in civil engineering: excitation sources and temperature effects. *Smart Materials and Structures*, **10** (3), 518–527.

45 Forde, M.C. (2004) Sonic and ultrasonic NDT methods in the assessment of concrete structures, in *Proc. First Workshop COST 534 "NDT Assessment and New Systems in Pre-Stressed Concrete Structures"* (ed. B. Elsener), 13 October, Zurich, pp. 55–69.

46 de Wit, M. (2004) The use of acoustic monitoring to manage concrete structures, in *Proc. First Workshop COST 534 "NDT Assessment and New Systems in Pre-Stressed Concrete Structures"* (ed. B. Elsener), 13 October, Zurich, pp. 25–30.

47 Fricker, S., Vogel, T., Ungricht, H., and Hunkeler, F. (2010) Akustische Überwachung einer stark geschädigten Spannbetonbrücke und Zustandserfassung beim Abbruch, Bericht VSS Nr. 643, 2010, download at http://www.ifb.ethz.ch/corrosion/research/reports.

48 Proverbio, E. (2009) New methods for the assessment and monitoring of prestressed concrete structures, in *COST 534, New Materials, Systems, Methods and Concepts for Prestressed Concrete Structures*, Final Report (eds R. Polder, M.C. Alonso, D. Cleland, B. Elsener, E. Proverbio, Ø. Vennesland, and A. Raharinaivo), European Commission, Cost Office, TNO Delft, pp. 117–240.

18
Principles and Methods for Repair

Reinforced concrete structures damaged by concrete degradation or steel corrosion need maintenance or repair interventions aimed at restoring the safety and serviceability of the structure to the required performance level and, in the meantime, providing a reasonable residual service life to the structure [1]. The recent Model Code of the International Federation for Structural Concrete (*fib*) proposes maintenance procedures to be included both within *proactive* and *reactive* approaches [2]. The former are based on the application of measures that prevent any damage before it becomes critical for the expected performance of the structure. These are usually defined at the time of design or construction and may be selected on the basis of a planned periodic inspection and systematic monitoring of critical deterioration processes. This type of strategy has been considered in Chapter 11 among the possible design options for the design of durability of new structures. Conversely, reactive approaches consist in the application of nonpreviously planned remedial measures or interventions intended to stop ongoing degradation that may damage or has already damaged the structure. A reactive approach is the only possibility for existing structures, which were not subjected to specific durability design at the time of building and nowadays are in the propagation stage of corrosion. Many buildings or infrastructures are in this condition and require maintenance interventions in order to prolong their life and guarantee reasonable serviceability performance. In the case of buildings or constructions belonging to the cultural heritage preservation requirements should also be met [3]. Various repair options are available for the rehabilitation of structures damaged by corrosion of steel reinforcement. In recent years, some recommendations and standards have been developed by different organizations, to help designers in the choice of the repair strategy [1, 2, 4, 5]. Many standards, however, refer to properties related to the materials used in the repair rather than to the whole performance of the repaired structure. This chapter deals with remedial techniques for structures that are subjected to carbonation-induced or chloride-induced corrosion and follows the approach proposed by the RILEM 124-SRC recommendation [1]. Details of conventional repair and electrochemical maintenance methods are reported in Chapters 19 and 20. The repair of prestressed concrete structures is outside the scope of this book.

Corrosion of Steel in Concrete: Prevention, Diagnosis, Repair, Second Edition. Luca Bertolini,
Bernhard Elsener, Pietro Pedeferri, Elena Redaelli, and Rob Polder.
© 2013 Wiley-VCH Verlag GmbH & Co. KGaA. Published 2013 by Wiley-VCH Verlag GmbH & Co. KGaA.

18.1
Approach to Repair

RILEM 124-SRC suggests considering three phases in the definition of a repair work. A first phase is aimed at defining the *repair option*, that is, at deciding whether to take on actions to control the corrosion process or adopt other approaches. If intervention in the corrosion process has been chosen, a second phase will be dedicated to the selection of a *repair principle* that can avoid further corrosion. Finally, the last phase is the choice of the *repair method* and the design of the details of its application.

18.1.1
Repair Options

Several factors may influence the decision of the approach for intervention. Figures 18.1 and 18.2 show the main steps for the decision on the repair option suggested by RILEM committee 124-SRC [1].

A preliminary assessment of the condition of the structure is required. Inspection techniques described in Chapter 16 may be used to diagnose the causes of deterioration and the extent of damage. Degradation processes that may indirectly contribute to corrosion of the reinforcement (freeze–thaw, sulfate attack, etc.) should also be considered. Distinction should be made between different parts of the structure where:

a) reinforcement is still passive, that is, corrosion has not initiated since carbonation or chloride penetration has not yet reached the steel surface;

b) reinforcement is corroding but the propagation is in the early stages, for example, concrete cover is not cracked and reduction in cross section of rebars is negligible;

c) corrosion of steel has already led to loss of serviceability of the structure, for example, due to cracking, spalling or delamination of the concrete cover and/or more than insignificant loss of rebar cross section.

In order to define a proper repair option, the future evolution of the damage should also be estimated in situations *a* and *b*, and the length of time required to reach the critical condition *c* should be evaluated (Chapter 11). This time should be compared with the design life of the repair, which can be the remaining period of use of the structure or the time expected before a new planned repair action. It is then possible to distinguish between the conditions where damage already exists, those where it is expected, and those where damage is not likely during the design life of the repair. Any future development of damage has to be considered in relation to the required residual service life of the repaired structure. The definition of the design life of the repair may be difficult in the case of buildings or structures with architectural, historical or cultural relevance. In this case, *a priori*, a very long residual life would be required, but this would often be associated with heavy

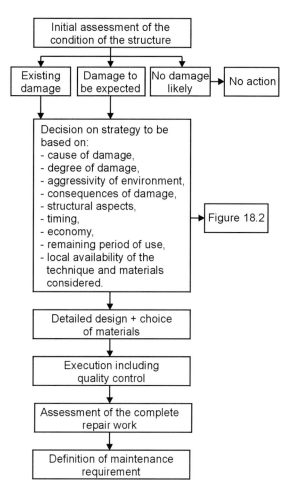

Figure 18.1 Steps to be taken in the repair process, according to RILEM 124-SRC [1].

repair works that may themselves alter both the original materials and the aspect of the building. In this case a compromise should be reached between extending as long as possible the service life and reducing as much as possible the invasiveness of the intervention [3].

When damage is not likely, no action is required, while in the case of already present or expected damage an option from those shown in Figure 18.2 has to be selected. A decision may be made to replace or partially reconstruct components that have undergone severe damage, for example, in the case of particularly exposed elements that can be easily replaced. More frequently, interventions aimed at stopping corrosion or reducing the corrosion rate in the existing structure are adopted, possibly after adding reinforcement (in the case of significant cross section loss) and repair and/or replacement of the damaged concrete. In some

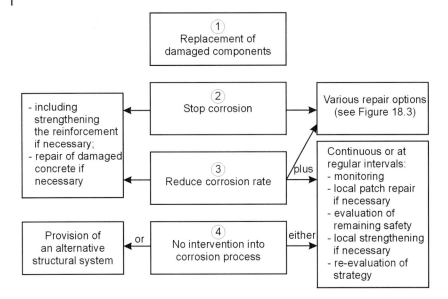

Figure 18.2 Basic strategies for intervention, according to RILEM 124-SRC [1].

cases, especially if the extent of damage is limited or the remaining period of use is short, it may be decided not to intervene in the corrosion process, but to keep the conditions of corrosion and serviceability under control by some form of monitoring. In other cases, an alternative structural system may be provided or the function of the structure may be downgraded, if structural requirements are not fulfilled.

The decision on the repair option has to be taken starting from the extent and the cause of damage, its evolution in time, the intended use and importance of the structure and the consequences of degradation for its structural safety and serviceability (Figure 18.1). The cost of the repair methods, the availability and the experience with their use in the place where the structure is located, and the time available for repair also have to be considered. Several questions should be answered during this step, which go far beyond corrosion aspects. Table 18.1 gives an example of the most frequent questions to be considered.

Although this chapter only deals with methods aimed at providing protection to the reinforcement, safety aspects should be taken into serious consideration. The risks to health and safety from falling debris or local failure due to loss of material, and the effects of deterioration upon the structural stability must be assessed. If this analysis leads to the conclusion that the structure has to be considered unsafe, appropriate actions must be specified to restore safety requirements before other protection or repair work is undertaken, taking into account any additional risk that may arise from the repair work itself. Such action may include local protection or repairs, the installation of supports or other temporary works, or partial or even complete demolition [5].

Table 18.1 Key factors in the decision of the repair strategy, according to RILEM 124-SRC [1].

Cause and extent of damage
What is the extent of depassivation?
Has the damage been caused by carbonation, chlorides or by other aggressive factors?
Has the damage been caused mainly by an extremely aggressive environment or by bad design, detailing or execution?
Has the damage been caused by an unexpected environment or use?
What are the dominant transportation mechanisms for the aggressive agents into and within the concrete (e.g., absorption or diffusion)?

The consequences of the damage
Does the damage influence structural safety or just the appearance of the structure?
Does the damage influence the fitness of the structure for its intended use?
In cases of damage influencing the structural safety, would in the worst instance a global or a local failure occur? What would be the consequence of a failure? Would the public be at risk from falling debris?

The appropriate time for intervention
Is immediate action necessary to avoid a sudden failure or to avoid further significant deterioration?
Is the rate of damage so low that repair could be postponed or not attempted?
Should early, preventive measures be taken to avoid damage in the future?
Is it advisable only to change the environment, and to repair the damage at a later stage (e.g., after the structure is in equilibrium under the new environ¬mental conditions), or not to attempt repair after all?

The remaining time in use
Is the remaining time in use sufficiently short for repair or maintenance not to be attempted?
Does the decision on the type of intervention fix the remaining time in use?
Economic aspects, that is, the most cost-effective solution including such aspects as the costs of maintenance.
May intervention be deferred to a later stage?
Is the structure accessible to the maintenance needed?
Are the costs of maintenance acceptable?

Practicability of repair work
Can the requirements be fulfilled under the given site conditions, for example, the required surface condition of steel for coating?
Are the appropriate techniques and materials available? Is sufficient access possible to all areas needing repair? Is pollution by sandblasting or water jetting acceptable?

18.1.2
Basic Repair Principles

The corrosion rate of steel can be controlled throughout the residual service life of the structure using several techniques that rely on four different principles, as shown in Figure 18.3. As illustrated in Chapter 7, in order to reduce the corrosion rate it is possible to stop the anodic process or the transport of current within the

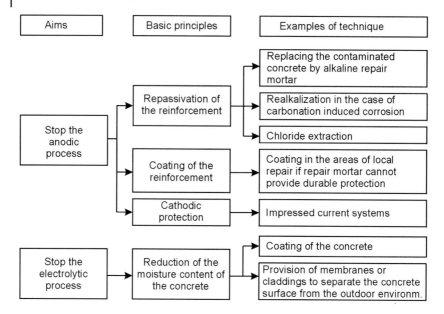

Figure 18.3 Principles of repair to stop corrosion of reinforcement, according to RILEM 124-SRC [1].

concrete (electrolytic process). The anodic process can be stopped by restoring passivity on the steel surface, by coating the rebars, or by application of cathodic protection. Alternatively, the electrolytic process can be stopped by reducing the moisture content of concrete. For structures exposed to the atmosphere, there are no reliable techniques that can stop the cathodic process; this would only be possible if the structure could be kept permanently and constantly saturated with water. In the past, attempts have been made to stop the cathodic process by applying a coating to the concrete surface aimed at excluding oxygen access. In many cases this approach has failed. Failure is caused by the fact that even the densest coating allows some oxygen transport; the oxygen already present in the concrete may be sufficient to sustain corrosion and for new cracks to be formed, allowing more oxygen to enter. Furthermore, in general it is not possible to apply a coating to the whole surface (i.e., on all sides) of a concrete element or structure. Consequently, there will always be a path for access of oxygen.

The principle adopted to control corrosion has to be clearly defined in the design phase of the repair work. Similarly, a specific repair method suitable to achieve such principle has to be found. Several methods of repair have been developed since the 1970s. Some of them have been used successfully in many countries; others have only local traditions; some are still in the experimental stage; others have been through the experimental phase and demonstrated their usefulness on many real structures but cannot be considered routine because their long-term behavior is not yet known. The next sections summarize the most common repair methods for carbonated or chloride-contaminated structures, and outline the minimum requirements necessary to satisfy the repair principle they are based on.

Normally only one method should be adopted and all the conditions that are necessary for it to be effective must be fulfilled. Although, in principle, more than one method can also be used simultaneously, it should be clear that all the requirements of at least one method must be carefully observed (RILEM 124-SRC calls this the "main method"). Additional methods must not interfere negatively with the main method and cannot, in any case, be used as an excuse for avoiding the fulfilment of requirements of the main method. In fact, any combination of different methods cannot be sufficient to guarantee reliable protection of the reinforcement if none of the methods has been applied properly.

18.2
Overview of Repair Methods for Carbonated Structures

Figure 18.4 shows the methods used for repair of carbonated structures, which are based on repassivation of steel, limitation of moisture content of concrete, and coating of the reinforcement.

18.2.1
Methods Based on Repassivation

Methods based on the principle of repassivation of rebars must insure that the reinforcement becomes passive and further depassivation is prevented during the remaining service life of the structure. For structures suffering carbonation-induced corrosion, this means that alkaline conditions have to be restored around the steel rebars. Therefore, the reinforcement will be protected even if the moisture content of the concrete should be high.

Conventional Repair The most utilized method consists in the removal of carbonated concrete and its replacement with alkaline mortar or concrete. This method is convenient when the corrosion attack is limited to zones of small extent (e.g., when the thickness of the concrete cover is reduced locally). In that case it is usually called *patch repair*. Conversely, it may be rather expensive when repair is required on large surfaces. In fact, concrete should be removed in all the zones where carbonation and subsequent corrosion of steel are *expected* to damage the structure within the design life of the repair. Even structurally sound concrete must be removed where the corrosion rate of the embedded steel is expected to be high enough to produce unacceptable cracking of concrete within this period. Furthermore, concrete removal may be required also in zones where carbonation has not reached the reinforcement yet, if carbonation to the depth of the steel and subsequent corrosion damage to the structure are to be expected.

After concrete has been removed, loose rust has to be removed from the surface of the rebars and a new alkaline material is applied. The properties of the alkaline material and the thickness of the new cover have to be adequate to guarantee the

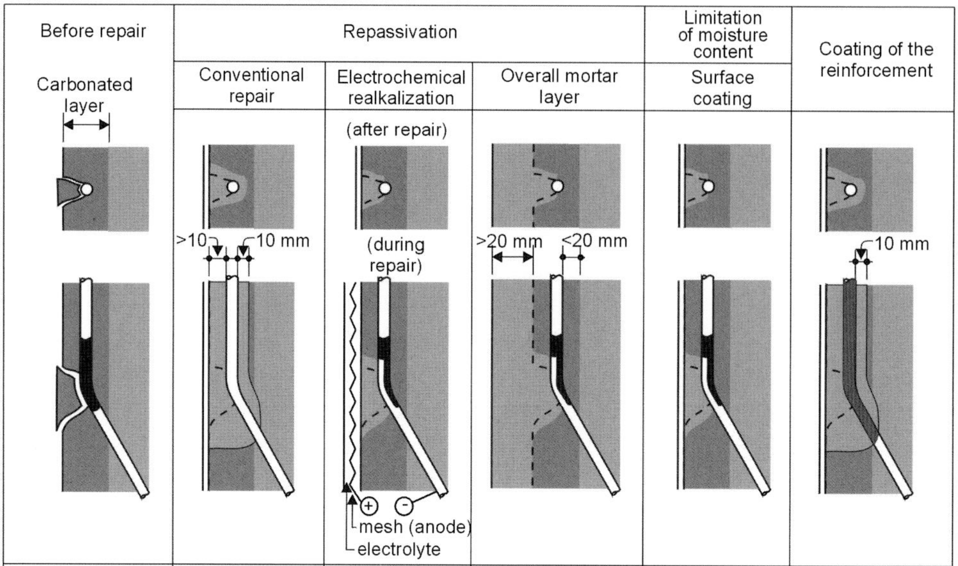

Before repair	Repassivation			Limitation of moisture content	Coating of the reinforcement
Carbonated layer	Conventional repair	Electrochemical realkalization	Overall mortar layer	Surface coating	
Concrete removal:	All areas expected to be damaged	Only cracked or spalled areas	Only cracked or spalled areas	Only cracked or spalled areas	All areas expected to be damaged
Cleaning of steel:	Loose rust	Loose rust	Loose rust	Loose rust	SA 2½
Coating of steel:	Not allowed	Not allowed	Not allowed	Not necessary	Required
Alkaline mortar:	Required	Required	Required	Not required	Not required
Surface treatment of concrete:	Not required (advisable)	Generally not required	Not required	Required	Not required (advisable)

Figure 18.4 Summary of widely used methods for the repair of structures affected by carbonation-induced corrosion (modified from [1]).

design life of the repair (Chapter 19). A coating may be applied on the repaired surface and the nonrepaired surface to increase the resistance to carbonation (Chapter 14). The beneficial effect of this coating can be considered both in the evaluation of the quantity of concrete to be removed and in the zones where the new material has been applied (e.g., if the cover thickness itself is not sufficient to prevent damage in the repair design life). It should, however, be taken into consideration that the effectiveness of coatings is limited in time and a strategy for their maintenance is required.

When the extent of carbonation is large, it may be necessary to remove a large amount of mechanically sound concrete. In order to avoid this expensive and undesirable operation, other methods have been developed that are aimed to realkalize the original concrete.

Repassivation with Alkaline Concrete or Mortar The application of a sufficiently thick (>20 mm) cement-based layer of concrete or mortar over the surface of carbonated concrete may induce realkalization of the underlying layer. Only cracked or delaminated concrete has to be removed, while mechanically sound concrete, even if it is carbonated up to the reinforcement and thus in contact with corroding steel, will not be removed. The method relies on the diffusion of hydroxyl ions (OH⁻) from the new external alkaline layer towards the carbonated concrete substrate. This may occur in wet environments or in the presence of wetting–drying cycles (i.e., the worst conditions for corrosion), possibly leading to repassivation of the reinforcement. This method is mainly applied in Germany [6]. This method should not be used if carbonation has penetrated behind the reinforcement more than 20 mm.

The beneficial effects of this type of repair are durable if the external layer offers an effective barrier to penetration of carbon dioxide. The thickness and permeability of this layer should therefore be sufficient to prevent its carbonation during the complete design life of the repair. In fact, the original concrete underlying this layer cannot be counted on to resist carbonation.

Electrochemical Realkalization This method is based on the application of direct current to the reinforcement from a temporary anode placed on the surface of the concrete. The treatment lasts several days to a few weeks. The realkalization of concrete takes place both from the surface of the concrete (due to the ingress of the alkaline solution used as anolyte) and from the surface of the steel (due to the cathodic process that takes place at the steel surface). This technique is described in Section 20.4.2.

Cathodic Protection The technique of cathodic protection, which is usually applied for chloride-induced corrosion, has also been applied to carbonated structures in the presence of small amounts of chlorides. It requires the permanent application of a small direct current to the steel and it can lead to repassivation of the reinforcement because of the realkalization of the concrete around the steel [7]. This technique is described in Section 20.3.

18.2.2
Reduction of the Moisture Content of the Concrete

In Chapter 5 it was shown that, at least in the absence of chloride contamination, the corrosion rate of steel in carbonated concrete is negligible when the concrete is dry (for instance when it is exposed to environments with relative humidity lower than 70%, Figure 5.9). In many exposure conditions where the concrete is dry, even if carbonation has reached the reinforcement, the corrosion rate is very low. This occurs for instance inside buildings where, in the absence of specific causes of wetting (such as capillary suction from the soil or condensation), the corrosion rate is usually low enough to permit a significant service life without any repair.

In environments with higher humidity or in the presence of wetting–drying cycles in the concrete, the application of a surface treatment that avoids absorption of water from the environment (Chapter 14) may lead to a reduction in the moisture content of the concrete and thus in the corrosion rate of the reinforcement.

This is usually achieved through hydrophobic treatments, impermeable coatings or cladding systems. Unexpected humidity that may enter the concrete, for example, through capillary suction, should be prevented. The efficiency of the treatment can be monitored by visual observation of the condition of the surface coating, by measuring the internal relative humidity of concrete, or with electrochemical measurements (Chapter 16) that measure the resistivity of concrete or the corrosion conditions of the embedded steel. The effectiveness will decrease in time and, when necessary, the coating must be maintained or even replaced. This method should not be used if the carbonated concrete is also contaminated by chlorides, since the corrosion rate may be high even if the moisture content of concrete is relatively low (Section 5.3.2). Concrete surface coatings may also be used to restrict penetration of carbonation or chlorides in zones where corrosion has not yet initiated (Chapter 14).

18.2.3
Coating of the Reinforcement

The anodic process could be stopped by applying a coating to the reinforcement that acts as a physical barrier between the steel and the repair mortar. For this purpose only organic coatings, preferably epoxy based, should be used. Protection is entirely based on the barrier between the reinforcement and the mortar, and passivation of steel cannot be achieved because contact with alkaline repair material is prevented. This method should be used to protect depassivated areas of the reinforcement only as a last resort, that is, when other techniques are not applicable and only for small specific applications [1]. It may be used, for instance, when the thickness of the concrete cover is very low and it is impossible to increase it, so that the repair material cannot provide durable protection to the embedded steel.

Concrete must be removed from all areas that are damaged or expected to be damaged within the design life of the repair. In order to reduce the amount of concrete that must be removed, a coating may be applied on the concrete surface so that penetration of carbonation may be kept under control after the intervention. The surface of the steel to be coated must be very carefully cleaned from corrosion products, bringing the metal to blast-cleaning grade SA $2^1/_2$.

It was suggested that such coatings operate by increasing the electrical resistance between the steel and the concrete; both epoxy and polymer-cement-based materials seem to have this effect, reducing local corrosion current exchange [8].

18.3
Overview of Repair Methods for Chloride-Contaminated Structures

Stopping or reducing the corrosion rate in reinforced concrete structures damaged by chloride-induced corrosion is more difficult than in carbonated concrete. For instance, distinguishing between "aggressive" and "protective" concrete, that is,

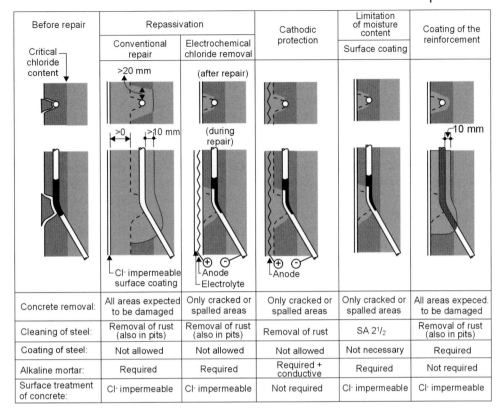

Figure 18.5 Summary of most widely used methods for the repair of structures affected by chloride-induced corrosion (modified from [1]).

concrete containing chloride above and below the critical threshold, is difficult, since the chloride threshold depends on the concrete composition and the exposure conditions (Chapter 6).

The most frequently used methods for repair of chloride-contaminated structures are shown in Figure 18.5. When the chloride level is high, conventional repair methods usually do not give a durable result. In these cases, it is necessary to consider other methods such as cathodic protection or electrochemical chloride removal.

18.3.1
Methods Based on Repassivation

In chloride-contaminated structures, repassivation of steel can be achieved by replacing the contaminated concrete with chloride-free material (*conventional repair*), by removing chlorides from the concrete (*electrochemical chloride removal*) or by means of *cathodic protection* (Chapter 20).

Repassivation with Alkaline Mortar or Concrete Repassivation of steel can be obtained by replacing the chloride-contaminated concrete with chloride-free and alkaline mortar or concrete. Because of the mechanism of chloride-induced corrosion, it is not sufficient to repair the concrete in the area where the reinforcement is depassivated. The concrete has to be removed in all areas where the chloride threshold has reached the depth of the reinforcement or is expected to reach it during the design life of the repair. In fact, the concrete that surrounds the zones of corrosion usually has a chloride content higher than the chloride threshold, even though the steel remains passive because it is protected by the corroding site. In fact, a macrocell forms (Figure 18.6a) that provides cathodic polarization to adjacent steel and thus prevents the initiation of corrosion (Section 7.3). If only concrete near the corroded reinforcement is replaced, the attack may start in the areas near to those repaired because they do no longer benefit from the cathodic polarization and, moreover, pitting corrosion may even be stimulated by anodic polarization from the repassivated steel in the repaired zone (Figure 18.6b). Sometimes, electrically insulating coatings are applied to the steel surface in the repaired area, but this does not solve the problem, as shown in Figure 18.6c. The use of sacrificial anodes embedded in the repair patches has been proposed as a means of preventing corrosion of the reinforcing bars surrounding the repair

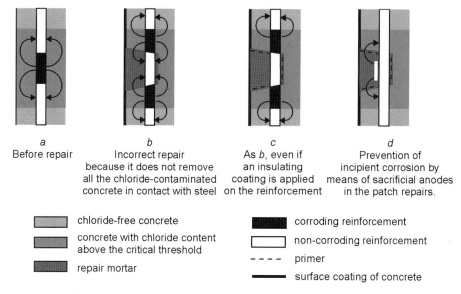

Figure 18.6 Schematic representation of the consequences of local repair on a structure contaminated by chlorides. Arrows indicate the flow of positive current due to corrosion or sacrificial protection.

[9]. A special anode consisting of zinc encased in a high alkalinity mortar saturated with lithium hydroxide was developed to provide continuing protection to the surrounding bars even when a chloride content higher than the critical threshold is left in the original concrete (Figure 18.6d); this is a sort of localized cathodic prevention (Chapter 20).

Besides removing all chloride-contaminated concrete, the surface of the reinforcement must be carefully cleaned to remove all chloride-contaminated rust around it, including that inside the corrosion pits.

For the replacement of the original concrete, an alkaline material with high resistance to chloride penetration should be applied with a sufficient cover thickness to prevent corrosion initiation during the design life of the repair (Chapter 19). Effects of possible diffusion of chlorides into the repair mortar from the concrete substrate should also be considered. A chloride-resistant coating or mortar layer can also be applied on the concrete surface to limit further chloride penetration (Chapter 14).

Electrochemical Chloride Removal (ECR) This method is based on the application of direct current to the reinforcement from a temporary anode placed on the surface of the concrete. The treatment may last up to several weeks. Chloride ions are removed from the concrete since they migrate towards the surface of concrete due to the applied current. This technique is described in Section 20.4.1.

18.3.2
Cathodic Protection

This technique requires the permanent application of a small direct current to protect the steel. It can also lead to repassivation of the reinforcement if it lowers the steel potential below the repassivation potential (Section 7.3). If it is applied properly, cathodic protection is able to stop corrosion for any level of chloride contamination of concrete or exposure condition of the structure. This technique is described in Section 20.3.

18.3.3
Other Methods

Because of the high penetration rate of pitting corrosion and the uncertainty due to structural consequences of localized attack, the methods aimed at repassivation of steel should be preferred for chloride-contaminated structures. Only if the chloride content in concrete is low and the penetration of chloride is limited in extent, other repair techniques can be taken into consideration.

Hydrophobic Treatment In principle, reduction in the moisture content of concrete can reduce the corrosion rate also in chloride-contaminated concrete, and thus hydrophobic treatment may be used to control the corrosion rate. Nevertheless, there is insufficient knowledge about the critical moisture content in the case of chloride-induced corrosion. The greater the content of chlorides and their hygroscopicity, the lower the moisture content must be. The literature is conflicting; in some cases hydrophobic treatments seem to slow down chloride-induced corrosion, in other cases they clearly do not [10]. However, since no precise data are available in this instance, an evaluation must be made of the advisability of applying hydrophobic treatments and impermeable coatings that prevent concrete from coming into direct contact with water (but allow aqueous vapor to pass so that the concrete will dry out), on a case-by-case basis.

Coating of the Reinforcement Protection of steel by means of an organic coating creating a barrier between the reinforcement and the repair mortar is very difficult in chloride-contaminated concrete (it should be considered only when it is not possible to apply any other repair technique [1]). It is essential to remove concrete around reinforcement that is corroding or will presumably corrode during the design life of repair. Also, removal of chlorides from pits is necessary. To reduce the amount of concrete that must be removed, this technique usually necessitates the use of a surface coating so that, after the intervention, chloride penetration remains under control. In order to insure adhesion to the reinforcement, the surface to be treated must be carefully prepared, with grit-blasting grade SA $2^1/_2$. In practice, this operation may be quite difficult to perform. Particularly crucial are the backwards sides of the reinforcement, corners, overlapping parts, etc. Therefore, before proceeding with repair based on coating the reinforcement, it is necessary to make sure that sufficient cleaning of the reinforcement has been achieved.

Migrating Inhibitors Substances have been developed that, once applied on the concrete surface, are claimed to migrate through the concrete cover. The effect of such treatment occurs only if the inhibitor blend reaches unaltered the steel surface. The properties and the effectiveness of migrating inhibitors are discussed in Chapter 13.

18.4
Design, Requirements, Execution and Control of Repair Works

Any repair work should be designed in order to define the requirements for the materials and the techniques used. First, the condition of the structure should be defined by means of inspection or monitoring (Chapters 16 and 17), in order to identify causes of damage (carbonation, chlorides, etc.) and their extent. Once the repair technique has been selected, details of its application should be defined. This phase will be described in Chapter 19 for conventional repair techniques and in Chapter 20 for electrochemical techniques.

As in the case of new structures, performance requirements for the materials to be used should be defined (e.g., resistance to carbonation or chloride penetration, electrical resistivity, etc.). Quality control tests should be selected to check that materials used during the execution fulfill the design requirements. Unfortunately, often repair works on existing structures fail to provide long-term protection due to incorrect design or execution [11–13]. For instance, from a survey it was reported that almost 50% of past repairs and interventions showed signs of deterioration (continued corrosion, cracking, disbonding or spalling of concrete) within five years [14, 15]. This was attributed to: wrong diagnosis of the cause of the initial damage or deterioration of the structure (16%), inappropriate design of the intervention works (38%), inappropriate specification or choice of the materials used (15%), poor workmanship (19%) and other factors (12%). As for the durability design of new structures, also in the case of repair the traditional prescriptive methods (e.g., based only on the properties of the materials) should be replaced by performance-based approaches that deal with the achievement of performance criteria of the repaired structure and reliable evaluation techniques should be found to verify the achievement of design requirements.

References

1 RILEM Technical Committee 124-SRC, P. Schiessl (ed.) (1994) Draft recommendation for repair strategies for concrete structures damaged by reinforcement corrosion. *Materials and Structures*, **27**, 415–436.

2 *fib*, International Federation for Structural Concrete (2010) Model code 2010–First complete draft–vol. 2, chapter 9 "Conservation", Bulletin 56, Lausanne.

3 Bertolini, L., Carsana, M., Gastaldi, M., Lollini, F., and Redaelli, E. (2011) Corrosion assessment and restoration strategies of reinforced concrete buildings of the cultural heritage. *Materials and Corrosion*, **62** (2), 146–154.

4 EN 1504-1 (2005) Products and Systems for the Protection and Repair of Concrete Structures. Definitions, Requirements, Quality Control and Evaluation of Conformity. Part 1: Definitions, European Committee for Standardization.

5 EN 1504-9 (2008) Products and Systems for the Protection and Repair of Concrete Structures. Definitions, Requirements, Quality Control and Evaluation of Conformity. Part 9: General Principles for the Use of Products and Systems, European Committee for Standardization.

6 Bier, T.A., Kropp, J., and Hilsdorf, H.K. (1987) Carbonation and realkalinization of concrete and hydrated cement paste, in *Proc. of 1st Int. RILEM Congress From Materials Science to Construction Materials Engineering, vol. 3, Durability of Construction Materials* (ed. J.C. Maso), Chapman & Hall, p. 927.

7 Bertolini, L., Pedeferri, P., Redaelli, E., and Pastore, T. (2003) Repassivation of steel in carbonated concrete induced by cathodic protection. *Materials and Corrosion*, **54** (3), 163–175.

8 Mattila, J. (2003) Durability of patch repairs, in *COST Action 521, Corrosion of Steel in Reinforced Concrete Structures*, Final Report (eds R. Cigna, C. Andrade, U. Nürnberger, R. Polder, R. Weydert, and E. Seitz), European Communities, Luxembourg, Publication EUR 20599, pp. 183–196.

9 Sergi, G. and Page, C.L. (2000) Sacrificial anodes for cathodic protection of reinforcing steel around patch repairs applied to chloride-contaminated concrete, in *Corrosion of Reinforcement in Concrete. Corrosion Mechanisms and Corrosion Protection*, European Federation of Corrosion Publication number 31 (eds J. Mietz, R. Polder, and B. Elsener), The Institute of Materials, London, pp. 93–100.

10 Polder, R.B., Borsje, H., and de Vries, J. (2000) Corrosion protection of reinforcement by hydrophobic treatment

of concrete, in *Corrosion of Reinforcement in Concrete. Corrosion Mechanisms and Corrosion Protection*, European Federation of Corrosion Publication number 31 (eds J. Mietz, R. Polder, and B. Elsener), The Institute of Materials, London, pp. 73–84.

11 Snover, R.M., Vaysburd, A.M., and Bissonnette, B. (2011) Concrete repair specifications: guidance or confusion? *Concrete International*, **33** (12), 57–63.

12 Matthews, S. (2007) CONREPNET: performance-based approach to the remediation of reinforced concrete structures: achieving durable repaired concrete structures. *Journal of Building Appraisal*, **3** (1), 6–20.

13 Matthews, S., Sarkkinen, M., and Morlidge, J. (eds) (2007) *Achieving Durable Repaired Concrete Structures – Adopting a Performance-Based Intervention Strategy*, BREpress.

14 Tilly, G.P. and Jacobs, J. (2007) *Concrete Repairs – Performance in Service and Current Practice*, BRE Electronic Publications.

15 Tilly, G. (2011) Durability of concrete repairs, in *Concrete Repair. A Practical Guide* (ed. M.G. Grantham), Spon Press/Taylor and Francis, pp. 231–247.

19
Conventional Repair

The term conventional repair is used in this book to indicate a repair work made on a damaged reinforced concrete structure, which is aimed at restoring protection to the reinforcement by means of replacement of nonprotective concrete with a suitable cementitious material. The durability of the repair work is due to the achievement and maintenance of passivity of the reinforcement by the contact with the protective repair material. The repair work can be divided in the following steps: (i) assessment of the condition of the structure; (ii) removal of concrete in well-defined parts of the structure and for specific depths; (iii) cleaning of the exposed rebars; (iv) application of a suitable repair material to provide an adequate cover to the reinforcement [1–9]. Each of these steps must be carried out properly in order to guarantee the effectiveness of the whole repair work. Recent developments in the field of concrete repair tend toward a performance-based approach similar to that used for new concrete structures [10, 11]. Additional protection measures can be used to increase the durability of the repair, but they should not interfere with the protection provided by the alkalinity of the repair material. Strengthening may also be required to restore the structural safety of the structure. Quality control should be performed onsite to check the fulfilment of requirements in terms of materials properties and repair work performance.

This chapter illustrates the procedure that should be followed in the design of the conventional repair to guarantee that it is effective in the aims of providing protection to the structure and preventing further corrosion damage during the remaining service life.

19.1
Assessment of the Condition of the Structure

As illustrated in Chapter 18, before making a decision on the use of conventional repair, the causes of deterioration and the condition of the structure have to be clearly assessed (otherwise, any repair work should only be considered a waste of time and money). Moreover, the remaining service life and/or the management strategy of the owner must be known, which define the required life of the repair. If carbonation or chloride contamination are the causes of degradation and the

Corrosion of Steel in Concrete: Prevention, Diagnosis, Repair, Second Edition. Luca Bertolini, Bernhard Elsener, Pietro Pedeferri, Elena Redaelli, and Rob Polder.
© 2013 Wiley-VCH Verlag GmbH & Co. KGaA. Published 2013 by Wiley-VCH Verlag GmbH & Co. KGaA.

decision of using conventional repair has been made, a more detailed analysis is required. A survey of the structure has to be carried out with the aim of identifying the extent of the areas where concrete has to be removed and the depth that should be removed. A clearly described working plan of the repair should be made.

A sufficient number of tests should be carried out in order to evaluate the depth of carbonation and/or the chloride profiles (Chapter 16) in different parts of the structure where the conditions of exposure or the properties of the concrete may have led to different penetration. The actual thickness of the concrete cover should also be detected in representative areas of the structure, taking into account its variability due to lack of quality control during the construction phase (often for structures built in the past, significant differences can be found in different parts). The designer should be aware that protection can only be guaranteed if all of the nonprotective concrete in contact with the rebars is removed. On the other hand, it is in general desirable to limit concrete removal as much as possible since it is a slow, costly, noisy, and dusty operation. The actual number of analyses required to outline the condition of the structure can only be judged on the basis of experience, depending on many factors related to the specific structure (dimensions, importance, possible risks related to safety, etc.). Although a smaller number of analyses may reduce the cost of the assessment, it may also lead to underestimation or overestimation of the quantity of concrete to be removed and thus to risks of failure of the repair or to much higher costs for concrete removal.

Even if the principal cause of corrosion is carbonation or chloride penetration, construction defects (voids, honeycombs, early cracks, etc.) or other processes of deterioration (sulfate, freeze–thaw, dynamic loading, etc.) can contribute to the damage of the structure. These should be detected in order to take appropriate measures in the repair work. Furthermore, the condition of the structure with regard to stability must be investigated in order to define if strengthening is required.

19.2
Removal of Concrete

19.2.1
Definition of Concrete to be Removed

To provide protection to the reinforcement after the repair work, concrete has to be removed not only in all zones where it is weak, cracked or damaged. Also, removal of structurally sound concrete is often necessary if carbonation or chloride contamination is expected to damage the structure (Section 18.3.1). Unfortunately, this requirement is often not taken into serious consideration in many repair specifications. It is therefore useful to go into details and give some examples. A single bar will be initially considered to illustrate the procedure to be followed for carbonation-induced or chloride-induced corrosion. Variability throughout the structure will be discussed afterwards. The proposed depth of removal in the following examples is only intended to satisfy requirements relating to corrosion protection. If other requirements are necessary, for example, to repair construction defects or mechanical damages, this should be considered as a minimum value. It should be emphasized that the consequences of concrete removal for the structural performance and the stability of the structure, even during the repair work, should be carefully evaluated.

Carbonation-Induced Corrosion If carbonation has been identified as the cause of corrosion of steel reinforcement and other deterioration processes can be neglected, the evaluation of the depth of concrete to be removed can be carried out as shown in Figure 19.1. The present condition of the structure has to be expressed by means of:

- the thickness of the concrete cover (x): this is usually evaluated on the most external reinforcement (e.g., stirrups),
- the carbonation depth d_0,

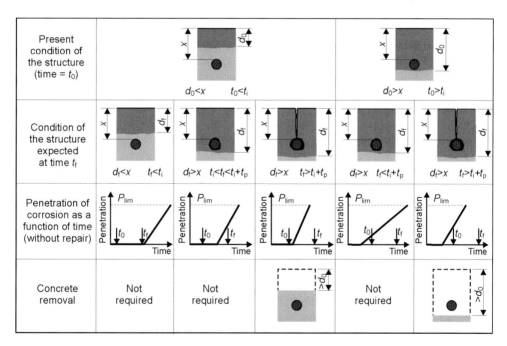

Figure 19.1 Definition of the depth of concrete removal for structures subjected to carbonation-induced corrosion as a function of the present condition of the structure (time t_0) and the estimated evolution of the damage at the time t_f when use of the structure is expected to finish (or new repair is planned). t_i and t_p are the duration of the periods of initiation and propagation of corrosion, x is the thickness of the concrete cover, d_t is the depth of carbonation at time t.

- the possible presence of small amounts of chlorides (both mixed-in or penetrated from the environment) that, even though they cannot cause pitting corrosion, may increase the corrosion rate of steel once the concrete is carbonated (Figure 5.10).

Three cases can then occur: (i) the depth of carbonation is lower than the concrete cover; (ii) the carbonation depth is higher than the concrete cover, but corrosion has not yet cracked the concrete cover; and (iii) carbonation has reached and passed the reinforcement and corrosion has damaged the concrete cover.

In the third case, which is not shown in Figure 19.1, it is obvious that concrete has to be removed beyond the reinforcement. In the first two cases, a further analysis is required in order to evaluate the future evolution of corrosion within the intended service life of the repair and to assess the expected condition of the structure at the end of this period (i.e., at time t_f equal to the time of repair t_0 plus the intended life of repair) following the hypothesis that no further repair work is made. In fact, since for carbonation-induced corrosion the propagation period may be long, for example, if the concrete is dry, to estimate the condition of the structure at time t_f, both the initiation time (t_i) and the propagation time (t_p) have to be considered according to Tuutti's model (Figure 4.1). This means that both penetration of carbon dioxide and subsequent corrosion of steel have to be considered. Concrete should be removed only if damage is expected before time t_f, namely if $t_f > t_i + t_p$, as shown in Figure 19.1.

To evaluate the penetration of carbonation, the square-root model can be considered (Section 5.2.1). In most cases it is reasonable to assume that the conditions of exposure of the structure will not change in time and thus the carbonation coefficient K can be calculated from the present carbonation depth: $K = d_0/(t_0)^{1/2}$. It is therefore possible to describe the carbonation depth in time as: $d = K \cdot (t)^{1/2}$ and thus to evaluate the time for initiation of corrosion: $t_i = (K/x)^2$. If this time is higher than t_f carbonation is not expected to reach the reinforcement within the required residual life of the structure, and thus neither concrete removal nor repair is required.

When $t_i < t_f$ carbonation is expected to reach the reinforcement and thus consequences of corrosion after time t_f should be considered. The time of propagation can be estimated as the time required to obtain a maximum penetration for corrosion attack that is expected to crack the concrete (often, a P_{max} value of $100\,\mu m$ is considered, Section 11.2.2). Estimation of the corrosion rate is thus required. The most critical factors are the humidity of concrete and its possible chloride contamination (Section 5.3). A surface treatment (Chapter 14) may be considered to slow down the rate of corrosion and further carbonation. If the concrete is already carbonated up to the reinforcement at time t_0 (case ii), electrochemical measurement (mapping) of steel potential, concrete resistivity, and polarization resistance may help in evaluating the moisture conditions of the concrete and the corrosion rate of the steel (Chapter 16).

Chloride-Induced Corrosion If chlorides have been identified as the cause of corrosion of reinforcement and other deterioration processes can be neglected

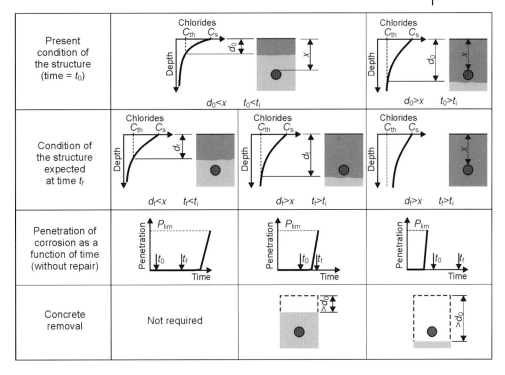

Figure 19.2 Definition of the depth of concrete removal for structures subjected to chloride-induced corrosion as a function of the present condition of the structure (time t_0) and the estimated evolution of the chloride penetration damage at the time t_f when use of the structure is expected to finish (or new repair is planned). t_i is the duration of the period of initiation of corrosion (the period of propagation is neglected), x is the thickness of the concrete cover, C_s is the surface chloride content, C_{th} is the chloride threshold value, d_t is the depth at which the chloride threshold is reached at time t.

(including carbonation), the evaluation of the depth of concrete to be removed can be carried out as shown in Figure 19.2. The present condition of the structure has to be expressed by means of:

- the thickness of the concrete cover (x);
- the chloride profile measured in the concrete cover;
- the chloride threshold for corrosion initiation (C_{th}).

While the thickness of the concrete cover and the chloride profile can be measured, the chloride threshold can only be estimated. This is a critical point for successful repair of chloride-contaminated structures. Because of the many influencing factors (Section 6.2.1) the evaluation of the chloride threshold is often rather difficult, especially if the actual concrete composition is not known. From a safe point of view a value of 0.4% by mass of cement may be usually considered, but even lower values may be possible for certain concretes and exposure conditions (Chapter 6).

In general, for chloride-induced corrosion the propagation period is neglected since consequences of corrosion are difficult to assess because of the local and penetrating nature of pitting corrosion and the usually high corrosion rate. Two cases can therefore be considered (Figure 19.2): (i) the chloride threshold (C_{th}) is found at depth d_0 lower than the concrete cover thickness (x); (ii) the chloride content at the depth of the reinforcement is higher than the chloride threshold. In the second case corrosion has already initiated and thus concrete has to be removed beyond the reinforcement. It is important to recall that concrete has to be removed everywhere the chloride threshold has been reached at the steel surface, even if corrosion has not initiated (Section 18.3.1).

In case i, future penetration of chlorides has to be evaluated in order to assess if the chloride threshold will be reached at the surface of the reinforcement before time t_f. If this is expected, the concrete with a chloride content higher than the critical value has to be removed (and replaced with a chloride-free material that prevents further penetration of chloride). Equation (6.1) may be used to evaluate the future penetration of chlorides. By fitting the present chloride profile, the surface content C_s and the apparent diffusion coefficient D_{app} may be calculated at time t_0. Since these parameters are evaluated on the actual structure and usually after a long time of service, it is often reasonable to assume that they will not change significantly in the near future (unless the conditions of exposure of the structure will change). Equation (6.1) can then be used to plot the expected chloride profile at time t_f.

Variability The examples of Figures 19.1 and 19.2 summarize the process of evaluation of the depth of concrete removal when a single rebar is considered and measurements of carbonation, chloride, and cover thickness are available locally. In a real structure carbonation and chloride penetration may vary due to spatial variation of exposure conditions (microclimate) and of concrete properties (e.g., cracking or bad compaction). The cover depth may also be very variable, as was shown in Figure 16.3.

Since an accurate analysis covering the complete structure is not generally possible, the evaluation should be carried out on parts of the structure that may be supposed to be homogeneous with regard to both concrete composition and exposure conditions, for example, as established by visual inspection. An adequate amount of sampling should then be carried out within each "homogeneous" area to assess the variability of the different parameters and thus define the depth of concrete removal. In principle, a statistical approach would be possible where all factors are described in terms of probability distributions and a depth of concrete removal is calculated that can guarantee a given probability of success of the repair. Nevertheless, this is in practice difficult to do, especially because the number of analyses of carbonation depth or chloride penetration that can be carried out is often limited and data are usually insufficient for a statistical approach. A statistical model for evaluating the extent and future evolution of corrosion damage due to carbonation has been proposed by Mattila [12]. An experienced and skilled engineer, however, should be able to determine the depth of removal on the basis of onsite observations and of a limited number of representative analyses. The accuracy of the estimation should also be adapted to the importance of the structure, the risks connected with failure of the repair and the required residual service life. Further tests can also be prescribed during the repair work to verify the preliminary assessment, for example, in the case of carbonation the actual depth of penetration can be easily verified by phenolphthalein tests on the freshly broken concrete and the amount of concrete to be removed can be adjusted accordingly.

19.2.2
Techniques for Concrete Removal

Many techniques are used to remove concrete for repair works, which vary from mild surface removal to destructive methods such as blasting techniques [5, 7, 9]. When conventional repair is used, however, only some methods are suitable to achieve the following requirements:

- concrete removal should be selective, that is, limited to the designated area and depth;
- damage of the original concrete that is intended to remain in place should be minimized to maintain the structural integrity and guarantee sufficient bond for the repair material;
- the reinforcement should not be damaged;
- the surface of concrete left in place should be rough and clean (otherwise further treatment is required).

Concrete removal is easy where concrete is cracked or weak, while it may be difficult where it is mechanically sound, especially if it should be removed also beyond the reinforcing bars.

The choice of the most suitable technique depends on the depth of concrete that should be removed. If only surface removal is required (i.e., only a thin surface layer has to be removed), milling methods (such as a scarifier) or abrading methods (such as grit blasting) can be used. When concrete removal has to extend to the concrete cover or even beyond the reinforcement, it is required to operate between and around the reinforcing bars without damaging them, cracking the concrete substrate or compromising the bond between concrete and the reinforcement in zones where concrete will not be removed. This task is usually performed with the use of pneumatic breakers, or with high-pressure water jetting (hydrodemolition) [9].

The first method is the most common. The rate of concrete removal depends on the size of the breaker and the skill of the operator. Massive breakers may proceed quicker, but it is difficult to control the microcracking they produce on the surface of the residual concrete and thus the result is more influenced by the quality of the workmanship. Hydrodemolition consists in the application of a water jet with very high pressure (70–500 MPa) that destroys the cement matrix of concrete and liberates aggregate particles. Aggregates are not broken and the final surface is irregular and thus suitable for application of the repair mortar (after removal of residual coarse aggregate particles and of the slurry produced by hydrodemolition). Hydrodemolition is usually convenient when concrete has to be removed on large surfaces and for large depths. The equipment required to generate and control the high pressure water jet is complex and expensive. Nevertheless, the productivity may be very high and the system may be programmed in order to remove the required depth. Furthermore, this method allows preservation and cleaning of the steel reinforcement and minimization of the damage to concrete remaining in place.

19.2.3
Surface Preparation

The surface of the concrete substrate has to be prepared to provide sufficient bond of the repair material. Factors that may affect the bond are the strength and integrity of the substrate, the cleanliness of the surface, and the roughness. The surface should be rough and dust or incoherent residues should be removed (for instance by grit blasting or waterblasting). This operation is usually not necessary if hydrodemolition has been used. If the cementitious repair material is applied directly on the surface of the concrete, the surface should be saturated by water in order to avoid absorption of water from the substrate and subsequent plastic shrinkage and incomplete hydration of the repair material, which will result in loss of bond.

Bonding agents may also be applied in some cases on the concrete surface in order to enhance adhesion of the repair mortar. These may be cementitious systems (cement paste or fine mortars), polymer latex or epoxy systems [7]. Epoxy systems should be used carefully since they create a moisture barrier between the substrate and the repair material, which can result in failure of the repair if moisture is trapped in the concrete.

19.3
Preparation of Reinforcement

The surface of the reinforcement that has been exposed by concrete removal should be cleaned by grit blasting or mechanical methods. Loose rust or mortar that could compromise bonding with the repair material should be removed without damaging the steel bars. If the concrete, and thus also the corrosion products, are free of chlorides the removal of adherent rust spots is not required since small residues of rust will not prevent repassivation when steel is in contact with the alkaline repair material. Conversely, cleaning of the reinforcement surface requires special attention when the concrete is contaminated by chlorides. It is not enough to eliminate only the loose rust, as in the case of carbonation. Corrosion products may contain significant amounts of chlorides that will attract moisture (hygroscopicity), causing reactivation of corrosion. Consequently, chloride-containing corrosion products need to be removed completely, including those in hidden parts of the reinforcement or inside pits. Very often this can only be achieved with high-pressure waterblasting.

19.4 Application of Repair Material

19.4.1 Requirements

A repair material suitable to protect the reinforcement during the required design life of the repair should be applied. Many requirements should be considered in order to achieve this goal.

Alkalinity and Resistance to Carbonation and Chloride Penetration Since the protection of the reinforcement relies on the long-term contact with alkaline and chloride-free environment, the repair materials should be:

- cement-based mortar or concrete (polymer concrete or mortar, obtained by embedding aggregates in a polymer matrix do not provide alkalinity[1]);
- resistant to carbonation;
- resistant to chloride penetration (if the structure is subjected to chloride penetration).

The repair material should also be able to resist other types of attack that could occur in the specific environment (sulfate attack, freeze–thaw attack, etc.).

Cover Thickness The thickness of the cover produced by the repair material should be designed, as in the case of new structures, to be sufficient to protect the reinforcement for the required time. Therefore, it depends on the resistance of the repair material to carbonation and chloride ingress (Section 19.4.3), on the aggressiveness of the environment and on the design life of the repair. Often for geometrical and aesthetical reasons the original thickness of the concrete cover is reconstructed; however, if this is not sufficient to obtain the required durability, it should be increased or additional protection should be used (Section 19.5).

Rheology and Application Method The properties required of the repair material in the fresh state depend on the thickness of the new cover and the method used for the application [5, 6]. If the thickness of the cover is high (e.g., higher than 50 mm), formworks may be used and the repair material should have a flowable or pumpable consistence and should be self-compacting, to be able to fill the space

---

1) Nonalkaline (polymer) mortars have been used for repairs; however, they do not promote repassivation and they are only based on physical effects. Epoxy mortars applied as patch repairs were in the past unsuccessful in stopping corrosion, in particular where chloride had caused corrosion, as was shown in many cases. This type of mortar will not be addressed.

inside the form without segregating. When a lower thickness is required or a form cannot be used, the material can be applied by troweling or can be sprayed or pneumatically conveyed at high velocity onto the surface (shotcrete process). In this case the repair material should not be flowable, but once it has been put in place it should not move until it sets, even on vertical surfaces or soffits (i.e., it should have a thixotropic behavior). Other techniques such as preplaced-aggregate concrete, dry packing or injection grouting, may also be used under particular circumstances [5, 13].

Bond to the Substrate and Dimensional Stability The repair should have a good bond to the concrete substrate. This bond depends on the surface preparation and the use of a bonding agent (Section 19.2.3), but it is also influenced by the properties of the repair material and the compaction during its application.

Dimensional stability of the mortar with regard to moisture or temperature variations is also important. As far as moisture is concerned, drying shrinkage should be of particular concern. The fresh repair material that is applied on the hardened concrete substrate will experience drying shrinkage due to loss of water present in its pores. This shrinkage is constrained since the repair material is bonded to the original concrete, which has negligible shrinkage, and tensile stresses may generate that tend to crack the repair material. Furthermore, shear stresses are built up at the bonding plane between the repair material and the concrete, which can compromise the adhesion. Therefore, the drying shrinkage of the repair material should be as low as possible.

Thermal variations may also lead to differential deformation between the repair material and the concrete substrate, especially in large and thick patches. To limit the risk of loss of bond, the thermal expansion coefficients of the two materials should be similar.

Mechanical Properties The compressive strength of repair materials is often rather high, even if it is not required for structural reasons, because it is a consequence of the low water/cement ratio required for durability reasons. A low modulus of elasticity is beneficial with regard to the deformations due to moisture or temperature changes; in fact, stresses generated by constrained shrinkage decrease as the modulus decreases and thus the risk of loss of bond is lower. However, if the repair material is intended to sustain stresses parallel to the bond plane, the modulus should match that of the original concrete. If the modulus of the repair material is significantly lower than that of the original concrete, loads will only be partially transferred to the repair material, which will therefore give a small contribution to the strength of the structure. Nevertheless, the contribution to the load-bearing capacity of repairs is generally neglected.

19.4.2
Repair Materials

Although ordinary portland cement mortars (or concretes) can be used as materials for conventional repair, different types of additions or admixtures may improve their performance [14, 15]. The water/cement ratio is usually low (e.g., 0.4 or even lower) to guarantee resistance to carbonation and chloride penetration and strength. Pozzolanic materials, such as fly ash or silica fume, are often added to improve the durability (Section 1.3.1). Superplasticizing admixtures (Section 12.1.4) may be used to obtain adequate workability, especially when self-leveling properties are required. Addition of fibers may improve the resistance to cracking of the repair mortar and may change the rheological behavior of the fresh material (usually they reduce the slump, but they may also contribute to obtain thixotropic behavior). Quick-setting admixtures may also be used.

Because of the high cement content, the drying shrinkage of these materials may be rather high and cracking or loss of bond has to be prevented. Shrinkage-compensating mortars can be obtained using expansive cements or adding expansive agents (e.g., calcium oxide, anhydrous calcium sulfoaluminates, etc.) in a proper amount to ordinary cement [7, 14]. The expansion that takes place in the plastic mortar or in the hardened mortar is intended to compensate for the expected tensile stresses due to plastic and drying shrinkage and to maintain a tight bond to the substrate. For the shrinkage-compensating mortar to be effective, however, two requisites should be fulfilled. First, the expansion should be constrained so that compressive stress can build up in the repair material; subsequent drying shrinkage will simply relieve the compressive stresses. Secondly, wet curing is essential for expansion to take place.

Polymer-modified mortars can be obtained by replacing part of the mixing water with a synthetic latex (e.g., styrene butadiene or acrylate) to the mix. Although the binder is still cementitious, and thus alkalinity is guaranteed, the latex may improve the workability, the waterproofness, the carbonation and chloride resistance, the tensile and flexural strength of the repair mortar [14]. It can also reduce the modulus of elasticity, increase the bond to the substrate, reduce the rate of drying out and thus the rate of shrinkage.

The selection of cement used in the repair material may also be influenced by other requirements, such as early strength or the necessity to match the color of the original concrete [5–7].

The definition of mix proportions that can fulfill the different properties required for the repair material may be rather difficult. A wide variety of commercial premixed products of proprietary compositions are available that can suit different requirements. Specifications for the repair mortars should, however, be clear regarding performance requirements in order to allow the selection of an appropriate material.

19.4.3
Specifications and Tests

Specifications for the selection of the repair material should be based on tests that can be carried out prior to the application. Each test should be suitable to assess the performance of the materials with regard to the specific property that it is intended to evaluate. Several standards cover tests to be used for the evaluation of workability or mechanical properties (compressive, tensile or flexural strength, elastic modulus, adhesion) of the repair material. In addition to mechanical and physical tests, also durability requirements have been introduced in recent standards: for instance, according to EN 1504-3 [3], the carbonation depth evaluated through an accelerated test on the repair material should not exceed the carbonation depth measured on a reference concrete in the same conditions. Recently, accelerated methods for testing carbonation and chloride resistance in repair mortars have been standardized [16, 17]. However, there is still a lack of well-established procedures concerning durability performance [18]. This is due to the difficulty in reproducing the long-term performance of the material with short-term tests. Parameters obtained from accelerated tests, such as the coefficient of carbonation K or the apparent diffusion coefficient D_{app}, can be hardly converted in values to be used in models for service life prediction. Even the simple comparison of different materials tested in the same condition may be misleading, because if the increased severity of the accelerated test introduces phenomena that do not occur in real conditions, even the ranking of the materials may be different.

19.5
Additional Protection

Under particular circumstances, the long-term protection given by the repair may be improved by additional measures that are aimed at prolonging the passivity of the reinforcement. In any case, the main objective of the repair work remains passivation of the reinforcement due to the alkalinity of the repair mortar (otherwise other repair methods illustrated in Chapter 18 should be considered).

Corrosion Inhibitors Corrosion inhibitors may in principle improve the protection of the steel reinforcement. Mixed-in inhibitors added to repair mortar may increase the chloride threshold or delay the chloride penetration in the repair material. Migrating inhibitors applied on the surface of the original concrete are intended to reduce or stop corrosion and thus make concrete removal and its replacement unnecessary. The role of inhibitors and their effectiveness are discussed in Chapter 13. It is, however, useful to recall that if a corrosion inhibitor is used, the concentration that is needed in the vicinity of the reinforcement should be specified and suitable means for demonstrating that such conditions are actually achieved and maintained for an adequately long time should be proposed [1].

Surface Treatment of Concrete Surface treatment of concrete may be used in association with conventional repair to achieve the required service life of the repair. A specific type of treatment may be used to delay penetration of carbonation or chlorides or to decrease the moisture content (Chapter 14) either in the original concrete or in the repair mortar. The effect of this treatment can be taken into consideration in the evaluation of the residual service life of the structure both in the repaired and unrepaired zones. This can lead to a reduction in the extent of the areas to be repaired or in the thickness of the repair material.

Coating of Rebars Proprietary products are often applied on the surface of the rebars, to promote adhesion to the repair mortar and to improve the corrosion resistance (often inhibiting properties are also claimed, and these are called "anti-corrosion coatings" [12]). The use of surface coatings on the reinforcement should be carefully evaluated. As far as corrosion of steel reinforcement is concerned they should not be recommended, since the repair mortar is the material intended to protect the reinforcement. Only if it is impossible to provide adequate thickness of repair material cover, so it is locally not possible to provide the required long-term protection, may a coating that acts as a physical barrier be useful. In that case, however, it should be clearly understood that the repair principle is locally changed and requirements illustrated in Sections 18.2.3 or 18.3.3 should be fulfilled [12].

19.6 Strengthening

Evaluation of the safety and stability of the structure and interventions when strengthening is required are outside the scope of this book. Nevertheless, it is useful to recall the main methods of strengthening of the structure that can be applied before, during or after repair work. In fact, although the repair of corrosion damage and strengthening of the structure are different in scope, in the methods and in the competence required, they should be considered together in the design of the rehabilitation work.

Strengthening is aimed at restoring the capacity of weakened components or elements to their original design capacity [5]. Supplemental members can be provided for temporary (e.g., during the repair work) or permanent support of the damaged structural members. If concrete that will not be removed and replaced is cracked, cracks must be sealed using cementitious or polymeric grouts. During the repair work, reinforcing bars may be substituted or additional bars may be added where corrosion has led to unacceptable reduction of the cross section (sufficient length of overlap should be provided). The repair material itself may be used to increase the cross section of structural elements and thus the cover thickness may be increased for structural reasons. New reinforcement may also be required in the repaired area to provide anchorage to the substrate, to control cracking or to counteract expansion of shrinkage-compensating mortar.

If concrete removal is not required or supplementary reinforcing bars cannot be used, external reinforcement can be applied. For instance, steel bars may be encased in a shotcrete layer or steel plates may be bonded onto the concrete surface. Nowadays, mainly fiber-reinforced polymers (FRP) are used, which are composite materials with glass, aramide or carbon fibers embedded in a polymeric matrix (usually an epoxy system). FRP are available in the form of laminates or sheets that are bonded to the concrete surface using an epoxy adhesive [19]. They are typically applied to improve the flexural and shear strength or to provide confinement to concrete subjected to compression. The low weight of these composite materials and the high design flexibility make this technology quite attractive, although experience on its long-term performance is limited. The durability of the complex system consisting of the combination of composite, adhesive and reinforced concrete has to be studied yet. The effects of temperature and moisture, including their cycles, on the long-term behavior of the structural adhesive should be clarified, in relation to the exposure condition of the structure.

References

1. RILEM Technical Committee 124-SRC, P. Schiessl (ed.) (1994) Draft recommendation for repair strategies for concrete structures damaged by reinforcement corrosion. *Materials and Structures*, **27** (7), 415–436.
2. EN 1504-9 (2008) Products and Systems for the Protection and Repair of Reinforced Concrete Structures. Definitions, Requirements, Quality Control and Evaluation of Conformity. Part 9: General Principles for the Use of Products and Systems, European Committee for Standardization.
3. EN 1504-3 (2005) Products and Systems for the Protection and Repair of Reinforced Concrete Structures. Definitions, Requirements, Quality Control and Evaluation of Conformity. Part 3: Structural and Non-Structural Repair, European Committee for Standardization.
4. EN 1504-10 (2003) Products and Systems for the Protection and Repair of Reinforced Concrete Structures. Definitions, Requirements, Quality Control and Evaluation of Conformity. Part 10: Site Application of Products and Systems and Quality Control of the Works, European Committee for Standardization.
5. ACI 546R-04 (2004) Concrete Repair Guide, American Concrete Institute.
6. Cambell-Allen, D. and Ropr, H. (1991) *Concrete Structures: Materials, Maintenance and Repair*, Longman Scientific and Technical.
7. Mailvaganam, N. (1992) *Repair and Protection of Concrete Structures*, CRC Press Inc., Boca Raton, FL.
8. BRE, Digest 444-3 (2000) *Corrosion of Steel in Concrete – Part 3: Protection and Remediation*, Building Research Establishment, Garston.
9. Babaei, K., Clear, K.C., and Weyers, R.E. (1996) *Workbook for Workshop of SHRP Research Products Related to Methodology for Concrete Removal, Protection and Rehabilitation*, Wilbur Smith Associates, Falls Church.
10. Matthews, S. (2007) CONREPNET: performance-based approach to the remediation of reinforced concrete structures: achieving durable repaired concrete structures. *Journal of Building Appraisal*, **3** (1), 6–20.
11. Matthews, S., Sarkkinen, M., and Morlidge, J. (eds) (2007) *Achieving Durable Repaired Concrete Structures – Adopting a Performance-Based Intervention Strategy*, BREpress.
12. Mattila, J. (2003) Durability of patch repairs, in *COST Action 521, Corrosion of Steel in Reinforced Concrete Structures*, Final Report, Publication EUR 20599 (eds R. Cigna, C. Andrade, U. Nürnberger, R. Polder, R. Weydert, and E. Seitz), European Communities, Luxembourg, pp. 183–196.
13. Taylor, G. (2011) Sprayed concrete for repairing concrete structures, in *Concrete Repair. A Practical Guide* (ed. M.G. Grantham), Spon Press/Taylor and Francis, pp. 212–230.
14. Rixom, R. and Mailvaganam, N. (1999) *Chemical Admixtures for Concrete*, 3rd edn, E & Fn Spon.
15. Neville, A.M. (1995) *Properties of Concrete*, 4th edn, Longman Group Limited, Harlow.

16 EN 13295 (2004) Products and Systems for the Protection and Repair of Concrete Structures. Tests Methods. Determination of Resistance to Carbonation, European Committee for Standardization.

17 EN 13396 (2004) Products and Systems for the Protection and Repair of Concrete Structures. Tests Methods. Measurement of Chloride Ion Ingress, European Committee for Standardization.

18 Frederiksen, J.M. (ed.) (1996) HETEK – Chloride Penetration into Concrete. State of the Art. Transport Processes, Corrosion Initiation, Test Methods and Prediction Models, The Road Directorate, Report No. 53, Copenhagen.

19 Nanni, A. (1997) CFRP strengthening. *Concrete International*, **19** (6), 19–23.

20
Electrochemical Techniques

Electrochemical techniques applied for controlling corrosion of steel in concrete are: cathodic protection, cathodic prevention, electrochemical chloride removal, and electrochemical realkalization.

Cathodic protection (*CP*) is applied to structures already affected by corrosion, mainly induced by chlorides; the steel is subjected to cathodic polarization, that is, its potential is brought to values more negative than the free corrosion value, so that the corrosion rate is reduced. Corrosion can actually be stopped if a potential more negative than the repassivation potential (E_{pro}, Figure 7.9) is reached.

Cathodic prevention (*CPre*) is applied to new structures that will presumably be contaminated by chlorides; the passive reinforcement is cathodically polarized with the aim of increasing the critical chloride content necessary to initiate pitting attack in such a way that it will not be reached within the service life of the structure.

Electrochemical realkalization (*RE*) and electrochemical chloride removal (*CE*) can be applied to structures in which corrosion has not or has already initiated. They are techniques aimed at modifying the composition of concrete that is carbonated or contains chlorides, in order to restore its original protective characteristics.

Due to increasing field experience, the confidence in the electrochemical techniques is growing. They share one advantage: only detached concrete needs to be removed and repaired. As the electrochemical treatment takes over the protection of the steel, there is no need to remove carbonated or chloride-contaminated but mechanically sound concrete.

Both the beneficial effects on the steel as well as those modifying the protective properties of the concrete are obtained by forcing a direct current to circulate between an anode, placed on the external surface of the structure, and the reinforcement (Figure 20.1). The current density imposed varies from $1-2\,mA/m^2$ for cathodic prevention, to $5-20\,mA/m^2$ for cathodic protection, up to $1000-2000\,mA/m^2$ for electrochemical realkalization and chloride removal.

Once applied, cathodic protection and cathodic prevention should operate for the remaining service life of the structure. Consequently, the design should take into account the durability of the installation, that is, its ability to provide sufficient

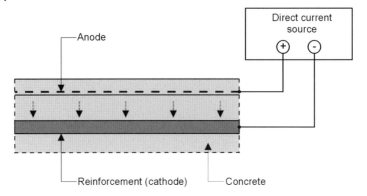

Figure 20.1 Schematic representation of application of electrochemical techniques.

protection current for a long period. Normally, the quality of protection offered by *CP* and *CPre* is checked regularly using electrical measurements.

On the other hand, electrochemical realkalization and chloride removal need only to be applied for a limited period of time (of the order of a few weeks and a few months, respectively). For these methods, the design of the treatment should take into account that a protected situation is established at the end of the treatment and that this will last as long as needed. Here, the durability of the final situation is the central issue. Monitoring of the protection during further service of the structure is recommended.

20.1
Development of the Techniques

20.1.1
Cathodic Protection

Cathodic protection of reinforced concrete structures exposed to the atmosphere was applied for the first time to bridge decks contaminated by deicing salts by Stratfull in California in 1973 [1, 2].[1] In the years following, design and protection criteria were elaborated, as well as power supply and monitoring systems completely different from those used for cathodic protection of buried steel structures or structures operating in seawater. Above all, it was proved that cathodic protec-

1) R. F. Stratfull may be considered the pioneer of cathodic protection in concrete. In 1956, he was the first to publish laboratory results demonstrating the applicability of this method in a high-resistivity material like concrete. Practical applications, however, had to wait nearly twenty years: cathodic protection experts were sceptical about such a strange electrolyte.

tion was a reliable repair technique even in the presence of high chloride contents, where traditional methods of rehabilitation are ineffective or very costly.

In the 1980s, anodes based on titanium mesh activated with special noble-metal oxides (especially ruthenium and iridium) and conductive paints were developed. The technique, by now used outside North America, is applied to not only bridge decks, but also to bridge substructures (girders, cross beams, piers), and marine structures, parking garages, industrial, office and residential buildings, etc. So far, applications were based on impressed current (*ICCP*).

In the late 1990s, sacrificial cathodic protection (*SCP*) of concrete reinforcement was introduced. In *SCP*, dissolution of a less noble metal such as zinc or aluminum connected to the reinforcement provides the current instead of an external power source.

Up to now several million square meters of concrete have already been treated worldwide with impressed current systems that utilize a variety of anodes.

The developments of the technique were reviewed in the 1990s [3–6]. Many papers have reported on case studies and practical experience [7–17]. European and North American standards are available [18, 19].

Until recently, cathodic protection was not considered generally advisable for prestressed concrete structures, in order to avoid the risk of hydrogen embrittlement in high-strength steel. Before 2000, only a few cases exist where reinforcing steel was protected in structures that also contain post-tensioning steel [7]. Once a clear criterion for safe application to prestressed structures was issued in the European Standard, this has now become more widespread. In any case, *CP* of prestressed structures requires special monitoring to prevent overprotection and related damage to the prestressing steel.

More recently, application to underground and submerged structures including those using sacrificial anodes has grown and experience has been described [20]. This document provides useful design information for such applications and collects further experience with *CP* on above ground structures. A new version of the European Standard, which has also now become an international standard, includes underground/submerged applications [21].

Cathodic protection also proved to be effective in repassivating steel in carbonated concrete. This type of cathodic protection was called "continuous realkalization", since the continuous application of low currents leads to protection effects similar to those achieved by the technique of electrochemical realkalization [22]. In general, cathodic protection is not applied to carbonated structures, because other methods of protection (conventional repair, see Chapter 19) tend to be more economical in most cases. Despite this, the technique was utilized in Italy and in the Netherlands to protect parts of buildings and structures during the 1960s and 1970s that were characterized by carbonated concrete with small quantities of chlorides (added as accelerating admixtures or contaminants of the mixing water at the time of construction, or penetrated from marine atmosphere). *CP* was chosen here because previous traditional methods of repair had not provided long-lasting results.

20.1.2
Cathodic Prevention

Cathodic protection of new reinforced and prestressed structures exposed to the atmosphere was applied for the first time in Italy: in 1989 to two highway bridge decks and in the following four years to many other bridges, viaducts and infrastructural works for a total surface of more than 100 000 m^2. Pedeferri [23] presented this technique as "a method of preventive maintenance of new structures that are expected to become affected by chloride contamination in the future". To underline that this type of cathodic protection has "aims, operating conditions, throwing power, effects, especially important those regarding hydrogen embrittlement, many engineering and economic aspects connected with design, construction, monitoring and maintenance different from the ones of the usual cathodic protection", he proposed to name it *cathodic prevention* [24].

The technique is based on the fact that the chloride threshold increases as the potential of steel decreases. In practice, application of very low current densities (<2 mA/m^2) can bring the potential to values in which steel operates in conditions of "imperfect passivity" so that initiation of pitting is suppressed even if high levels of chlorides build up at the surface of the steel by penetrating through the cover concrete [25].

A European standard is available [19, 21] and the development of the technique has been reviewed [3–6, 26, 27].

A variation of this technique has been applied in conjunction with conventional patch repair of chloride-contaminated structures in order to avoid the initiation of incipient pitting around the repaired zones, by utilizing sacrificial anodes embedded near the periphery of the repair patches [28–30]. Cathodic prevention is now being used in several countries.[2]

20.1.3
Electrochemical Chloride Removal

The possibility to remove chloride ions from concrete by an electrochemical process was first studied in the USA in the 1970s (see [33] for references), using a liquid electrolyte and very high DC voltages (up to 220 V). In Europe, chloride extraction was patented by the Norwegian company Noteby in 1986, using a water-retaining substance, such as paper fiber pulp or retarded shotcrete [34], wetted with calcium hydroxide or tap water and moderate voltages, normally less than 40 V. Since the beginning of the 1990s, research papers were published on this

---

2) Page [31] writes: "Although cathodic prevention is now being used in several countries, some researchers think there is still a need to convince some engineers of its appropriateness as the following statement, reproduced from the most recent edition of a widely-used reference book on concrete would appear to indicate: 'Cathodic protections have shown to be effective in some applications, but its use in a new structure is an admission of defeat in that a particular reinforced concrete structure is manifestly not durable.'" [32].

method, see for instance [35]. There is no standard for electrochemical chloride extraction, but a Technical Specification was published in 2011 [36]. A state-of-the-art report was published by the European Federation of Corrosion [33] and technical information was provided from European Commission supported COST Materials Actions, 509 and 521 [27, 37]. Because the original concrete surface is left unchanged after the treatment, CE may be particularly suited for structures with special architectural values such as monuments.

20.1.4
Electrochemical Realkalization

Reinstalling the passivation of steel in carbonated concrete by electrochemical realkalization with impressed current was introduced by Noteby in the late 1980s [38]. The original system used a surface-mounted steel mesh anode and sprayed paper pulp wetted by sodium carbonate solution as the electrolyte. Later, titanium mesh anodes and liquid electrolytes were introduced. In various countries, particularly in Norway and the United Kingdom, hundreds of thousands of square meters have been treated. A variant of this technique was recently developed by Ciment d'Obourg and Freyssinet, wherein the current is obtained from a sacrificial anode [39, 40]. Because after the treatment the original concrete surface is left unchanged, realkalization may be particularly suited for structures with special architectural values such as monuments. Examples of application to historical buildings are reported in refs. [41, 42].

A state-of-the-art document on realkalization (and chloride extraction) was published by the European Federation of Corrosion [33] and technical information was provided from two European COST Actions [27, 37]. Presently, there is no standard but a technical specification was published in 2004 [43].

20.2
Effects of the Circulation of Current

Circulation of current between the anode and the reinforcement has several consequences. Besides lowering of the steel potential (which was discussed in Section 7.5), chemical reactions on the surfaces of the reinforcement and the anode and ionic migration within the concrete take place.

20.2.1
Beneficial Effects

Reactions on the Steel Surface Oxygen reduction takes place on the surface of the reinforcement: $O_2 + 2H_2O + 4e \rightarrow 4OH^-$ and, if very negative potentials are reached, hydrogen evolution also occurs: $2H_2O + 2e \rightarrow 2OH^- + H_2$. The two reactions produce hydroxyl ions and thus alkalinity in the concrete closest to the steel surface.

Migration Circulation of current in concrete is produced by migration of ions present in the pore solution: positive ions Na^+, K^+ move towards the steel, negative ones, OH^- and Cl^-, in the opposite direction. The flow of each ion is proportional to its transport number and thus to its mobility and concentration (Section 2.5).

Hydroxyl ions carry most of the circulating current because of their high mobility and their generally high concentration. Consequently, most of the alkalinity produced by the cathodic reaction is removed from the reinforcement surface. On the other hand, the fraction of current carried by chlorides is modest and it decreases with decreasing concentration. High current densities are therefore required to remove large quantities of chlorides from the cathodic region or to completely prevent their ingress into the concrete of new structures. At medium to low current densities typical for cathodic protection, there will nevertheless be a certain reduction in the level of these anions on the cathodic surface.

Chemical modifications resulting from the cathodic reaction and from ionic transport produce, respectively, increases in the OH^- concentration and decreases in the chloride level in the cathodic region, as illustrated in Figure 20.2. Their amount grows along with current density and time until a steady state is reached.

These chemical modifications are at the basis of chloride removal and concrete realkalization, but are also a decisive factor for the long-term operation of cathodic protection and prevention.

20.2.2
Side Effects

The circulation of current may have undesirable side effects: it may lead to hydrogen embrittlement of high-strength steel, stimulate alkali aggregate reactions,

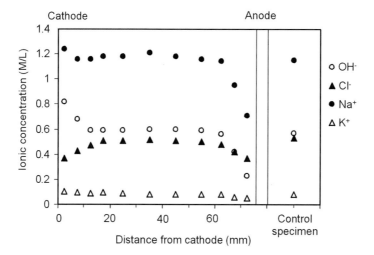

Figure 20.2 Profiles of ionic concentration measured between an anode and a cathode embedded in cement paste after one month of circulating current (average current density 350 mA/m^2) obtained by maintaining a potential of the cathode at −850 mV SCE [28].

reduce bond between reinforcement and concrete [33] or lead to acidification in the anodic region.

Hydrogen Embrittlement If the imposed current is such that the steel potential becomes more negative than −1000 mV SCE, hydrogen evolution takes place on the surface of the steel and thus high-strength prestressing steel (but not ordinary reinforcing steel) may be subject to hydrogen embrittlement (Chapter 10) [44]. Taking into account a safety margin of 100 mV, the European Standard limits the polarization of prestressing steel to −900 mV SCE [18, 21]. This clear statement made in the European Standard has opened the way for safe application of CP to prestressed structures. Later work in COST 534 has further underpinned how and why such applications are possible and several examples are given [45]. Subsequently, more and more cases of CP of prestressed structures have followed (e.g., [46]).

Realkalization and electrochemical chloride removal bring the potential to about −1100 mV SCE, at which value hydrogen will be formed. These processes may lead to critical conditions and thus their application on prestressed structures must be considered carefully. Presently, their application to pretensioned structures is not recommended [45]. The situation with regard to cathodic protection and prevention is more complex, and will be dealt with later.

Alkali Aggregate Reaction The increase in alkalinity produced at the cathode can cause damage if the concrete contains aggregates potentially susceptible to alkali silica reaction (Section 3.4). In practice, this may happen only for current densities well over $20\,mA/m^2$ [47]. Therefore, this may regard electrochemical chloride removal or (less likely) electrochemical realkalization, which in fact cause large increases in pH [48–50] at the surfaces of the reinforcement, but not cathodic protection, and much less, cathodic prevention.

Loss of Bond Strength At very negative potentials, a loss of bond strength between reinforcement and concrete may be observed. There are many uncertainties regarding this phenomenon, but it cannot be excluded that this happens in the long-run if the potential falls below −1100 mV SCE. Therefore, this value may be taken as the lower limit for cathodic protection (even though this is probably a conservative figure). Application of strong negative polarization due to high current density for relatively short periods is involved in chloride extraction and realkalization, which will be dealt with in the separate sections concerning these techniques.

Anodic Acidification At the anode surface, the anodic process of oxygen evolution takes place: $2H_2O \rightarrow O_2 + 4H^+ + 4e$. In the presence of chlorides, even chlorine develops: $2Cl^- \rightarrow Cl_2 + 2e$. Such processes may directly or indirectly produce acidity and may thus lead to destruction of the cement paste in contact with the anode [48]. Experience shows that such deterioration is negligible for activated titanium mesh anodes if the anodic current density does not exceed $100\,mA/m^2$ (or values 3–4 times greater for brief periods). Design of the anodes for cathodic prevention and cathodic protection must respect these limits.

A few studies have tried to quantify acid formation at the anode. Early work used relatively high current densities for CP [51]. From tests up to one year it was found by spraying color-pH indicators that acidification occurred, but only over distances of about one millimeter from the anode. The pH was reduced from 13 down to 8 after seven months at current densities of about $100\,mA/m^2$ (anode surface) and down to 4 at densities considerably higher (up to $2\,A/m^2$ anode surface). The amount of acid formed by the anodic process can be calculated by Faraday's law. The amount of acid observed in these experiments can be estimated. It appears that only about 10% of the acid theoretically produced actually attacks the cement paste. This is because most of the hydroxyl ions produced at the cathode migrate to the anode and neutralize the acid formed. Later work studied acid production in CP installations in the field and long-term laboratory tests. Samples were studied of conductive coatings that had served as anode material in several CP systems in the field for up to about 9 years [52]. The current densities in these systems were relatively low, about $1\,mA/m^2$ (anode surface), however, the $100\,mV$ depolarization criterion (see later, Section 20.3.1) was systematically met. Cores were taken and thin sections were prepared and studied using light microscopy and scanning electron microscopy (SEM). No significant dissolution due to acid formation could be observed at the coating/concrete interface. Also, when migration of hydroxyl ions from the cathode was taken into account, the amount of dissolution was less than what was calculated from the current flow. It was suggested that another mechanism mitigates the acid formation near the anode. Further work studied samples from CP and $CPre$ tests in the laboratory using activated titanium mesh anodes [53]. These tests involved current densities of about $20\,mA/m^2$ for up to 6 years. Light microscopy and element mapping were used. Again less acid formation was observed than calculated, also when hydroxyl ion migration was taken into account. Various alkaline substances (NaOH, KOH, Ca(OH)$_2$) could be involved in these processes, which are insufficiently understood [53]. Whatever the mechanisms, it appears that acid formation at the anode, at least at moderate current densities, is not critical for the service life of CP systems. However, it may be relevant for (unintended) local high current densities (see Section 20.3.7).

Anodic processes may cause premature failure of oxidizable anode materials, however. A CP system based on a carbon-filled polymer cable anode functioned properly until 6 to 8 years of service. Later, it became increasingly difficult to achieve the criterion of $100\,mV$ depolarization. Detailed examinations after 15 years showed that the carbon had dissolved from the outer layers of the cable and the polymer had become brittle. This caused high resistance build-up in the circuit and decreasing current density [17]. In another case using the same anode, however, the material itself was found to be in good condition after 12 years. This was probably related to lower operation current densities. In this case, the system required maintenance in that the power sources, the connections and the reference electrodes had failed and needed to be replaced [54].

In the case of realkalization and chloride extraction, acidification is not relevant because the anode is placed outside the structure in an appropriate (alkaline) electrolyte and so any acid formation would only affect the surface of the concrete.

20.2.3
How Various Techniques Work

Electrochemical techniques are applied in order to avoid corrosion: by stopping it, by containing it within reasonable limits or by preventing it. The way to do this, as already stated, is to have a direct current flow from the concrete to the reinforcement. The mechanism by which the circulation of current leads to avoiding corrosion is different in different techniques, as shown in Figure 20.3. Further details are discussed in the following sections.

20.3
Cathodic Protection and Cathodic Prevention

20.3.1
Cathodic Protection of Steel in Chloride-Contaminated Concrete

To achieve cathodic protection of steel reinforcement embedded in chloride-contaminated concrete it is not necessary to establish immunity conditions, that is, lowering the potential below the equilibrium potential E_{eq} given by Nernst's law. These immunity conditions are normally required on steel in active condition as on steel structures in soil or immersed in seawater, where potentials more negative than −850 mV CSE or than −950 mV CSE in the presence of sulfate-reducing bacteria should be imposed (Figure 20.3a). Conversely, the target of cathodic protection in concrete structures is to reduce the corrosion rate by taking the steel into the passivity range or by reducing the macrocell activity on its surface, and this can be done with a small reduction in potential and a small current (Figure 20.3b).

The conditions for pitting corrosion initiation and those for repassivation of steel in chloride-contaminated concrete were pointed out in Chapter 7.

Figure 20.4 shows Pedeferri's diagram that summarizes the different domains of potentials and chloride contents where pitting can initiate and propagate. The boundaries between these domains are indicative, since they depend on several factors such as pH and temperature. Domain *A* (corrosion zone) indicates the conditions that cause initiation and stable propagation of pits. Domain *B* ("imperfect" passivity zone) indicates the conditions that do not allow the initiation of new pits but that *do* allow the propagation of pre-existing ones. Domain *C* ("perfect" passivity zone) represents the conditions that do not allow either the initiation or the propagation of pits; and domain *D* (immunity and also hydrogen evolution zone) where corrosion cannot take place for thermodynamic reasons and where hydrogen evolution and consequently hydrogen embrittlement of high-strength steel can take place; finally domain *E* where loss of bond between steel and concrete may occur.

The path of the potential of steel reinforcement in a concrete structure exposed to chlorides and then protected by a cathodic protection system is shown in Figure 20.4. The initial condition is represented by point *1* where the chloride content is

(a) Cathodic protection of steel in soil or in seawater (immunity protection)

(b) Cathodic protection of steel in chloride containing concrete

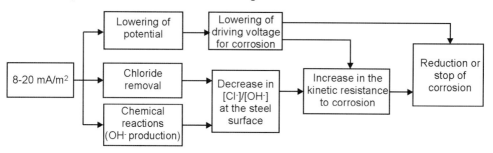

(c) Cathodic protection of steel in carbonated concrete

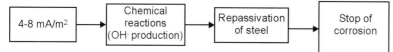

(d) Cathodic prevention of steel in concrete in contact with chloride environments

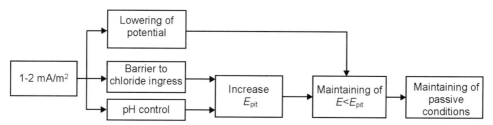

(e) Electrochemical realkalization or electrochemical chloride removal

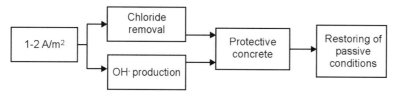

Figure 20.3 Mechanisms by which different electrochemical techniques control the corrosion rate [3].

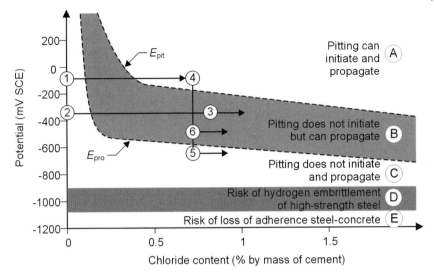

Figure 20.4 Pedeferri's diagram showing evolution paths of potential and chloride content on the rebar surface of an aerial construction during its service life for: cathodic prevention (1-2-3); CP restoring passivity (4-5); CP reducing corrosion rate (4-6). By definition, cathodic prevention is applied *before* corrosion has initiated, CP *after* corrosion has initiated [24].

nil and the steel is passive. By increasing the chloride content in time, the working point shifts to 4, within the corrosion region. Corrosion of the steel occurs rapidly by pitting or macrocell mechanisms. Applying cathodic protection leads to 5 so that the passivity is restored or to 6 without fully restoring passivity. In all cases the corrosion rate is reduced.

Initial CP current densities in the range of 5–15 mA/m² of reinforcing steel surface area are generally needed for protecting reinforced concrete structures exposed to the atmosphere. Much lower current densities are required under conditions that reduce the access of oxygen towards the surface of the steel, such as in water-saturated concrete. For elements operating under water, current densities typically in the range 0.2 to 2 mA/m² are sufficient [20].

The experience on bridge decks shows that, in the cases in which the cathodic protection path runs according to 4–5 of Figure 20.4, the current required to maintain protection conditions (verified by the so-called four-hour 100 mV potential decay empirical criterion[3]) decreases with time, even after months or years

---

3) The "100 mV decay" criterion, used for verifying the protection conditions in reinforced concrete structures, is based on the measurement of potential decay after switching off the current. The structure is considered protected if the difference between the instant-off potential and the potential measured after four hours (or up to 24 h) is higher than 100 mV. This criterion is adopted by NACE [19] and CEN standards [18, 21]. For longer periods of depolarization than 24 h, 150 mV is the minimum required [18, 21].

from start up. This happens because the cathodic current can bring about repassivation of steel in active zones. When passivity is established on the entire surface of the steel, the current required to maintain passivity is reduced to a few mA/m^2 (e.g., 2–5 mA/m^2). If the CP path runs according to 4–6, the current density to fulfill the protection criterion remains high and does not decrease with the time, since passivity is not obtained.

The development of passivity is also favored by the decrease of the $[Cl^-]/[OH^-]$ ratio on the steel surface produced by the cathodic reaction (i.e., production of OH^-) and by the migration of ionic species inside the concrete (i.e., removal of Cl^- from the steel surface). These beneficial effects do not cease immediately if the current is interrupted but can last for months and, under some situations, give rise to the possibility of intermittent application of cathodic protection [55] as well as the application of initial prepolarization at high currents in order to achieve passivity or a more persistent protection condition. Recent field data suggest that even over 24 months after stopping CP current, some protection of steel may remain despite high chloride contents [16].

20.3.2
Cathodic Prevention

Cathodic protection is also applied to noncorroding structures in order to prevent corrosion otherwise expected during the service life. This technique, called "cathodic prevention", is based on the strong influence of the potential on the critical chloride content (Figure 20.3d). In Figure 20.4 a typical evolution path (*1–2–3*, in terms of potential and chloride content) is shown for a structure to which cathodic prevention is applied. At usual current densities in the range 1–2 mA/m^2, a decrease in potential of at least 100–200 mV is produced leading to an increase in the critical chloride content higher than one order of magnitude.

This is illustrated in Figure 20.5, showing the results of application of cathodic prevention to slabs subjected to ponding with a NaCl solution [56]. After about 700 days, initiation of rebar corrosion was detected in the control slab (in the free corrosion condition) at spots where the chloride content at the steel surface had reached more than about 1% by mass of cement. From about that point in time, the slab receiving a very low current density of 0.4 mA/m^2 showed a similar (instant-off) potential as the control slab and its four-hour decay values became lower than 100 mV. The slab receiving 0.8 mA/m^2 showed lower decays from about 24 months on, but generally still above 100 mV. Its instant-off potential was about 50–100 mV lower than in the control slab. Further work on these slabs including destructive analysis showed that after five years, strong local corrosion occurred in the control slab; limited but significant corrosion had occurred with 0.4 and 0.8 mA/m^2 and no corrosion at all with 1.7 mA/m^2 or higher [25].

On the basis of these results, the increase of the chloride threshold brought about by cathodic prevention in practical applications is expected to be sufficient to avoid corrosion initiation throughout the entire service life, even where the

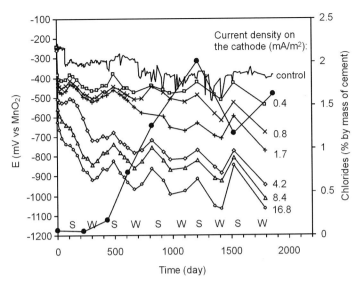

Figure 20.5 Instant-off potential of noncorroding steel cathodically polarized with current densities typical of cathodic prevention and chloride content in the control slab (solid circles). Specimens exposed to saturated NaCl solution alternating ponding, consisting of wetting for 1 week followed by drying for 2 weeks. Control slab started corroding after about 700 days (S = summer, W = winter) [56].

exposure is very aggressive and the concrete itself is unable to sufficiently resist chloride penetration.

20.3.3
Cathodic Protection in Carbonated Concrete

Figure 20.3c shows the effect of application of cathodic protection to carbonated concrete. The applied cathodic current density, even if it brings about only a small lowering of the steel potential, can produce enough alkalinity to restore the pH to values higher than 12 on the reinforcement surface and thus promote passivation. The effectiveness of cathodic protection in carbonated concrete was studied with specimens with alkaline concrete, carbonated concrete and carbonated concrete with 0.4% chloride by cement mass that were tested at current densities of 10, 5, and $2\,mA/m^2$ (of steel surface) [57]. Carbonated concrete specimens polarized at $10\,mA/m^2$ showed that, although initially protection was not achieved since the four-hour decay was slightly lower than 100 mV, after about four months of polarization, the protection criterion was fulfilled and higher values, in the range 200–300 mV of the four-hour potential decay were measured (Figure 20.6). The same results were obtained on carbonated and slightly chloride-contaminated concrete.

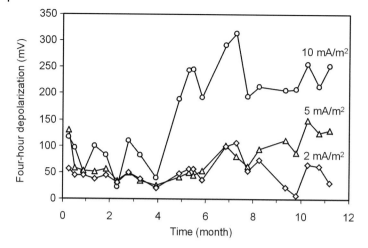

Figure 20.6 Four-hour depolarization of steel in carbonated concrete as a function of time and cathodic current density [57].

The variations in the pH of concrete around the cathode were monitored by means of two pH-sensitive probes (activated titanium wires) placed respectively 1 mm and 3 mm far from the steel surface. During the first 4 months, the potential difference between the two electrodes for all specimens was negligible, showing that the pH was low (carbonated) throughout the concrete.

After five months, in the carbonated specimens protected at $10\,mA/m^2$, the potential difference between the two activated titanium electrodes increased significantly and reached a value up to 200 mV, suggesting that the alkalinity produced by the cathodic reaction at the steel surface induced localized realkalization of the concrete in the vicinity of the rebar. For the other specimens (i.e., in carbonated concrete at 2 and $5\,mA/m^2$) realkalization of concrete and repassivation of steel did not occur even after five years of testing [57]. In the same study, it was found that the application of a start-up current density of $70\,mA/m^2$ for 1 month proved to be an effective way for achieving repassivation of steel in carbonated concrete. The tests showed that, once repassivation is induced, even a current density of $5\,mA/m^2$ is sufficient to guarantee protection.

Recently, the role of precorrosion of steel reinforcement on the effectiveness of CP in carbonated concrete has been investigated [58]. Tests were carried out on reinforced specimens made with concretes of various composition (in terms of water/cement ratio, cement type, and chloride content in small amount); some specimens were precorroded after carbonation and before the application of CP, by exposure to a wet environment that promoted the propagation of corrosion. After precorrosion steel reinforcement was heavily corroded, although not enough to crack the concrete cover. The effectiveness of CP with an applied current of $10\,mA/m^2$ (for given concrete type and exposure condition) was lower on precorroded specimens compared to nonprecorroded specimens, as indicated by lower values of four-hour depolarization even after a year. These results show that, although CP is a repair technique, and as such is usually applied after corrosion initiation due to carbonation, its effectiveness may be affected by the presence of corrosion products on the steel surface, which suggests that the sooner it is applied, the more effective it will be.

20.3.4
Throwing Power

High-strength steel utilized in prestressed structures (but not ordinary steel utilized for reinforced concrete) can be subjected to hydrogen embrittlement if its potential is brought to values at which hydrogen evolution can take place (Chapter 10). However, the risk of hydrogen embrittlement of prestressed steels is low, provided the potential is more positive than −900 mV SCE [59]. Cathodic protection applied to corroding steel works mainly in point 5 of Figure 20.4, that is, relatively close to the hydrogen evolution zone. Consequently, in order to operate cathodic protection properly, it is necessary to consider the throwing power.

The throwing power of cathodic polarization depends on the corrosion condition of the steel. In cathodic protection, which is applied to corroding reinforcement, a uniform current distribution is difficult to achieve because of the high electrical resistivity of concrete, the generally small distance between anode and (outer) reinforcement, the low polarizability of corroding steel and the complex geometry of reinforcement. To achieve a protection condition on inner active rebars, higher current densities are required on the outer rebars, with the risk of overprotection (the accepted thresholds for overprotection are −900 mV SCE for high-strength steel and −1100 mV SCE for ordinary steel). The highest throwing power is achieved when these potential thresholds are obtained on the outer rebars. Respecting these thresholds, sufficient protection is obtained (by operating in zone C or in the lower part of zone B of Figure 20.4) for the case of actively corroding steel at about 0.25 m from the first mat for ordinary steel and 0.15 m for high-strength steel [60]. The throwing power is much higher in the case of passive steel, where protection is achieved (by operating in zone B of Figure 20.4) even at a depth of 0.80 m from the first mat for ordinary steel and 0.6 m for high-strength steel [61].

The higher throwing power and the less negative potentials needed allow safe application of cathodic prevention with regard to avoiding the risk of hydrogen embrittlement of prestressing steel [62].

20.3.5
The Anode System

Titanium activated with oxides of different metals, in particular ruthenium or iridium, used in the form of mesh, wire or strip, is the most reliable and widely used type of anode [63]. It has good mechanical properties and can easily be adapted to the entire surface of the structure in order to obtain a good distribution of current. It is usually coated with a overlay of mortar but can also be embedded directly into the concrete. It can deliver current densities up to $100\,mA/m^2$ (of anode material) over long periods, with short-term maximum levels of even 300–$400\,mA/m^2$. Laboratory tests and field experience indicate that the service life can range from 20 to over 100 years (if the quality of concrete and overlay are adequate, naturally).

A second type of widely used anode is based on a conductive coating with carbon powder in an organic matrix. It has the following advantages: it does not require an overlay, it can easily be applied to structures of any form, and it does not present problems of additional weight to the structure or limitations of dimension; the cost of installation can be lower, down to about half that of activated titanium mesh systems. On the other hand, it cannot deliver current densities above $20\,mA/m^2$ (anode surface equals concrete surface) over long periods (and maximum levels of $35\,mA/m^2$). Conductive coating service life may be up to 15 years, but less in humid climates where there may be loss of adhesion and effects of premature deterioration (see Section 20.3.7).

Anodes of conductive asphalt that have been operating for over 20 years in the protection of bridge slabs in North America must be excluded from many applications due to problems of weight and limitations of dimension, and because they cannot be applied to vertical surfaces or to roofs. However, systems that are operated well (not abandoned) appear to function well after such long periods.

New types of anodes (cementitious overlays with bare or nickel-coated short carbon fibers, long carbon fibers mesh, sprayed titanium coatings, ceramic conducting oxides, etc.) have been used for some time. However, the small number of systems or the duration of use does not allow definition of their reliability over long periods, as damage of the cement matrix may occur due to acidification [64].

Thermally sprayed coatings of zinc or aluminum (or their alloys) are also used as sacrificial anodes. Practical experience exists in the USA, among others on bridges. Another sacrificial system is based on a zinc foil attached to the concrete surface via an ion-conducting gel adhesive [65–67]. Sacrificial CP systems have obvious advantages of simplicity and ease of application; the main questions concern their effective service life.

20.3.6
Practical Aspects

Practical aspects of a CP system should be taken into account when deciding to apply CP and during the design process. The design should consider the anode system, the power source and other electrical parts, the monitoring system, the execution and the operation and maintenance of the installation. Carrying out and evaluating trials can be part of the design and decision process.

Design The design of a CP system should consider the type and location of anodes in order to achieve sufficient and durable protection. Generally, the anodes are placed on the concrete surface or inside the structure in such a way that uniform current distribution can be achieved. A number of examples are given in

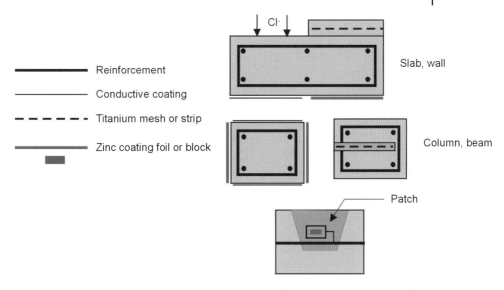

Figure 20.7 Examples of anode layouts with respect to a concrete cross section.

Figure 20.7 [7, 29, 68]. In specific cases, the use of localized anodes (either galvanic or with impressed current) may be considered, for instance to reduce the invasiveness of the intervention on the surface of the concrete. In such circumstances, the distribution of current to the reinforcement should be estimated either with a trial or with numerical modeling, in order to guarantee protection to all reinforcement and avoid overprotection to reinforcement close to the anodes. A practical example of a trial on a real element is reported in [69].

Anode System The anode system, which consists of the anode material, a primary anode and for some types an overlay, must supply the required current for the anticipated service life and distribute it to the reinforcement that needs to be protected. Anode materials and current density aspects have been dealt with in a previous section. The general requirements of an anode system are: it has to adhere to the concrete surface; it should be suitable for application to the surface needing protection (top, bottom, horizontal, vertical, flat, curved), it should be durable and have low installation cost; it should produce acceptable weight addition and change of the appearance and dimensions of the structure. If an overlay is used, it should have durable bond to the substrate concrete, sufficient mechanical strength and electrical characteristics equal to those of base concrete (ionic conductivity).

Power System The current densities needed to reduce or stop corrosion in atmospheric structures will be generally less than $20\,mA/m^2$ of steel surface area. The required current density decreases with time; typical long-term values for activated titanium systems are lower than $10\,mA/m^2$, for conductive coating systems $2\,mA/m^2$ or less. The current density needed for immersed structures to obtain sufficient polarization is indicatively in the range $0.2–2\,mA/m^2$ of rebar area. Power is usually delivered by transformer/rectifiers. They transform mains (AC) voltage to low-voltage direct current. A *CP* system can be run either at constant voltage or constant current. Power units should be able to keep the current or the voltage (whichever is intended) constant at the preset level over long periods of time. In situations where mains power is not available, solar or wind-powered systems can be used with a normal battery backup. In recent years an increase in solar-powered systems has occurred.

Electrical Connections Electrical connections should be made of durable materials and especially in the anode circuit, should be isolated well against corrosion that may occur due to anodic polarization. Cables are usually normal isolated low-voltage wires. They should have sufficient diameter for transporting the current with minimal voltage drop. Cables should be color coded or labeled to facilitate identification in junction boxes, wire bundles, and at the central cabinet containing the power sources. Usually multiple connections and cables are provided to anode and cathode circuits to ensure uninterrupted current flow in case one connection is lost, for example, by damage to the installation due to vandalism.

Zones In cases where parts of the structure need different current densities, have different resistivities or simply are very large, *CP* systems are divided in electrically separated zones. A zone then is a separate part of the anode system with its own power source and current or voltage control and monitoring devices. The reinforcement can be continuous over more than one zone. Different resistivities may be present if parts of the concrete are exposed to different moisture conditions (sheltered versus exposed to rain; marine tidal/splash/atmospheric) or differ by composition (portland cement versus blended cement).

In practice, it has proven useful to divide a *CP* system into smaller parts ("sections") during the execution phase. Small sections allow for checking the absence of short circuits by electrical resistance measurements. The method is based on the principle that a continuous metallic system has a resistance of (much) less than $1\,\Omega$. In the absence of short circuits, the resistance of the concrete between anode and cathode substantially contributes to the total resistance. If a section is small enough, the total resistance is well over $1\,\Omega$, clearly demonstrating the absence of short circuits.

Repair Materials Repair materials to be used in areas where current is flowing should have similar electrical resistivity as the parent concrete in order not to disturb uniform current flow. Repair mortars should be tested for resistivity for several months to obtain mature values, preferably in a climate that simulates the exposure of the structure [7]. A difference of a factor two between mortar and concrete resistivity is considered acceptable [18, 21]. It appears that most cementitious mortars or modified mortars with limited polymer content are in the practical range of resistivities for concrete. Polymer mortars (e.g., epoxy mortars) do not conduct protection current sufficiently; they should not be used in association with CP. The resistivity of the concrete of the structure can be measured during the preliminary investigation [70] or can be obtained from reference tables [37] (see Chapter 16).

The Monitoring System   To determine the performance of the CP system and to ensure that protection conditions are reached and overprotection is avoided, monitoring systems have to be installed. Such systems are based mainly on potential measurements of the reinforcing steel with respect to embedded reference electrodes. In some cases measurement of the current picked up by probes, or exchanged in macrocells are also performed. Reference electrodes and current pick-up probes have to be embedded in the concrete in the most critical areas or where the control of the potential is most important. True reference electrodes for permanent embedment are: silver/silver chloride (Ag/AgCl/KCl-0.5M) and manganese dioxide (Mn/MnO_2/KOH-0.5M). In practice, so-called decay probes are used widely based on activated titanium and graphite. The European Standard recognizes both true reference electrodes and decay probes [18, 21].

The most widely used criteria are based on potential variation, such as the so-called 100 mV decay or depolarization criterion: protection conditions are reached if a potential decay of at least 100 mV over a period between 4 and 24 h from instant-off potential is measured [18, 19, 21, 71]. The instant-off potential is measured between 0.1 and 1 s after switching the protection current off, in order to remove ohmic drop from the potential measured. In some cases, for example, in splash or submerged zones, criteria based on absolute potential measurements can be more convenient: for example, protection or prevention conditions are reached when the potential is more negative than −720 mV with respect to Ag/AgCl [18, 21, 72]. To avoid overprotection, the potential should not be more negative than −1100 mV with respect to Ag/AgCl for plain reinforcing steel or −900 mV for prestressing steel.

The need of monitoring sacrificial CP systems is under discussion. Some authors state that sacrificial CP does not need monitoring, because the current is self-regulating. Others state that all CP installations are self-regulating to a certain extent and that verification is needed to make sure the system is not outside its region of ability to provide protection current where and as much as it is needed. Because most sacrificial CP systems involve direct metallic contact between the steel and the sacrificial anode system, monitoring then requires special provisions. A possibility is to make parts of the system without anode-to-steel connections and to provide an external circuit that is normally connected, but may be switched open to allow current and depolarization measurements. Those parts should be representative of the whole system.

CP systems are increasingly being provided with data loggers and wireless telephone connections for remote sensing and control from a computer in the office of the responsible party. Remote control allows frequent monitoring without actually visiting the structure.

Trials In several cases it has proved to be useful to carry out a trial, that is, a small part of a structure to which *CP* is applied in order to test various aspects of the proposed system(s) [69, 73]. Usually, a trial area is between 10 and 100 m^2. Various anode layouts or different anode systems can be tested. Monitoring probes and other control systems can be tested as well. Seasonal variations can be established. That is why a trial should last at least several seasons, preferably one year.

Execution In the execution phase, the *CP* system is applied to the structure following the design. During the work, the complete concrete surface is checked for cracking, delaminations, cover depth, steel continuity, and the presence of metal objects that might cause short circuits. If necessary, continuity is provided and metal objects are connected to the reinforcement. Subsequently the cracked and spalled areas are removed and repaired. The reference electrodes and other monitoring probes are embedded. Then the anode is applied, with an overlay or top coat as designed. All electrical connections are made and the power source is installed. After sufficient curing of materials, the installation is first checked for proper electrical functioning of all parts and absence of short circuits and then commissioned. For a period of several weeks to months, the functioning is tested frequently and adjustments to the voltage or current are made. If this has been carried out successfully, the normal operation phase begins.

A practical aspect to be considered for the execution phase is the availability of the structure. For example, parts of a bridge deck will be closed to traffic during removal of asphalt, surface preparation, installing of the anodes and applying the overlay. The execution plan has to take into account the diversion of traffic. When work is carried out on the galleries of a building, inhabitants should be provided means to access their houses.

Operation and Maintenance In the operation phase, a *CP* system must be kept working properly. This is achieved by checking voltage, current, and depolarization (or other criteria) regularly, usually two to four times per year and more frequently when a remote control system is present. Once a year the installation is visually inspected for cracks, rust spots, loss of adhesion, cable defects, etc., which should all be corrected or further investigated. Usually, the responsibility for checks and maintenance is organized under a maintenance contract between the owner and the contractor. Normally, the contractor gives a ten-year guarantee that no corrosion and damage to the concrete will occur, provided he is allowed to carry out the control measurements. Increasingly, maintenance contracts up to 25 years are asked for by clients and provided by *CP* companies.

20.3.7
Service Life

Because installing *CP* on a concrete structure represents a significant investment, the question may be asked how long such a system will work. This issue was studied from a practical point of view for *CP* systems in the Netherlands [74]. Information on the performance of concrete *CP* systems was asked from *CP* and repair companies who had installed them. In 2011, information on performance was obtained for 105 structures with *CP*, out of which 50 had been operating for ten years or longer. About two thirds of these required minor interventions, for example, replacement of components. The following analysis assumes that all systems have been designed and executed properly and applies only when proper maintenance is carried out [18, 21].

Working lives of *CP* systems of ten to twenty years without major intervention have occurred; corrosion and related damage to concrete have been absent in all documented cases. Necessary interventions were mainly related to defective details such as local leakage and poor electrical isolation.

Survival analysis of the 105 documented cases suggests that (minor) interventions are increasingly necessary with increasing age. The results indicate that there is a 10% probability that a *CP* system needs maintenance at an age of about 7 years or less; and a 50% probability that maintenance is needed at an age of 15 years or less.

Complete replacement of the anode was carried out in only two cases, one conductive coating anode (on a bridge) and one titanium anode (of a nonmesh type in a building). Some more systems showed poor coating condition, but corrosion protection was still adequate, corresponding to findings in the United Kingdom [16].

Conductive coatings have shown local deterioration in limited numbers, mainly related to water leakage, which caused the need for local repairs of the coating (ten cases); however, corrosion protection may be provided for another several years even if their condition is (visually) poor. Anodes based on activated titanium have shown long working lives, up to more than 20 years. Replacement of a titanium based system was carried out in only one case, which was a nonmesh type. Replacement of primary anodes and anode-copper connections was necessary in a number of cases, in particular with older systems; it appears that the working life of such critical details has been improved.

Power units and reference electrodes have been replaced in some cases. Simple power units and reference electrodes are relatively inexpensive today, so the cost of these actions is limited.

Based on time-to-failure rates for individual components obtained from survival analysis, the life-cycle costs of a *CP* system was predicted. For an example case it was shown that the cost of replacement of components is relatively small compared to the usual cost of inspection and electrical (depolarization) checks.

Service life may be a critical issue for sacrificial anode systems, where the anode itself is consumed by the oxidation reaction. This is usually taken into account in the design stage of anode dimensioning, however long-term data on the performance of sacrificial anodes are few. Recently, it was reported that sacrificial anodes embedded in patch-repaired concrete showed good performance for a period of at least ten years [30].

20.3.8
Numerical Modeling

Present-day computer power and improved understanding of electrochemical processes have stimulated numerical modeling of CP in concrete. Generally, the intention is to predict system behavior in the design stage. Ideally, critical details can be reviewed and the design can be modified to reach optimal protection where it is needed and to avoid overprotection, in particular in the case of prestressed structures. The economic and technical benefits are minimal use of materials for sufficient corrosion protection and system performance.

For numerical modeling of CP systems a finite element model is made of a typical part of the structure to be protected, which ideally includes all steel bars with the right dimensions at the right place, polarization processes at the anode and at the steel and ohmic drop in the concrete. Using such a model, current densities and potentials, both absolute potentials and potential shift (polarization), can be calculated at every point in the modeled structure. By equating the calculated polarization with depolarization, the usual criterion for quality of protection, such a model can be used to verify the spread of protection for various anode configurations and to predict local polarizations that can be tested experimentally.

The modeling is based on solving Laplace's equation in the (ohmic) concrete electrolyte space with boundary conditions that represent steel polarization (usually based on Butler–Volmer-type equations) and some form of anode polarization. It was shown that the general modeling concept works [75]. However, calculations using active steel with a typical corrosion current density ($10\,mA/m^2$), produced much too high CP current densities compared to practical systems. The explanation is that the beneficial effects of CP on corrosion current density [3] are neglected, one of which is pH increase at the steel due to generation of hydroxyl ions. A model for the beneficial effects of current flow was proposed and validated [76, 77], suggesting that the transition from active to passive behavior occurred in a relatively short time (days to weeks). Nevertheless, such transient behavior has not yet been included in models. Consequently, at present the modeling of impressed current CP basically involves passive steel. Numerical modeling has been developed to assess the possibilities and limitations of protecting deeper layers of steel with respect to the anode [78]. Another branch of CP modeling is the application to sacrificial systems. Based on experimental work, the spread of protection (current and polarization) from a zinc anode has been modeled for steel in concrete that extends from below to above seawater with the zinc anode in the seawater [79, 80], with good results.

20.4
Electrochemical Chloride Extraction and Realkalization

Electrochemical chloride extraction and electrochemical realkalization have been proposed as an alternative to traditional repair methods for the rehabilitation of chloride-contaminated or carbonated structures. Due to increasing field experience and laboratory studies, the confidence in these methods is growing [27, 33, 37, 81–86].

20.4.1
Electrochemical Chloride Extraction

For electrochemical chloride extraction (abbreviated CE, also called electrochemical chloride removal, or desalination), a direct current is applied between the reinforcement (cathode) and an anode that is placed temporarily on the outer surface of the concrete. The anode is an activated titanium wire mesh or a reinforcing steel mesh. The anode is surrounded by tap water or saturated calcium hydroxide solution in ponds (upper, horizontal surfaces) or tanks (vertical or overhead surfaces) or as a paste that can be sprayed onto all types of surface. The applied current forces chloride ions to migrate from the reinforcement to the anode. Due to a relatively high current density of 1 to $2\,A/m^2$ (of steel surface), relatively large amounts of chloride can be removed from the concrete within a relatively short time, usually 6 to 10 weeks. After that, the anode, the electrolyte and the incorporated chloride ions are removed from the structure. The principle layout and electrode reactions involved are indicated in Figure 20.8.

The strong polarization associated with such high current densities brings the potential of the steel below $-1000\,mV$ SCE and the electrolysis of water produces a significant amount of hydrogen gas at the steel surface. The hydroxyl that is produced simultaneously increases the pH at the steel surface and together with the reduction of the chloride concentration in the pore water near the cathode, will repassivate the steel in a short time. During the treatment, migration takes away

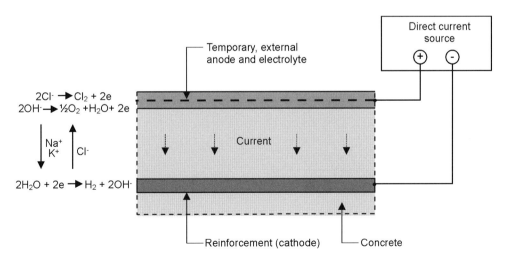

Figure 20.8 Principal reactions involved in chloride extraction.

the chloride ions from the concrete cover zone to the anode system, where it is oxidized to chlorine gas (depending on the electrolyte pH) or removed together with the electrolyte solution. The strongly negative potential and the evolution of hydrogen poses the risk of hydrogen embrittlement of high-strength steel. Consequently, chloride extraction should not be carried out on prestressed concrete, unless convincingly it has been demonstrated that in the particular situation at hand the risk is negligible.

Until recently, CE always involved impressed current. However, sacrificial anodes have been used as well [40, 87]. With sacrificial systems, a zinc or aluminum anode is connected to the reinforcement, providing current by galvanic dissolution. No power source is needed, but the treatment time is longer than with impressed current CE.

Since 2011, a CEN Technical Specification exists [36] for impressed current chloride extraction of atmospherically exposed reinforced concrete structures. For post-tensioned structures, verification of the risk of hydrogen embrittlement is necessary. It is not applicable to pretensioned structures or where epoxy-coated or galvanized bars have been used. Indicative criteria given for termination of the treatment are: a remaining acceptable chloride profile (in the range of 0.4% chloride by mass of cement and not over 0.8%), a total charge of 1000 to 2000 Ah/m^2 of steel surface and mapping of rest potentials that should indicate elimination of macrocells, to be done 4 to 6 months after the treatment.

Treatment Effectiveness From many studies it is clear that large amounts of chloride can be removed from concrete and that the steel will be repassivated, even inside corrosion pits [88]. The quantity of chloride removed depends on the initial chloride content and on the total charge circulated (which is the integral of the current imposed and the duration of the treatment). For practical reasons, the treatment duration should be as short as possible. An upper limit for the current density should be respected in order to avoid deterioration of the concrete, which occurred in early tests above $10 A/m^2$. The effect on steel/concrete bond is discussed below. Limit values between 1 and $5 A/m^2$ are given in the literature [89]. Because the current distribution may be nonuniform, a safety margin should be imposed on the average "design" current density. For systems with a galvanic anode, the current density decreases spontaneously with time, so it is not a process parameter. The design in such cases must be based on the total charge, which means that the current must be monitored and integrated and thus the total time needed cannot be known beforehand.

Only a fraction of the total current is actually removing chloride ions, corresponding to the chloride ion transport number (Chapter 2.5). The remaining part of the current is transporting other ions, mainly hydroxyl and to a lesser extent alkali-metal ions. As the chloride concentration in the pore solution is high in the beginning, but decreases upon removal, its transport number decreases and consequently the efficiency is higher in the early stages but it decreases going to later stages. The removal efficiency depends on the initial chloride content, the distribution of chlorides in the concrete, the rate of release of bound chloride, the presence of other ions and on the geometry of the reinforcing bars (which determines the current distribution). The removal efficiency has been reported from laboratory studies to be between 0.1 and 0.5 [88, 90, 91]. In practical cases it is probably between 0.1 and 0.3 [92]. On heavily chloride-contaminated structures an average removal efficiency of 0.5 has been reported [81], reaching maximum values of 0.7 [81, 93]. Experiments with chloride removal of cores from a marine structure showed that the efficiency using sodium carbonate solution as the electrolyte was very low. Apparently, the introduction of sodium and carbonate ions and the associated increase of pH strongly reduced the chloride ion transport number [94].

Predictions of chloride removal can be made using physicochemical models [95, 96] and estimated transport numbers. Literature data indicate that between about 50% [33, 97] and 90% [88] of the total chloride content can be removed. These figures, however, apply to penetrated chloride. A special case is concrete with chloride mixed-in in the fresh concrete. In several laboratory studies it was shown that mixed-in chloride is removed with low efficiency. In practice, it is unlikely that sufficient chloride can be removed within a practical time span.

As mentioned above, the total treatment time required varies from a few weeks to a few months. If the maximum current density is respected, the removal efficiency and the required final result (in terms of the acceptable level of remaining chloride) determine the duration. As the steel is repassivated quickly during the treatment, the main consideration should be the durability, that is, the period after the end of the treatment during which corrosion most likely will be absent. Repeated treatment may be needed if results of a first treatment period are insufficient [81].

All the data from laboratory and field studies reported above on treatment efficiency rely on measurements of the total chloride content in concrete. From fundamentals of ion transport in the pore solution (Chapter 2.5) it is obvious that only free chloride ions dissolved in the pore solution can be removed by applying an electric DC field. Recent laboratory studies using concrete blocks instrumented with sensors measuring the free chloride content in the pore solution [98] have shed a new light on the chloride-removal process [99, 100]: it was found that the free chloride ions present in the pore solution are removed completely within a few days (Figure 20.9). Due to the complete removal of free chloride during the treatment, bound chloride is dissolved until the equilibrium between bound and free chlorides is re-established. The rate of release of bound chloride is slow compared to the rate of chloride removal and thus the *CE* process quickly becomes inefficient. Current-off periods allow the system to re-establish the equilibrium between bound and free chlorides. Subsequently, the process is highly efficient again (Figure 20.9). Current-off periods greatly reduce the total amount of (inefficient) current flow. A further interesting result is that the free chloride content in the pore solution measured by the chloride sensor increases immediately upon switching on the electrical DC field (Figure 20.9). This can be explained by chloride ions physically adsorbed in the pore system of concrete (Figure 20.10). The results show that the *CE* process with constant application of the electrical field for weeks is inefficient and only hydroxyl ions are transported. Short treatment times with

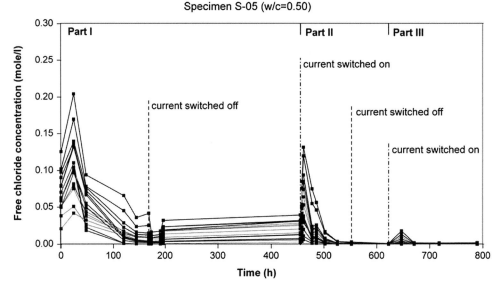

Figure 20.9 Change in the free chloride content in concrete ($w/c = 0.5$) with time. Note the rapid decrease within few days and the immediate increase after switching on the current [100].

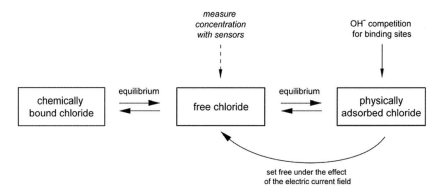

Figure 20.10 Equilibria between bound, adsorbed and free chloride in the CE process [100].

longer current-off periods reduce the charge required for the desalination significantly and make the process more efficient [99, 100]. This is also of interest with regard to the risk of undesirable side effects that currently limit the use of CE.

Durability after Chloride Extraction After the treatment, significant amounts of chloride (total chloride content) may be left in the concrete. In addition, the remaining chloride is distributed nonhomogeneously. Usually a profile remains with higher values near the concrete surface and lower values near the rebar. Also, because the removal current density is higher near the rebars, the remaining chloride content near the reinforcing bars is lower than in between the bars. For a durable result of the treatment, the remaining chloride should not initiate corrosion for a reasonable period of time. Unless the exposure to chloride is taken away, at the end of the treatment additional measures should be taken to avoid future chloride penetration. This makes accepting the final result of a particular case of chloride extraction a difficult matter. If it is guaranteed that no new chloride would penetrate, a safe upper limit for accepting the remaining chloride would be 0.4% by mass of cement *throughout the complete cross section*. In practical cases, this requirement can be hard to meet.

A solution to this problem can be found considering that in the course of time, the remaining chloride would redistribute by diffusion, going from regions with high contents (from zones between the rebars) to regions with lower contents, such as near the rebars [88]. Consequently, situations with a higher *average* chloride content than 0.4% may be tolerated if it can be shown that during a reasonable period of time, the chloride content *at the rebar* will not become higher than 0.4%.

Following this approach, a procedure was proposed assuming that the remaining chloride is redistributed by diffusion from the profile present immediately after CE, to a uniform distribution, as observed in specimens studied one year after treatment [101]. Diffusion coefficients for portland and blast-furnace slag cement concrete were taken from a study of 16 years submersion of concrete in seawater [102]. Furthermore, it was assumed that no additional chloride penetrates. This can be achieved by applying a coating or hydrophobic treatment to the concrete surface after CE (Chapter 14). Disregarding differences between the remaining chloride contents in the mesh and near the rebars, the calculation model may be one dimensional. For a remaining chloride profile with the highest values near the concrete surface (similar to a penetration profile) and assuming no new chloride entering, it can be calculated at what minimum depth the chloride content will not exceed 0.4% by mass of cement in 10 years. This depth is termed the "durable cover depth": steel at this depth will not develop corrosion in a period of 10 years. In many cases, the chloride content at this depth does not exceed 0.4% for much longer times due to diffusion of chloride to deeper parts. If significantly more chloride remains in the mesh than near the rebars, two-dimensional calculations are necessary [101].

From the profiles of remaining chloride in specimens subjected to chloride extraction in the laboratory with various amounts of charge, the relationship between the "durable cover depth" values and the charge was established [88]. Before CE testing, the 144 specimens studied were subjected to chloride ponding that resulted in the penetration of about 2.5% chloride by mass of cement in the outermost 15 mm and about 0.6% from 15 to 30 mm depth. The results of the CE tests were interpreted as follows. It was considered that it should take at least 10 years after treatment before the chloride content at the rebar surface would exceed 0.4% by mass of cement. If no new chloride would penetrate, it was concluded that [88]:

- for portland cement concrete, at least 2400 A h/m² (concrete surface) is required to obtain a sufficiently low remaining chloride profile, independent of the cover depth above a minimum of 15 mm; this is equivalent to 100 days at 1 A/m²;
- for blast furnace slag cement concrete, the required charge depends on the cover depth; for cover depths of 15 to 20 mm, 2230 A h/m² is required; for cover depths of 20–25 mm, 1130 A h/m² is needed; for cover depths over 25 mm, about 1000 A h/m² is required (equivalent to 42 days at 1 A/m²).

In general, these required amounts of charge correspond to those mentioned by the European Technical Specification [36].

Only few long-term results from field applications of CE regarding the durability of the treatment are documented. On one of the earliest field application of CE (1992), an abutment of an underpass [81], condition assessment by half-cell potential mapping has been performed several times. The last measurements from 2005 showed that all the reinforcement was completely repassivated and remained fully passive 15 years after the treatment [93]. At least in this case, it is clear that CE may have a long and durable effect.

Trials The effectiveness of chloride extraction depends on characteristics of individual structures, such as the concrete composition, the actual chloride penetration profile and the depth of cover. So, it may be useful to carry out a trial on an area (about 1 to 10 m²), which should be representative of the structure to be treated and should last at least 4 to 8 weeks. The results of such a trial in terms of the chloride profile before, during and after chloride extraction gives an indication of the duration required and can be used to show that chloride-extraction treatment of the particular structure will be effective under field conditions. Trials are most certainly recommended if prestressed structures are to be treated with chloride extraction. Careful monitoring of the potential of the prestressing steel should be carried out to establish the risk of hydrogen embrittlement. As a safe criterion, the potential should not become more negative than −900 mV SCE, as applies for cathodic protection [18, 21].

Monitoring of the Process During the treatment, a form of monitoring of its progress is necessary. Primary data are the voltage applied and current flowing during the treatment. Taking cores and determining the remaining chloride profile has been found to be a reliable method [81, 93, 97]. For electrolyte solutions confined in tanks, monitoring the chloride in the electrolyte is another method. Provided that the electrolyte pH is high enough to prevent chlorine gas evolution, the amount of chloride in the electrolyte is equal to the amount of chloride removed from the concrete [91, 94]. Under that condition, the chloride in the electrolyte can be reliably used to monitor the progress.

Monitoring after Treatment After the application of chloride extraction, nondestructive measurements may be carried out for controlling the durability of the structure, for example, by measuring the half-cell potential of reinforcement.

However, it should be realized that due to strong polarization, immediately after chloride extraction steel potentials are very negative. They become more positive with time. Consequently, potentials are not stable during the first months. Where potential mapping before treatment has shown the presence of potential gradients over the surface, successful treatment may be expected to flatten out the gradients [81, 97, 103]. In general, potentials should shift towards positive values in a few months, showing reduction of corrosion activity and reinstatement of passivity. When at some point in time potentials become more negative again, this is a sign that repeated treatment may be needed. Embedding reference electrodes and measuring regularly may be used to monitor the durability of a structure treated with chloride extraction (Chapter 16).

Side Effects During chloride extraction, hydroxyl ions are formed around the reinforcing steel, locally increasing the pH and sodium and potassium ions are enriched around the steel. These changes might stimulate alkali silica reaction (ASR, Section 3.4). In several studies, the possibility of ASR was checked as a side effect of chloride extraction [27, 43, 49, 104, 105]. The aggregates used were reactive and the alkali content of the cement was just below the critical values. The results obtained with noncarbonated concrete showed that, under the worst conditions, chloride extraction induced concrete expansion, but no cracking was observed. In [49], a pessimum effect was observed: at low and high charge less expansion was observed, but at intermediate charge, high expansion and cracking were found.

For practical application of chloride extraction on structures with potentially sensitive aggregate, it is recommended to assess this in a stepwise process, going to the next step only if the previous step confirms the possible risk of ASR:

- evaluate the potential reactivity of the aggregate by considering its geological source and practical experience in structures containing the aggregate;
- study the concrete using microscopy for actual presence of reactive material;
- perform chloride extraction trials, either on a small part of the structure or in the laboratory on cores taken from it and check treated parts by microscopy and expansion measurements.

If deleterious expansion is found in representative tests, chloride extraction should not be recommended.

Concern may be justified for high current density or very long chloride extraction treatment and loss of bond between steel and concrete. Steel/concrete bond was determined by pulling out ribbed bars from cylindrical concrete specimens after polarization [106]. Water/cement ratios were 0.48, 0.58 and 0.66. Polarization duration was 1, 3, and 5 months, at current densities of 4 and $12 \, A/m^2$ of the cathode (steel) surface. It was found that bond strength was reduced by 7% to 55%, increasing with higher total electrical charge passed and higher w/c of the concrete. Microhardness and analysis of sodium and potassium near the steel suggested that softening of the cement paste related to polarization caused the reduction of bond strength. The current densities studied are relevant for CE application. One month at $4\,A/m^2$ (or 4 months at $1\,A/m^2$) is equivalent to $2880\,A\,h/m^2$, which was found to cause 7% to 15% loss of bond strength. The results support the view expressed above that current densities significantly higher than $1\,A/m^2$ of steel surface and treatment times longer than 4 months should be avoided in order to avoid loss of bond between reinforcement and concrete. Similar results in terms of bond-strength reduction and accumulation of sodium and potassium in chloride extraction tests were found by Vennesland and coworkers [90].

Changes in the pore structure have been reported also at lower current densities than previously mentioned [107]. Portland cement concrete specimens ($w/c = 0.4$) were subjected to chloride-migration tests at about $0.75\,A/m^2$ for one to four weeks. Samples from near the cathode electrolyte chamber and near the anode surface were analyzed by mercury intrusion porosimetry. It appeared that after treatment for 3 and 4 weeks, the number of the smallest pores (diameter below 50 nm) increased near the cathode and to a lesser extent, also near the anode. The volume of pores with diameters between 50 nm and 5 μm decreased near the anode. After post-treatment storage for 8 months, pore sizes had redistributed, decreasing pores below 50 nm and increasing between 50 nm and 5 μm. Reductions were found near the anode in the amount of calcium hydroxide and ettringite. It appears that current flow influences the equilibria between liquid and solid phases in the concrete, changing the pore-size distribution. In other work, chemical changes were observed (i.e., liberation of sulfate ions), but no reduction of microhardness [48]. This subject requires further study. Reducing the total amount of charge passed by using an intermittent CE treatment with short current-on and long current-off periods will greatly reduce negative side effects [99, 100].

20.4.2
Electrochemical Realkalization

For electrochemical realkalization (abbreviated RE), a direct current is applied between the reinforcement (cathode) and an anode that is placed temporarily on the outer surface of the concrete. The application of this method is comparable to that of chloride extraction. The anode is an activated titanium mesh or a reinforcing steel mesh. The anode is surrounded by a sodium (or other alkali metal) carbonate solution of about 1 mole per liter in ponds (upper, horizontal surfaces) or tanks (vertical or overhead surfaces) or as a paste that can be sprayed onto all types of surface. Due to a relatively high current density of 1 to $2\,A/m^2$, a carbonated concrete cover of several centimeters can be realkalized within a relatively short time, usually a few days to a few weeks. After that, the anode and the electrolyte are removed from the structure. The layout and principle reactions involved are indicated in Figure 20.11.

As illustrated in the figure, hydroxyl ions are produced by electrolysis at the reinforcement, of which only a part migrates to the anode, the remaining part being balanced by sodium ions migrating in. In addition, carbonate ions penetrate

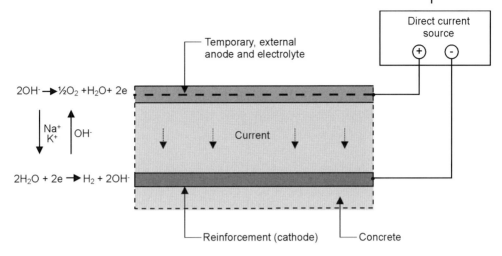

Figure 20.11 Principal reactions involved in electrochemical realkalization.

from the electrolyte into the concrete by electro-osmotic flow, diffusion, and capillary absorption. The high amount of hydroxyl ions formed produces pH values near or above 14 [94, 108]. After realkalization at least part of the hydroxyl ions will react with atmospheric carbon dioxide, further increasing the carbonate concentration, resulting in a drop of pH. Sodium carbonate of sufficient concentration stabilizes the pH between 10.5 and 11.

There has been some controversy about electro-osmosis. No evidence of its occurrence has been found in early laboratory tests or during testing of cores in a realkalization demonstration project [33, 108, 109]. On the other hand, laboratory experiments by Andrade and coworkers [110, 111] have confirmed the occurrence of electro-osmosis, supporting previous observations by Banfill [104, 112]. Electro-osmotic flow was observed in carbonated concrete, while it was not observed in uncarbonated concrete. This was confirmed by tests carried out on mortars that showed that the electro-osmotic flow of water was inverted in alkaline mortar compared to carbonated mortar [113]. Full understanding of the mechanism and the controlling factors in terms of practical characteristics of structures needs further study.

The development of realkalization is schematically shown in Figure 20.12. The left figure shows the extent of carbonation before treatment. The middle figure shows the presence of alkaline material after a short duration: around the reinforcement (due to electrolysis) and from the surface into the concrete (due to absorption, electro-osmosis and/or diffusion). The right-hand figure shows a more advanced state of realkalization, where the alkaline zone around the steel has become continuous with that penetrating from the surface. Since 2004, a CEN Technical Specification exists [43] for impressed current realkalization of carbonated concrete in reinforced structures. It is not applicable to post-tensioned or

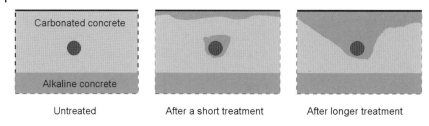

Figure 20.12 Development of realkalization detected with pH indicator [27, 108].

pretensioned structures or where epoxy-coated or galvanized bars have been used; or if chlorides contribute to corrosion.

End-Point Determination and Treatment Effectiveness An important issue is how to determine when the realkalization treatment can be stopped. As the amount of realkalization is related to the total charge passed, contractors tend to empirically apply values between 200 to 450 Ah/m². Such amounts of charge are obtained with a treatment duration between 8 and 18 days with a current density of 1 A/m² of steel surface. In any case, the total charge flow can only indicate the quality of the treatment in a general sense; however, it does not indicate steel repassivation.

The sodium content in cores taken during and after realkalization is often used as an additional criterion for the effectiveness of the treatment. However, the sodium content can vary strongly, depending among others on the cement type. Pollet and Dieryck [114] have observed no correlation between the sodium content and the amount of realkalization in their experiments. It appears that the sodium content does not provide a direct indication on the pH value around the steel nor on its repassivation. Van den Hondel and Polder [115] found some correlation between the sodium content and the amount of electrical charge passed, but the correlation was poor in the range of usually applied amounts of charge. They also observed significant amounts of sulfate ions to be present in the pore solution in realkalized blast furnace slag cement concrete [115]. This shows that the carbonate concentration can be lower than the value indicated by sodium analysis.

Using pH indicator liquids on cores taken during or after treatment provides another means of assessing the effect of realkalization. Phenolphthalein is a widely used pH indicator for measuring the carbonation depth in concrete, turning from colorless to pink as the pH rises over about 9 (Chapter 16). However, a pH (just) above 9 may leave the steel in an unpassivated condition. Thymolphthalein has a color change near pH 10 and thus appears more appropriate. To be visible on concrete, the solution should contain 1% thymolphthalein in a 70% ethanol solution [114]. In laboratory tests it was documented that phenolphthalein has another color change (as proposed by applicators of realkalization), namely at pH values between 13.5 and 14, from pink back to colorless. At amounts of charge normally used for realkalization, only a narrow zone around the rebars has such a high pH that this color change actually takes place. It is, however, a method that can be used to show that realkalization has progressed to an advanced stage. Sergi and coworkers [116] suggest to use a two-stage method applied to cored samples taken from concrete structures that have undergone realkalization. This two-stage method involves the use of acid/base indicators and galvanostatic polarization to determine the effectiveness of realkalization treatment. For realkalization process control, spraying phenolphthalein onto cores is the most widely used method in addition to monitoring the total amount of charge passed.

The criterion given in [43] for termination of the treatment is a total charge of 200 A h/m^2 of steel surface and the realkalization effectiveness must be demonstrated by phenolphthalein testing, showing a high pH in a zone of at least 10 mm around a rebar.

In addition to the methods outlined above for direct process control, there is a need to effectively assess the passivation of reinforcement brought about by realkalization. Pollet and Dieryck [114] and Elsener and coworkers [97] have studied electrochemical methods.

Half-cell potential measurement is the most frequently used technique onsite to evaluate corrosion of steel in concrete, giving an indication of the corrosion probability (Chapter 16). However, special precautions have to be respected when taking potential measurements after realkalization. According to references [101, 103, 114], the following development takes place. Immediately after switching off the current, potentials below −1000 mV SCE are measured and after 1 day the potentials are in the range of −600 to −800 mV SCE. The process of depolarization is quite slow. After about one month the potentials stabilize around −100 mV SCE. At longer times, a trend towards slightly more positive potentials was noted. Thus, meaningful half-cell potentials can be measured only one to several months after completion of the treatment. In addition to absolute potential values, evaluating potential gradients can be useful. Before treatment, strong potential gradients may be present that indicate corroding spots. The presence of a homogeneous potential field after the treatment is considered a strong indication of repassivation. In this context, it is interesting to note that the potential field over the treated area can be quite homogeneous already after 7 days.

Corrosion rate assessment by linear polarization resistance (LPR, Section 16.2.3) measurement after realkalization was also studied. Applying the usual criteria for LPR appeared not to be suitable for evaluating the corrosion state after realkalization [114]. Unusually low polarization resistances were found, suggesting high corrosion rates. The explanation may be that during such measurements, oxidation takes place of (low-valence) iron oxides present on the reinforcement after the treatment [117].

As was observed for *CP* in carbonated concrete (Section 20.3.3), also for electrochemical realkalization the effectiveness of the treatment may be affected by the corrosion conditions of reinforcement. Laboratory tests carried out on reinforced specimens made with concrete of various composition subjected to carbonation showed that the corrosion rate measured (with LPR) after the treatment was higher compared to that measured before the treatment if the reinforcement was precorroded, while nonprecorroded specimens showed a reduction in the corrosion rate [118]. This behavior requires further investigation in order to clarify if the increase in the corrosion rate is an effect connected with the use of the linear polarization technique, or a real worsening of the corrosion conditions of steel.

Influence of the Cement Type Most of the published data and empirical information on realkalization is based on tests on portland cement concrete. Effective realkalization of concrete made with blast furnace slag and fly ash cements seems much more difficult [108, 115]. After a total charge of 336 A h/m^2 of steel surface had passed (14 days at 1 A/m^2), nearly total realkalization was observed using phenolphthalein for portland cement (CEM I 42.5) concrete. However, after 672 A h/m^2 had passed (28 days) through concrete made with blast furnace slag cement (CEM III/A 42.5), only a small pink ring around the rebar was observed after spraying phenolphthalein. Similar effects were found with fly ash cement concrete. It appears that concrete made with slag cement or fly ash cement needs considerably more electrical charge than portland cement concrete before it has reached a high pH over a substantial part of the cross section. Consequently, the treatment time is longer for structures made with slag or fly ash cements. Van den Hondel and Polder [115] recommend applying 100 A h/m^2 of concrete surface for portland cement and 350 A h/m^2 for slag cement concrete (CEM III/B was studied). The former figure falls within the recommended value in the European Specification (200 A h/m^2 of steel surface) [43], the latter does not (see below).

Durability Electrochemical realkalization has been applied for over twenty years, mainly to structures made with portland cement concrete. Although well-documented case studies are rare, those available show that the corrosion protection obtained is durable [119]. Laboratory tests followed over several years support the durability of the effect on reinforcement corrosion, as long as sufficient electrical charge has been applied [114, 115]. As noted above, the amount of charge needed may depend on the cement type.

Elsner and coworkers [97] reported half-cell potential measurements made after six months and after more than one year on a test site. The potential field was homogeneous with values around −100 mV SCE, indicating good corrosion protection. Pollet and Dieryck [114] have studied the durability of realkalization of concrete specimens made with different cement types and different realkalization durations by following the corrosion potential during two years after realkalization and by measuring the extent of the highly alkaline zone with a pH indicator just after treatment and two years after treatment.

The two types of measurement indicate a reduction of the effect of realkalization with time. In the case of specimens made with blast furnace cement (CEM III/A 42.5) and fly ash cement (CEM II/A-M 42.5), this reduction leads to reactivation of corrosion. At the end of the realkalization, first potentials are very negative; then a change to less negative values of potential can be observed. In the case of blast furnace slag cement and fly ash cement, a third stage occurred, in which a subsequent change to more negative values was observed with time, indicating a higher probability of corrosion. Visual examination of rebars two years after the treatment and the decrease of pH observed using indicators corroborate the potential measurements.

The CEN Technical Specification states that the treatment shall be terminated when a total charge of 200 A h/m^2 has been delivered to the reinforcement [43]. Research carried out within COST 521 has shown, however, that such an amount of charge may be insufficient to achieve durable corrosion protection for concrete made with blended cements [109, 114, 115].

The reduction of pH after realkalization by reaction between hydroxyl ions and atmospheric carbon dioxide can be prevented by using a carbon-dioxide-resistant coating on the concrete surface (Chapter 14).

Side Effects The increased content of alkali ions and the production of hydroxyl ions might theoretically cause accelerated ASR in concrete with reactive aggregate. Following Miller [120], it is highly improbable that realkalization can have any pronounced effect on ASR and none have been reported from actual structures. Banfill and coworker [50, 104] found that realkalized concrete expanded about the same as the reference concrete, because the treatment did not raise the hydroxyl ions concentration to the threshold for expansion. In potentially critical cases, lithium-based electrolyte can be used to reduce the risk of ASR, as lithium does not cause expansive gel to be formed [116].

The amount of charge passed during realkalization treatment is well below empirical thresholds for loss of steel/concrete bond strength [37]. It seems highly unlikely that realkalization will negatively affect the bond between steel and concrete.

20.4.3
Practical Aspects

Practical aspects for chloride extraction and realkalization are similar [27]. The treatment has to be designed for achieving sufficient protection from immediately after treatment until some future point in time, at least for ten years. This means that the required treatment time and current density must be established beforehand, taking into account local variations of chloride penetration or carbonation and variations occurring during the treatment. Monitoring the process should be anticipated, for example, by indicating where control cores are going to be taken. Preparation includes cleaning of the surface, filling wide cracks, and preventing short circuits and points with excessively low surface-to-steel resistance (low cover). During the process, the current is monitored. Leaking of the electrolyte should be avoided and paste electrolytes should be kept wet. Calcium hydroxide or sodium/potassium carbonate solution concentrations should be maintained. Recent developments include using small tanks (each containing an anode and electrolyte) with individual current monitoring and control. Finally, the working area and the concrete surface are cleaned and, if planned, a surface treatment is applied.

References

1 Stratfull, R.F. (1957) The corrosion of steel in a reinforced concrete bridge. *Corrosion*, **13**, 173t.
2 Stratfull, R.F. (1974) Cathodic protection of a bridge deck. *Materials Performance*, **13** (4), 24.
3 Pedeferri, P. (1996) Cathodic protection and cathodic prevention. *Construction and Building Materials*, **10** (5), 391–402.
4 Page, C.L. (1997) Cathodic protection of reinforced concrete–principles and applications. in Proc. Int. Conf. on Repair of Concrete Structures, Svolvaer, Norway, 28–30 May, 123–132.
5 Bertolini, L., Bolzoni, F., Lazzari, L., Pastore, T., and Pedeferri, P. (1998) Cathodic protection and cathodic prevention in concrete: principles and applications. *Journal of Applied Electrochemistry*, **28** (12), 1321–1331.
6 Pedeferri, P. (1998) Principles of cathodic protection and cathodic prevention in atmospherically exposed concrete structures, in *The European Federation of Corrosion Publication Number 25* (eds C.M.J. Mietz, B. Elsener, and R. Polder), The Institute of Materials, London, pp. 161–171.
7 Polder, R.B. (1998) Cathodic protection of reinforced concrete structures in The Netherlands–experience and developments, in *Corrosion of Reinforcement in Concrete. Monitoring, Prevention and Rehabilitation*, The European Federation of Corrosion Publication number 25 (eds J. Mietz, B. Elsener, and R. Polder), The Institute of Materials, London, pp. 172–184.
8 Haldemann, C. and Schreyer, A. (1998) Ten years of cathodic protection in concrete in Switzerland, in *Corrosion of Reinforcement in Concrete. Monitoring, Prevention and Rehabilitation*, The European Federation of Corrosion Publication number 25 (eds J. Mietz, B. Elsener, and R. Polder), The Institute of Materials, London, pp. 184–197.
9 Virmani, Y.P. and Clemena, G.C. (1998) Corrosion Protection: Concrete Bridge, US Department of Transportation, Federal Highways Administration, Report FHWA-RD-98-088.
10 Chaudhary, Z. and Chadwick, J.R. (1998) Cathodic protection of buried reinforced concrete structures, in *Corrosion of Reinforcement in Concrete. Monitoring, Prevention and Rehabilitation*, The European Federation of Corrosion Publication number 25 (eds J. Mietz, B. Elsener, and R. Polder), The Institute of Materials, London, pp. 198–205.
11 Schuten, G., Leggedoor, J., and Polder, R.B. (2000) Cathodic protection of concrete ground floor elements with mixed in chloride, in *Corrosion of Reinforcement in Concrete. Corrosion Mechanisms and Corrosion Protection*, The European Federation of Corrosion Publication number 31 (eds J. Mietz, B. Elsener, and R. Polder), The Institute of Materials, London, pp. 85–92.
12 Grefstad, K. (2005) Cathodic protection applied on Norwegian concrete bridges. Experience and recommendations, Eurocorr05, Lisboa, 2005 (CD-ROM).
13 Nerland, O.C.N., Eri, J., Grefstad, K.A., and Vennesland, Ø. (2007) 18 years of cathodic protection of reinforced concrete structures in Norway–facts and figures from 162 installations, EUROCORR 2007, Freiburg im Breisgau, Germany, 9–13 September 2007 (CD-ROM).
14 Wenk, F. and Oberhänsli, D. (2007) Long-term experience with cathodic protection of reinforced concrete structures, EUROCORR 2007, Freiburg im Breisgau, Germany, 9–13 September 2007 (CD-ROM).
15 Tinnea, J.S. and Cryer, C.B. (2008) Corrosion control of Pacific coast reinforced concrete structures: a Summary of 25 years' experience. 1st Int. Conf. on Heritage and Construction in Coastal and Marine Environment, MEDACHS08, Portugal, 2008 (CD-ROM).
16 Christodoulou, C., Glass, G., Webb, J., Austin, S., and Goodier, C. (2010) Assessing the long term benefits of

Impressed Current Cathodic Protection. *Corrosion Science*, **52** (8), 2671–2679.
17 Mietz, J., Fischer, J., and Isecke, B. (2001) Cathodic protection of steel-reinforced concrete structures – results from 15 years' experience. *Materials Performance*, **40** (12), 22–26.
18 EN 12696 (2000) *Cathodic Protection of Steel in Concrete – Part I: Atmospherically Exposed Concrete*, European Committee for Standardization.
19 NACE, Standard RP0290-90 (1990) Cathodic protection of reinforcing steel in atmospherically exposed structures.
20 The Concrete Society (2011) Cathodic protection of steel in concrete, Technical Report No. 73.
21 EN-ISO 12696 (2012) Cathodic protection of steel in concrete, CEN/ISO.
22 Pedeferri, P. (1993) Cathodic prevention and protection of reinforced and prestressed concrete structures. *L'Edilizia*, **XII** (10), 69–81. (in Italian).
23 Pedeferri, P. (1989) Cathodic protection of reinforced concrete structures. in Proc. Int. Conf. on Internal and External Protection of Pipes, BHRA, Florence, October 23, 1989.
24 Pedeferri, P. (1992) Cathodic protection of new concrete construction. in Proc. Int. Conf. on Structural Improvement through Corrosion Protection of Reinforced Concrete, Institute of Corrosion, Conf. Docum. E7190, London, 2–3 June 1992.
25 Bertolini, L., Bolzoni, F., Gastaldi, M., Pastore, T., Pedeferri, P., and Redaelli, E. (2009) Effects of cathodic prevention on the chloride threshold for steel corrosion in concrete. *Electrochimica Acta*, **54** (5), 1452–1463.
26 Bertolini, L. Cathodic prevention. in Proc. COST 521, Workshop, 28–31 August, (eds D. Sloan and P.A.M. Basheer), The Queen's University Belfast, 2000.
27 COST 521 (2003) Corrosion of Steel in Reinforced Concrete Structures, Final Report, (eds R. Cigna, C. Andrade, U. Nürnberger, R. Polder, R. Weydert, and E. Seitz) European Commission, Directorate General for Research, EUR 20599, 2003.
28 Sergi, G. and Page, C.L. (1995) Advances in electrochemical rehabilitation techniques for reinforced concrete. in Proc. UK Corrosion 95, 21–23 November, 1995.
29 Sergi, G. and Page, C.L. (2000) Sacrificial anodes for cathodic prevention of reinforcing steel around patch repairs applied to chloride-contaminated concrete, in *Corrosion of Reinforcement in Concrete. Corrosion Mechanisms and Corrosion Protection*, The European Federation of Corrosion Publication number 31 (eds J. Mietz, B. Elsener, and R. Polder), The Institute of Materials, London, pp. 93–100.
30 Sergi, G. (2010) Ten-year results of galvanic sacrificial anodes in steel reinforced concrete. *Materials and Corrosion*, **62** (2), 98–104.
31 Page, C.L. (2002) Advances in understanding and techniques for controlling reinforcement corrosion. ICC 15th International Corrosion Congress, Granada, 22–27 September, 2002 (CD-ROM).
32 Neville, A.M. (1995) *Properties of Concrete*, 4th edn, Longman Group Limited, Harlow.
33 Mietz, J. (1998) Electrochemical Rehabilitation Methods for Reinforced Concrete Structures – A State of the Art Report, European Federation of Corrosion Publication number 24, IOM Communications, London.
34 Vennesland, O. and Opsahl, O. A. (1986) Noteby, European Patent Application No. 86 30 2888.2.
35 Tritthart, J. (1995) Changes in the composition of pore solution and solids during electrochemical chloride removal in contaminated concrete, in *Proc. 2nd CANMET/ACI International Symposium on Advances in Concrete Technology*, Las Vegas, SP154-8, pp. 127–143.
36 CEN/TS 14038-2 (2011) Electrochemical Realkalisation and Chloride Extraction for Reinforced Concrete. Part 2: Chloride Extraction, European Technical Specification.
37 COST 509 (1997) Corrosion and Protection of Metals in Contact with Concrete, Final report (eds R.N. Cox,

R. Cigna, O. Vennesland, and T. Valente), European Commission, Directorate General Science, Research and Development, Brussels, EUR 17608 EN.

38 (1987) Noteby, Norwegian Patent Application No. 875438.

39 Pollet, V., Guerin, R., Tourneur, C., Mahouche, H., and Raharinaivo, A. (1997) Concrete realkalisation using sacrificial anode, in *Proc. Eurocorr'97*, vol. 1, Trondheim, pp. 523–528.

40 Raharinaivo, A., Lenglet, J.C., Tourneur, C., Mahouche, H., and Pollet, V. (1998) Chloride removal and realkalisation of concrete by using galvanic anodes. Int. Conf. on Corrosion and Rehabilitation of Reinforced Concrete Structures, Federal Highway Administration, Orlando, 7–11 December 1998.

41 Bertolini, L., Carsana, M., and Redaelli, E. (2008) Conservation of historical reinforced concrete structures damaged by carbonation induced corrosion by means of electrochemical realkalisation. *Journal of Cultural Heritage*, **9** (4), 376–385.

42 Bertolini, L., Redaelli, E., Lattanzi, D., and Mapelli, M. (2010) Conservation techniques on the reinforced concrete pillars of the church of the San Carlo Borromeo hospital in Milan. *European Journal of Environmental and Civil Engineering*, **14** (4), 411–425.

43 CEN/TS 14038-1 (2004) Electrochemical Realkalisation and Chloride Extraction for Reinforced Concrete. Part 1: Realkalisation, European Technical Specification.

44 Page, C.L. (1992) Interfacial effects of electrochemical protection method applied to steel in chloride containing concrete, in *Proc. of the Int. RILEM/CSIRO/ACRA Conf. Rehabilitation of Concrete Structures* (eds D.W.S. Ho and F. Collins), Melbourne, 31 August–2 September, pp. 179–187.

45 COST 534 (2009) New Materials, Systems, Methods and Concepts for Prestressed Concrete Structures, Final Report (eds R. Polder, M.C. Alonso, D. Cleland, B. Elsener, E. Proverbio, Ø. Vennesland, and A. Raharinaivo) European Commission, Cost office.

46 van der Vaart, F., de Moor, W., and Polder, R.B. (2009) Electrical current protects reinforcement, Renovation of De Muinck Keizer bridge using cathodic protection. *Cement*, **1**, 62–66. (in Dutch).

47 Sergi, G., Page, C.L., and Thompson, D.M. (1991) Electrochemical induction of alkali-silica reaction in concrete. *Materials and Structures*, **24**, 359–361.

48 Bertolini, L., Yu, S.W., and Page, C.L. (1996) Effects of electrochemical chloride extraction on chemical and mechanical properties of hydrated cement paste. *Advances in Cement Research*, **8** (31), 93–100.

49 Page, C.L. and Yu, S.W. (1995) Potential effects of electrochemical desalination of concrete on alkali-silica reaction. *Magazine of Concrete Research*, **47** (170), 23–31.

50 Banfill, P.F.G. and Al-Kadhimi, T.K.H. (1998) The effects of re-alkalisation and chloride removal on alkali-silica expansion in concrete. Eurocorr '98, European Federation of Corrosion, Utrecht, 28 September–1 October 1998 (CD-ROM).

51 Mussinelli, G., Pedeferri, P., and Tettamanti, M. (1987) The effect of current density on anode behaviour and on concrete in the anode region. 2nd Int. Conf. on Deterioration and repair of reinforced concrete in the Arabian Gulf, Bahrain, 11–13 October 1987.

52 Polder, R.B., Peelen, W., Leggedoor, J., and Schuten, G. (2007) Microscopy study of the interface between concrete and conductive coating used as concrete CP anode, in *Corrosion of Reinforcement in Concrete. Mechanisms, Monitoring, Inhibitors and Rehabilitation Techniques*, European Federation of Corrosion Publication number 38 (eds M. Raupach, B. Elsener, R. Polder, and J. Mietz), Woodhead Publishing Limited, Cambridge, pp. 277–287.

53 Polder, R.B. and Peelen, W.H.A. (2011) Service life aspects of cathodic protection of concrete structures, in *Concrete Repair, A Practical Guide* (ed. M. Grantham), Taylor and Francis, Abingdon, pp. 248–261.

54 Schuten, G., Leggedoor, J., Polder, R.B., and Peelen, W. (2007) Renovation of the

cathodic protection system of a concrete bridge after 12 years of operation, in *Corrosion of Reinforcement in Concrete. Mechanisms, Monitoring, Inhibitors and Rehabilitation Techniques*, European Federation of Corrosion Publication number 38 (eds M. Raupach, B. Elsener, R. Polder, and J. Mietz), Woodhead Publishing Limited, Cambridge, pp. 300–306.

55 Glass, G.K., Hassanein, A.M., and Buenfeld, N.R. (2001) Cathodic protection afforded by an intermittent current applied to reinforced concrete. *Corrosion Science*, **43**, 1111–1131.

56 Bertolini, L., Bolzoni, F., Pastore, T., and Pedeferri, P. (1997) Three year tests on cathodic prevention of reinforced concrete structures. Int. Conf. Corrosion/97, paper 244, NACE, Houston, 1997.

57 Bertolini, L., Pedeferri, P., Redaelli, E., and Pastore, T. (2003) Repassivation of steel in carbonated concrete induced by cathodic protection. *Materials and Corrosion*, **54**, 163–175.

58 Redaelli, E. and Bertolini, L. (2011) Electrochemical repair techniques in carbonated concrete. Part II: cathodic protection. *Journal of Applied Electrochemistry*, **41** (7), 829–837.

59 Klisowski, S. and Hartt, W.H. (1996) Qualification of cathodic protection for corrosion control of pretensioned tendons in concrete, in *Corrosion of Reinforcement in Concrete Construction* (eds C.L. Page, P.B. Bamforth, and J.W. Figgs), The Royal Society of Chemistry, Cambridge, pp. 354–368.

60 Pastore, T., Pedeferri, P., Bertolini, L., and Bolzoni, F. (1992) Current distribution problems in the cathodic protection of reinforced concrete structures, in *Proc. Int. RILEM Conf. Rehabilitation of Concrete Structures* (eds D.W.S. Ho and F. Collins), Melbourne, pp. 189–200.

61 Bertolini, L., Bolzoni, F., Cigada, A., Pastore, T., and Pedeferri, P. (1993) Cathodic protection of new and old reinforced concrete structures. *Corrosion Science*, **35**, 1633–1639.

62 Bazzoni, A., Bazzoni, B., Lazzari, L., Bertolini, L., and Pedeferri, P. (1996) Field application of cathodic prevention on reinforced concrete structures, Corrosion/96, NACE, Houston, paper 312, 1996.

63 Mudd, C.J., Mussinelli, G., Tettamanti, M., and Pedeferri, P. (1988) Cathodic protection of steel in concrete. *Materials Performance*, **27** (9), 18.

64 Bertolini, L., Bolzoni, F., Pastore, T., and Pedeferri, P. (2004) Effectiveness of a conductive cementitious mortar anode for cathodic protection of steel in concrete. *Cement and Concrete Research*, **34** (4), 681–694.

65 Bennet, J.E. and Firlotte, C. (1996) A zinc/hydrogel system for cathodic protection of reinforced concrete structures, Corrosion/96, Nace, paper 316, 1996.

66 Peelen, W.H.A. and Polder, R.B. (2002) Throwing power of the zinc-hydrogel anode for sacrificial protection of steel in concrete. ICC 15th Int. Corrosion Congress, Granada, 22–27 September 2002 (CD-ROM).

67 Raupach, M. and Bruns, M. (2002) Effectiveness of a zinc-hydrogel anode for sacrificial cathodic protection or reinforced concrete structures. ICC 15th Int. Corrosion Congress, Granada, 22–27 September 2002 (CD-ROM).

68 Nerland, O.C. and Polder, R.B. (2002) Cathodic protection of RC structures–far more than bridge decks. ICC 15th Int. Corrosion Congress, Granada, 22–27 September 2002 (CD-ROM).

69 Redaelli, E., Lollini, F., and Bertolini, L. (2013) Throwing power of localised anodes for the cathodic protection of slender carbonated concrete elements in atmospheric conditions. *Construction and Building Materials*, **39**, 95–104.

70 Polder, R., Andrade, C., Elsener, B., Vennesland, Ø., Gulikers, J., Weydert, R., and Raupach, M. (2000) Test methods for on site measurement of resistivity of concrete. *Materials and Structures*, **33** (10), 603–611.

71 Bennet, J.E. and Mitchell, T.A. (1989) Depolarization testing of cathodically protected reinforcing steel in concrete, Corrosion/89, NACE, paper 373, 1989.

72 Linder, B. (1994) Cathodic protection criteria for passivated metals, in Proc. Corrosion and prevention '94, Glenelg, South Australia, 1994.

73 Broomfield, J.P. (2000) Cathodic protection of reinforced concrete structures and a case study of a multiple anode trial. *International Journal for Restoration of Buildings and Monuments*, **6** (6), 619–630.

74 Polder, R.B., Leegwater, G., Worm, D., and Courage, W. (2012) Working life modelling of cathodic protection systems for concrete structures – analysis of field data, in *Proc. International Conference on Concrete Repair, Rehabilitation and Retrofitting III, Cape Town* (eds M. Alexander et al.), Taylor & Francis, London, pp. 504–510.

75 Polder, R.B., Peelen, W.H.A., Lollini, F., Redaelli, E., and Bertolini, L. (2009) Numerical design for cathodic protection systems for concrete. *Materials and Corrosion*, **60** (2), 130–136.

76 Pacheco, J., Polder, R.B., Fraaij, A.L.A., and Mol, J.M.C. (2011) Short-term benefits of cathodic protection of steel in concrete, in *Proc. Concrete Solutions, Dresden, 26–28 September* (eds M. Grantham, V. Mechtcherine, and U. Schneck), Taylor and Francis, pp. 147–156.

77 Polder, R.B., Peelen, W.H.A., Stoop, B.J.T., and Neeft, E.A.C. (2011) Early stage beneficial effects of cathodic protection in concrete structures. *Materials and Corrosion*, **62** (2), 105–110.

78 Bruns, M. and Raupach, M. (2010) Protection of the opposite reinforcement layer of RC-structures by CP – results of numerical simulations. *Materials and Corrosion*, **61** (6), 505–511.

79 Bertolini, L., Gastaldi, M., Pedeferri, M., and Redaelli, E. (2002) Prevention of steel corrosion in concrete exposed to seawater with submerged sacrificial anodes. *Corrosion Science*, **44** (7), 1497–1513.

80 Bertolini, L. and Redaelli, E. (2009) Throwing power of cathodic prevention applied by means of sacrificial anodes to partially submerged marine reinforced concrete piles: results of numerical simulations. *Corrosion Science*, **51** (9), 2218–2230.

81 Elsener, B., Molina, M., and Böhni, H. (1993) Electrochemical removal of chlorides from reinforced concrete structures. *Corrosion Science*, **35** (5–8), 1563–1570.

82 Tritthart, J., Pettersson, K., and Sorensen, B. (1993) Electrochemical removal of chloride from hardened cement paste. *Cement and Concrete Research*, **23** (5), 1095–1104.

83 Mietz, J. and Isecke, B. (1994) Investigation on electrochemical realkalization for carbonated concrete, Corrosion/94, NACE, paper 297, 1994.

84 Kennedy, D., Miller, J.B., and Nustad, G.E. (1994) Review of chloride extraction and realkalisation of reinforced concrete, in *Corrosion and Corrosion Protection of Steel in Concrete* (ed. R.N. Swamy), Sheffield Academic Press, Sheffield, p. 1449.

85 Schmid, M. Electrochemical chloride removal: site example, *SIA tec21* **127** (31–32), 21–23, 2001. (in German).

86 Elsener, B. and Böhni, H. (1996) Electrochemical rehabilitation methods – progress and new results, in *Maintenance of Bridges – Actual Research Results*, SIA Documentation D0129, Schweizer Ingenieur und Architektenverein, Zürich, pp. 47–59. (in German).

87 Raharinaivo, A., Lenglet, J.C., Tourneur, C., Mahouche, H., and Pollet, V. (1998) Chloride removal from concrete by using sacrificial anode. Eurocorr '98, European Federation of Corrosion, Utrecht, 28 September–1 October 1998 (CD-ROM).

88 Polder, R.B. and van den Hondel, A.W.M. (2002) Laboratory investigation of electrochemical chloride extraction of concrete with penetrated chloride. *Heron*, **47** (3), 211–220.

89 Manning, D.G. (1990) Electrochemical removal of chloride ions from concrete. Elektrochemische Schutzverfahren für Stahlbetonbauwerke, Schweizerischer Ingenieur-und Architekten-Verein (SIA), Dokumentation D 065, Zürich, 61–67.

90 Vennesland, Ø., Humstad, E.P., Gautefall, O., and Nustad, G. (1996)

Electrochemical removal of chlorides from concrete – effect on bond strength and removal efficiency, in *Proc. 4th Int. Symp. Corrosion of Reinforcement in Concrete Construction* (eds C.L. Page, P.B. Bamforth, and J.W. Figg), Society of Chemical Industry, Cambridge, pp. 448–455.

91 Polder, R.B. (1996) Electrochemical chloride removal from reinforced concrete prisms containing chloride penetrated from sea water. *Construction and Building Materials*, **10** (1), 83–88.

92 Bennett, J.E. and Schue, T.J. (1990) Electrochemical chloride removal from concrete: a SHRP contract status report, NACE, Corrosion/90, paper 316, 1990.

93 Elsener, B. (2008) Long term durability of electrochemical chloride extraction. *Materials and Corrosion*, **59** (2), 91–97.

94 Polder, R.B., Walker, R.J., and Page, C.L. (1995) Electrochemical desalination of cores from a reinforced concrete coastal structure. *Magazine of Concrete Research*, **47** (173), 321–327.

95 Andrade, C., Diez, J.M., Alamán, A., and Alonso, C. (1995) Mathematical modeling of electrochemical chloride extraction from concrete. *Cement and Concrete Research*, **25** (4), 727–740.

96 Castellote, M., Andrade, C., and Alonso, C. (1999) Electrochemical chloride extraction: influence of testing conditions and mathematical modelling. *Advances in Cement Research*, **11** (2), 63–80.

97 Elsener, B., Zimmermann, L., Bürchler, D., and Böhni, H. (1998) Repair of reinforced concrete structures by electrochemical techniques – field experience, in *Corrosion of Steel in Concrete*, European Federation of Corrosion Publication number 25 (eds. J. Mietz, B. Elsener, and R. Polder), The Institute of Materials, London, pp. 125–140.

98 Elsener, B., Zimmermann, L., and Böhni, H. (2003) Nondestructive determination of the free chloride content in cement based materials. *Materials and Corrosion*, **54** (6), 440–446.

99 Angst, U. (2005) *Electrochemical chloride extraction – efficiency measured with chloride sensors and effect on the steel–concrete interface*, Master Thesis, Institute of Building Materials (IfB) ETH Zurich, January 2005.

100 Elsner, B. and Angst, U. (2007) Mechanism of electrochemical chloride removal. *Corrosion Science*, **49** (12), 4504–4522.

101 Stoop, B.T.J. and Polder, R.B. (1996) Redistribution of chloride after electrochemical chloride removal from reinforced concrete prisms, in *Proc. 4th Int. Symp. on Corrosion of Reinforcement in Concrete Construction* (eds C.L. Page, P.B. Bamforth, and J.W. Figg), Society of Chemical Industry, Cambridge, UK, 1–4 July, pp. 456–465.

102 Polder, R.B. and Larbi, J.A. (1995) Investigation of concrete exposed to North Sea water submersion for 16 years. *Heron*, **40** (1), 31–56.

103 Elsener, B. (2001) Half-cell potential mapping to assess repair work on RC structures. *Construction and Building Materials*, **15** (2–3), 133–139.

104 Banfill, P.F.G. (1999) Electrochemical re-alkalisation, chloride removal and alkali-silica reaction in concrete. Annual Progress Report 1998–1999, COST 521 Workshop, Annecy, 1999.

105 Raharinaivo, A., Grimaldi, G., and Marie-Victoire, E. (2002) Effectiveness of realkalisation and chloride removal. Final report COST 521 project F2, Luxembourg, 18–19 February 2002.

106 Chang, J.J. (2002) A study of bond degradation of rebar due to cathodic protection current. *Cement and Concrete Research*, **32** (4), 657–663.

107 Castellote, M., Andrade, C., and Alonso, C. (1999) Changes in concrete pore size distribution due to electrochemical chloride migration trials. *ACI Materials Journal*, **96** (3), 314–319.

108 Polder, R.B. and van den Hondel, H.J. (1992) Electrochemical realkalization and chloride removal of concrete, in *Proc. of the Int. RILEM/CSIRO/ACRA Conf. Rehabilitation of Concrete Structures* (eds D.W.S. Ho and F. Collins), Melbourne, 31 August– 2 September, pp. 135–147.

109 Pollet, V. (1999) Re-alkalisation: specification for the treatment

application and acceptance criteria. Annual Progress Report 1998–1999, COST 521 Workshop, Annecy, 1999.

110 Andrade, C., Castellote, M., and Alonso, C. (1998) An overview of electrochemical realkalisation and chloride extraction, in *Proc. of the 2nd Int. RILEM/CSIRO/ACRA Conf. Rehabilitation of Structures* (eds D.W.S. Ho, I. Godson, and F. Collins), Melbourne, pp. 1–12.

111 Andrade, C., Castellote, M., Sarría, J., and Alonso, C. (1999) Evolution of pore solution chemistry, electro-osmosis and rebar corrosion rate induced by realkalisation. *Materials and Structures*, **32** (6), 427–436.

112 Banfill, P.F.G. (1994) Features of the mechanism of realkalisation and desalination treatments for reinforced concrete, in *Proc. Int. Conf. on Corrosion and Corrosion Protection of Steel in Concrete* (ed. R.N. Swamy), Sheffield Academic Press, 24–29 July, pp. 1489–1498.

113 Bertolini, L., Coppola, L., Gastaldi, M., and Redaelli, E. (2009) Electroosmotic transport in porous construction materials and dehumidification of masonry. *Construction and Building Materials*, **23** (1), 254–263.

114 Pollet, V. and Dieryck, V. (2000) Re-alkalisation: specification for the treatment application and acceptance criteria. Annual Progress Report 1999–2000, COST 521 Workshop, Belfast, 2000.

115 van den Hondel, A.W.M. and Polder, R.B. (1998) Laboratory investigation of electrochemical realkalisation of reinforced concrete. Eurocorr '98, European Federation of Corrosion, Utrecht, 28 September–1 October 1998 (CD-Rom).

116 Sergi, G., Walker, R.J., and Page, C.L. (1996) Mechanisms and criteria for the realkalisation of concrete, in *Proc. 4th Int. Symp. Corrosion of Reinforcement in Concrete Construction* (eds C.L. Page, P.B. Bamforth, and J.W. Figg), Society of Chemical Industry, Cambridge, pp. 491–500.

117 Rossi, A., Puddu, G., and Elsener, B. (2001) The surface of iron and Fe10Cr alloys in alkaline media. in Proc. Eurocorr 2001, Riva del Garda 30 September–4 October 2001 (CD-ROM).

118 Redaelli, E. and Bertolini, L. (2011) Electrochemical repair techniques in carbonated concrete. Part I: electrochemical realkalisation. *Journal of Applied Electrochemistry*, **41** (7), 817–827.

119 Odden, L. (1994) The repassivating effect of electro-chemical realkalisation and chloride removal, in *Proc. Int. Conf. on Corrosion and Corrosion Protection of Steel in Concrete* (ed. R.N. Swamy), Sheffield Academic Press, 24–29 July, pp. 1473–1488.

120 Miller, J. (1998) The perception of the ASR problem with particular reference to electrochemical treatments of reinforced concrete, in *Corrosion of Steel in Concrete*, European Federation of Corrosion Publication number 25 (eds J. Mietz, B. Elsener, and R. Polder), The Institute of Materials, London, pp. 141–149.

Index

Page numbers in italics refer to figures and tables.

100 mV decay criterion   375

A
AC impedance   325
AC resistance   326
AC stray current   141, 149, 150
ACI 201 recommendation   60
ACI 308.1 recommendation   218
acid attack   54, 55
acid gases   79
acid rain   58
acoustic emission monitoring methods   328
active monitoring   325
admixed inhibitors   234
admixtures   205, 206
adsorbed water   26
adsorption inhibitors   228
aggregates   204, 205
air voids   3
alkali aggregate reaction   371
alkali content/alkalinity   62, 63, 79, 95, 96
– influence on alkali silica reaction   64, 65
– influence on carbonation   85
– influence on corrosion (aluminum)   75, 76
– influence on corrosion (steel)   71
alkali oxides   62
alkali silica reaction (ASR)   61–66, 399
alkanolamine-based inhibitors   232, 233, 235, 236
aluminum   75, 76, 380
amines   232
aminoalcohol   235–237
aminocarboxylate   235
ammonium attack   58
ammonium sulfate   60
anaerobic conditions   57, 58
anode systems   379–381

anode-ladder system   317
anodic acidification   371, 372
anodic polarization   116–119, 122, *123*, *125*, *127*
– macrocells   135, 137
anodic process   113, 115, 117, 126, 129, 338, 342
anodic resistance control   121
anodic stress corrosion cracking   156
apparent diffusion coefficient   31, 101, 102, 106–108, 190
Arrhenius equation   38
asset management   287
ASTM C876 standard   296–298
atmospherically exposed concrete   26, 37
atmospherically exposed structures
– carbonation-induced corrosion   86, 89
– chloride-induced corrosion   94, 100, 108, 109, *123*, *124*
– macrocells   129–131
atomic hydrogen, *see* hydrogen
AutoClam instrument   33
average corrosion rate   303

B
bacteria   56
Bakker's formula   87
binding   30–32
biogenic sulfuric acid (BSA) attack   56–58
biological attack   49
black rust   75
blast furnace slag   10
blast furnace slag cement/concrete, *see also* ground granulated blast furnace slag   10–13, *14*, *15*, *16*, *17*, 204, 223
– alkali content   63
– alkali silica reaction   62, 64–66

Corrosion of Steel in Concrete: Prevention, Diagnosis, Repair, Second Edition. Luca Bertolini, Bernhard Elsener, Pietro Pedeferri, Elena Redaelli, and Rob Polder.
© 2013 Wiley-VCH Verlag GmbH & Co. KGaA. Published 2013 by Wiley-VCH Verlag GmbH & Co. KGaA.

- biogenic sulfuric acid attack 57
- carbonation 80, 85
- chloride-induced corrosion 99, 100, 102, 105, *106*, 107, *108*, 109, 223
- pore solution 23, *24*, *25*
- pure water attack 58
- seawater attack 67
- transport processes *31*, *37*, 38, 41–44

blended cements/blended cement concrete 8–13
- alkali content 63
- alkali silica reaction 65
- carbonation 79, 80, 84, 85
- chloride-induced corrosion 99, 100, 108, 109
- pore solution 23
- sulfate attack 60
- transport processes 40

bonding agents 356
bound chloride 98–100
brucite (Mg(OH)$_2$) 67
buried structures 131–134

C

calcium aluminate (Ca$_3$Al) 1, 59, 67
calcium carbonate (CaCO$_3$) 55, 58
calcium chloride 23, 90
calcium ferroaluminate (Ca$_4$AlFe) 1
calcium hydroxide (Ca(OH)$_2$) 2, 55, 58, 59
- carbonation 79, 80
calcium (Ca^{2+}) ions 23, 24, 25, 36, 58
calcium nitrite 229–231
calcium oxide 85
calcium silicate hydrate gel (C–S–H) 1, 2
- carbonation 79, 80
calcium silicates (Ca$_2$Si, Ca$_3$Si) 1–3
calcium sulfoaluminate (CSA) cements 19
calomel reference electrode (SCE) 116
capillary absorption parameter 32, 33, 40, *41*
capillary action 21, 32, 33
capillary pores 3, 4, 7, 8
- water content 23–27, 51
carbon dioxide (CO$_2$) 31, 55, 58, 72, 79, 249
carbon dioxide (CO$_2$) emissions 11, 19
carbon steel 270, 273–275
carbonated concrete/structures
- cathodic protection 377–379
- hydrogen evolution 167
carbonation, *see also* carbonation-induced corrosion 72, 73, 79–85
- measurement 307, 308
carbonation coefficient 81

carbonation-induced corrosion 85–91, 172, *184*
- effect on service life 180
- electrochemical aspects 120–122
- repair 339–342, 351, 352
cathodic polarization 118–120, 127
- macrocells 135–137
cathodic prevention (CPre) 365, 366, 368, *372*, *374*, 376, 377
cathodic process 113, 115, 118, 119, 126, 129
cathodic protection (CP) 168, 169, 338, 341, 345, 365, 367, *372*, *374*
- carbonated concrete 377–379
- chloride-contaminated concrete 373–376
- design of system 380, 381
- maintenance 384
- monitoring 383
- service life of system 384, 385
CDF test 51
Cembureau method 35
cement, *see also* blended cements, portland cement 203, 204
cementitious coatings 258
chemical attack 49, 54–56
chemically combined water 26
chloride contamination, *see* chloride-contaminated concrete/structures
chloride content 98–100, *186*, 188
- measurement 323
chloride (Cl$^-$) ions, *see also* bound chloride
- concentration *24*, *25*
- diffusion 28–32, 101, 102
- diffusion coefficient 30, 31, 40–43, *222*
- electrochemical extraction/removal 345, 368, 369, 372, 387–394, 399
- influence on corrosion 72, 93, 122
- migration 36, 38
chloride migration coefficient 191
chloride-contaminated concrete/structures 90, 91, 187–189
- cathodic protection 373–376
- influence on hydrogen evolution 167, 168
- influence on macrocells 129–133, 137, 138
- influence on stray-current-induced corrosion 141, *144*, *146*, 147–149, *150*, 151
- measurement 308–310
chloride-induced corrosion (pitting corrosion) 73, 93–109, *184*
- detection 176
- effect on service life 180
- electrochemical aspects 122–126

– influence of macrocells   130
– influence of stray currents   149
– prevention
–– use of galvanized steel   279
–– use of inhibitors   233, 235
–– use of stainless steel   269–274
– repair   342–346, 352, 353
circulation of current   369, 373
classification, cements   13–15, *16–18*
cleaning, surfaces   356
clinker   1, 8
coated rebars   130
coating as a repair intervention   342, 346, 361
cold-drawn/cold-worked steel   164, 264, 266
combined chloride/resistivity sensors   318
compacting factor test   208
complex geometries   219
composite cement   15
compressive strength   7, 15, *18*, 40, 41, *42*, 53, 208–210
conductive coatings   380, 385
conventional repair   349–362
copper   77, 130, 131
corrosion inhibitors   227–240, 346, 360
corrosion potential   119, 121, 134, 292, 295
– measurement   322
corrosion products   74, 75, 88
corrosion rate   39, 73, 74, 114, 115, *116*, 119
– carbonation-induced corrosion   87–91, 121
– chloride-induced corrosion   96, 97, 108, 109
– control/reduction   337, 342
– influence of macrocells   130, 133, 134, 138, 281
– measurement   302, 303
cracks   173–175
creep   212
critical content (threshold value), chlorides   93, 94, 96–100, 102, 123, 187, 272, 353
critical size, cracks   160
critical stress   161
curative applications, corrosion inhibitors   229
curing, cement   4, 6, *8*, 214, 215, 217, 218
– influence on carbonation rate   84, 85
– influence on chloride-induced corrosion   105
– influence on compressive strength   209

– influence on concrete deterioration
–– freeze–thaw attack   53
–– sulfate attack   61
– influence on transport processes   38, *43*

D

Darcy's law   33, 34, 51
DC stray current   72, 141–149
deemed-to-satisfy rules   178
deicing agents   50, *54*, 93, 132, 188
delayed ettringite formation (DEF)   59–61
delayed fracture   157
depassivation   71
design details   219, *220*
designed concrete   213
detailed surveys   287
DIB test   165
diffusion   21, 27–32, 40–43
diffusion coefficient   31, *41*, 221
DIN 1048 standard   34
driving voltage, stray currents   142, 144–147, 151, 153
durability   177, 219
– chloride extraction   390–394
– realkalization   398, 399
DuraCrete method   179, 189, 195

E

elastic deformation   212
electrical conductivity/resistivity   12, 13, *14*, 37–40
– correlations   41
– influence on corrosion rate   87, *88*, 109, 114, 115
– measurement   40, 41, 298–302
electro-osmosis   395
electrochemical aspects, corrosion   113–127
– macrocells   134–137
electrochemical chloride extraction/removal (ECR)   345, 368, 369, 372, 387–394, 399
electrochemical inhibitor injection (EII)   238
electrochemical inspection techniques   291–307
electrochemical realkalization, *see* realkalization
electrochemical techniques, corrosion controlling   365–399
embedded sensors   316
EN 197-1 standard   13, 15, *16–18*, 55, 221
EN 206-1 standard   53, 55, *56*, 58, 60, 93, 183–189, 197, 208, 210, 212, 214, 216
EN 450-1 standard   204
EN 934-2 standard   206

EN 1008 standard   205
EN 1504-2 standard   243–245, *246*, 248–250, 253, 255
EN 1504-3 standard   360
EN 1990 standard   181
EN 10088-1 standard   267
EN 12390-8 standard   34
EN 12620 standard   204
EN 13670 standard   218
entrapped air   4, 207
– influence on chloride-induced corrosion   98
– influence on freeze–thaw attack   53
environmental aggressiveness   172, 173
epoxy-coated rebars   130, 199, 280–283
ettringite   59, 60, 80
Eurocode 2 standard   187, 264
exposure classes   188
external sulfate attack   59, 60

F

ferroaluminate of calcium (Ca_4AlFe)   1
Ferrogard 903 inhibitor   237
fib Model Code   179, 189–194, 195
fiber-reinforced polymers (FRPs)   263, 361, 362
Fick's first law   28, 29
Fick's second law   29–31, 101, 180, 250
field tests, inhibitors   238
film-forming inhibitors   228
FIP standard test   164, *165*
flow test   208
fly ash (pulverized fly ash, PFA)   9, 10
fly ash (PFA) cement/concrete   10–13, *16, 17*
– alkali content   63
– carbonation   85
– chloride-induced corrosion   99, 100, 102, 105, 107, 109
– deterioration
– – alkali silica reaction   62, 65, 66
– – seawater attack   67
– pore solution   23, *24, 25*
– transport processes   38, 42–44, 221
fracture surface   162
fracture toughness   161
freeze–thaw attack   50–53, *54, 55*
frost resistance   52, 53

G

galvanic coupling   77
galvanized steel   199, 276–280
gas permeability coefficient   35

gel pores   3, 51
global inspection methods   327
ground granulated blast furnace slag (GGBS)   10, 224
ground granulated blast furnace slag cement/concrete, *see also* blast furnace slag cement/concrete   11–13, 221
– alkali silica reaction   64, 66
– carbonation   85
– chloride-induced corrosion   *107*, 187
– pore solution   *24, 25*
– transport processes   *37*, 221
ground limestone   11, 13, 224
ground systems   133, 134
ground water   54, *56*, 59, 137
grouts   155, 156
gypsum   1, 59

H

half-cell potential   240, 291–298, 397
hardening accelerators   206
heat-induced internal sulfate attack   60
high-alumina cement (HAC)   18
– biogenic sulfuric acid attack   57
high-level monitoring   315
high-performance concrete (HPC)   219, 223
high-pressure water jetting   355
high-strength steel   150, 223, 276, 379
– stress corrosion cracking   157–165
high-volume fly ash cement/concrete   10
hot-rolled steel   164, 264
humidity effects, *see also* relative humidity
– carbonation rate   81–84
– carbonation-induced corrosion   88–90, 172
– chloride-induced corrosion   108, 109
– composition of pore solution   23, 26, *27*
– concrete deterioration
– – alkali silica reaction   65
– – chemical attack   59
– – sulfate attack   61
– corrosion rate   73, 74, 115, *116*
– electrical resistivity   39, 300
– transport processes   26, 28, 35, 38, 39, 42
hydration
– blended cements   10, 11, *12*
– calcium sulfoaluminate cements   19
– high alumina cement   18
– portland cement   1–3, 9, 10
hydraulic pressure mechanism   51
hydrochloric acid   55
hydrogen embrittlement   73, 157, 164, 168, 169, 371, 379
hydrogen evolution   157, 165–169, 379

hydrogen (H$^+$) ions   36
hydrogen sulfide   56
hydrogen-induced stress corrosion cracking (HI-SCC)   156, 157, 162–166
hydrophobic treatments   251–257, 345, 346
hydroxyl (OH$^-$) ions   23, *24*, *25*, 36, 54, 61, 63, 65, 97

I

ice overpressure mechanism   51
ICP technique   309
immersed/submerged structures
– carbonation-induced corrosion   172
– chloride-induced corrosion   94, 100, 103, 109, 123
– corrosion rate   119
– macrocells   131–133
immunity conditions   116
inhibitor/chloride ratio   239, 240
inhibitors, *see* corrosion inhibitors
initiation, corrosion   71
– carbonation-induced corrosion   85–87
– chloride-induced corrosion   96, 102, *103*
– influence of macrocells   131, 133, 138
– stray-current-induced corrosion   144, *146*, 147, 148, 152
initiation time   71, 72, 180, *188*, 221
inspection   151, 152, 176, 177, 287–310
instantaneous corrosion rate   303
interfacial transition zone   6
interlayer water   26
intermediate-level monitoring   315
internal sulfate attack   59–61
ions
– concentration in pore solution   23, *24–5*
– migration   21, 35–40, 370

L

law of five   171, 189
lead   76, 77
limestone aggregates   61
limestone, addition   11, 221
linear polarization resistance (LPR) measurements   240, 323, 397
lithium compounds   66
load and resistance factor design (LRFD) method   195
loads   181
loss of bond strength   371
low-alloyed steels   266
low-heat cements   15
low-level monitoring   315

M

macrocell corrosion   130
macrocell current   129–134, 240
– measurement   319–321
macrocell sensor systems   317
macrocells   73, 129–138
– electrochemical aspects   134–137
magnesium sulfate   60
magnetite scale   266
map cracking   64
marine environments   103, *104*
mechanical deterioration   49
metakaolin   11
metal-hexafluorosilicates   257
microclimate   81, 82
migrating inhibitors   232
migration (ions)   21, 35–40, 370
mineral additions   9, 221
mix design   212–215
mixed-in inhibitors   233, 234
mixing water   205
mobility, ions   36
modeling
– carbonation   79, 80
– carbonation-induced corrosion   86, 87
– cathodic protection systems   386
– chloride induced corrosion   101
– macrocells   137, 138
– service life   177–183
– transport processes   37
moisture content   22, *41*, 115
– influence on carbonation-induced corrosion   87, 120, 121
– influence on chloride-induced corrosion   100
– influence on corrosion rate (aluminum)   76
– influence on corrosion rate (steel)   115
– reduction   341, 342
monitoring   315–328
– cathodic protection   383
– chloride extraction   392, 393, 399
– realkalization   399
motorways   *105*, 106, 107
multiring-electrode sensors   318

N

natural pozzolana   9, 221
NDT monitoring methods   328
Nernst's law   166, 373
nickel   77
nitric acid   55

O

ohmic control   115
on-off technique   151
opal   64
organic acids   55
organic coatings   245–251, 342, 346
organic inhibitors   233, 238
organic sealers   257, 258
osmotic pressure   51
oxygen
– evolution   117, 144, 371
– reduction   113, 118, 369
– transport   323, 324
oxygen diffusion control   115

P

Parrott's formula   87
passivators   228
passive control   115
passive film (steel)   71–73
passive film (zinc)   278
passive monitoring   325
passivity conditions   117, 120, 123, 125
passivity range   117, 122
patch repair   130, 339, 340
Pedeferri's diagram   123, *124*, 125, 373
percolation theory   37
percolation threshold   8
performance-based design   177, 179–83
permeability   7, 8, 21, 58, 60
permeability coefficient, air   41
permeability coefficient, gas   35, *41*
permeability coefficient, oxygen   35
permeability coefficient, water   34, 35, 40, *41*
permeation   21, 33–35
pH levels, pore solution   23, 54
– indicator liquids   396, 397
– influence of carbonation   72, 79, 80, 120, 271
– influence on chloride-induced corrosion   96, 99, 100, 122
– influence on hydrogen evolution   166
– influence on passivity (aluminum)   75, 76
– influence on passivity (steel)   72
– measurement   323
phenolphthalein   80
physical deterioration, concrete   49
pitting corrosion, *see* chloride-induced corrosion
pitting potential   122
polarization   72
polarization curves   115, 116
polarization resistance method   304–307
polymer-modified mortars   359
pore blocking   257, 258
pore fillers   258
pore solution   22, 23, *24*, *25*, 71
porosity, cement   3–8, 10, 18
– influence of carbonation   80, 84
– influence on carbonation-induced corrosion   86, 120
– influence on corrosion rate   115
– influence on freeze–thaw attack   53
– influence on transport processes   29, 34
portland cement/portland cement concrete   1, 8, 11, 15
– alkali content   63
– alkali silica reaction   65, 66
– biogenic sulfuric acid attack   57
– carbonation   79, 80, *83*, 85
– carbonation-induced corrosion   86
– chloride-induced corrosion   99, 100, 107, 109
– freeze–thaw attack   50
– microstructure   *12*
– pore solution   23, *24*, *25*
– production process   1–3
– seawater attack   67
– transport processes   *37*, *38*, 40, 44
portland-composite cements   15
post-tensioning systems   325
post-tensioning tendons, *see* tendons
potassium hydroxide (KOH)   2
potassium (K$^+$) ions   23, *24*, *25*, 36, 61
pozzolana   9
pozzolanic cement/concrete   9–12, 15, 16–*17*, 204, 221
– alkali content   63
– alkali silica reaction   64
– chloride-induced corrosion   187
pozzolanic materials   9, 10, 224
pozzolanic reaction   9
preliminary surveys   287
prescribed concrete   213
prescriptive design   177, 178
prestressed concrete/structures   155–170
prestressing steel   155, 164, *165*, 169, 170, 264–266
preventative applications, corrosion inhibitors   229
preventative measures   197, 198
– alkali silica reaction   65, 66
– biogenic sulfuric acid attack   57, 58
– corrosion   197
– cost   198–200
– hydrogen embrittlement   168–170

– stray-current-induced corrosion   152, 153
– sulfate attack   60, 61
proactive conservation   176
proactive repair interventions   333
propagation phase   71, *72*, 73
protection, *see* preventative measures
protection potential   125
pulverized fly ash, *see* fly ash
pure water attack   58

Q

quality, concrete   173
quartz   64
quasi-immunity conditions   117
quenched steel   164, 264

R

rate of corrosion, *see* corrosion rate
reactive repair interventions   333
ready-mixed concrete   215, 216
realkalization (RE)   340, 341, 365, 369, 372, 394–399
recycled glass   11
reference electrodes   116, 292, 322
reinforcing bars (rebars)   263, 264
relative humidity (RH), *see also* humidity effects   321, 322
reliability index   182
removal, concrete   350–356
repair interventions, *see also* conventional repair   333–347
repair materials   357–360, 382, 383
repassivation
– carbonated structures   339–341
– chloride-contaminated structures   343–345
resistivity, *see* electrical conductivity
rice husk ash   11
RILEM recommendations   39, 62, 301, 333, 334, *337*

S

sacrificial cathodic protection (SCP)   367
San Bernardino curve   297
seawater   *54*, 93, 94, 100, 132, 133, 173, *184*
seawater attack   66, 67
self-compacting concrete (SCC)   219, 223–225
sensors   316, 325
service life   177–183
– cathodic protection systems   384, 385
serviceability limit states (SLS)   181
sewers   57

shotcrete   258
shrinkage
– concrete   212
– repair materials   359
silanes   251, 252
silica fume   9, 10
silica fume cement/concrete   10, 13, *16*, *17*, 204
– alkali content   63
– alkali silica reaction   65, 66
– chloride-induced corrosion   99, 100
– pore solution   23, *24*, *25*
– transport processes   *37*, 38
silicon dioxide ($Si(OH)_4$)   63
siloxanes   251, 252
slag, *see* blast furnace slag
slump test   208
sodium chloride   23
sodium content   396
sodium equivalent (Na_2O_{eq}) content   62–64
sodium hydroxide (NaOH)   2
sodium (Na^+) ions   23, *24*, *25*, 36, 61
sodium mono-fluoro phosphate (Na_2PO_3F, MFP)   231, 232, 234, 238
sodium silicate   257
soft water   58
soil conditions   *54*, *56*, 59, 133
sorptivity tests   32, 33
spacers   215
splash zones   *42*, *43*, 66, 67, 105, 107, *108*
stable propagation   124, 125
stagnation, water   219
stainless steel   130, 131, 266–276
standard hydrogen electrode (SHE)   116
stationary diffusion   28, 29
steel-fiber-reinforced concrete (SFRC)   151
Stern–Geary equation   306
stray-current-induced corrosion   141–153
strength classes   15, *18*, *55*, 61, 210, *222*
strengthening   361, 362
stress corrosion cracking (SCC)   156–162
stress intensity factor   159, 161
structural deterioration   49
structural health monitoring   327, 328
subcritical propagation   158
submerged structures, *see* immersed structures
sulfate attack   59–61
sulfate (SO_4^{2-}) ions   23, *24*, *25*, 36
sulfate-resistant cements   15, 60, 99
sulfur dioxide (SO_2)   79
sulfuric acid   55–58
superplasticizers   205, 206, 224

supplementary cementitious materials (SCMs), *see also* mineral additions   8, 221
surface content, chloride ions (C_s)   101–106
surface protection systems (SPSs)   243–260
surface treatment   360

T

temperature effects
– carbonation rates   84
– concrete deterioration
– – alkali silica reaction   65
– – freeze–thaw attack   50–53
– – sulfate attack   59
– corrosion rates   73
– – chloride-induced corrosion   109
– electrical resistivity   38, 39, 299
– transport processes   28, 36
tempered steel   164, 264
tendons   155, 264
tensile strength   62, 162, 164, 210, 211
thermal expansion coefficients   268
thermogravimetric analysis (TGA)   80
thickness, concrete   175, 176, 350
thickness, repair materials   357
thiobacilli   57
threshold value, *see* critical content
throwing power, cathodic polarization   379
tidal zones   66
time of wetness   88
titanium   77
titanium mesh   367
total chloride content   98, 309
transpassivity   117, 120
transport, inhibitors   236
transport numbers, ions   36
transport processes, *see also* capillary action, diffusion, migration, permeation   21, 22, 26, 27, 40
true cementitious coatings   258

tunnels   59, 83, 84, 89, 132
Tuutti's model   *72*, 86, 228

U

ultimate tensile stress   161

V

Valenta's equation   34
Vebe test   208
viscosity-modifying agents   206
visual inspection   288–291

W

waste water   56
water penetration   43, 44
water reducers   205, 206
water/binder ratio (w/b)   9, 61
water/cement (w/c) ratio   4–6, *7*, 8, 9, 18, 214
– influence on carbonation   84, 85
– influence on carbonation-induced corrosion   90
– influence on compressive strength   209, 210
– influence on concrete deterioration
– – freeze–thaw attack   52, 53
– – seawater attack   67
– – sulfate attack   60
– influence on transport processes   32, 34, 38, 39
– relationship to strength class   55
weldability, stainless steel   268
workability   207, 208

X

XPS technique   232

Z

zinc   277–279, 380
zones, anode systems   382